Heinz Schade, Klaus Neemann
Tensor Analysis

Also of Interest

Numerical Tensor Methods.
Tensor Trains in Mathematics and Computer Science
Ivan Oseledets, 2018
ISBN 978-3-11-046162-6, e-ISBN (PDF) 978-3-11-046163-3,
e-ISBN (EPUB) 978-3-11-046169-5

Irreducible Cartesian Tensors
Robert F. Snider, 2017
ISBN 978-3-11-056363-4, e-ISBN (PDF) 978-3-11-056486-0,
e-ISBN (EPUB) 978-3-11-056373-3

Groups and Manifolds.
Lectures for Physicists with Examples in Mathematica
Pietro Giuseppe Fré, Alexander Fedotov, 2017
ISBN 978-3-11-055119-8, e-ISBN (PDF) 978-3-11-055120-4,
e-ISBN (EPUB) 978-3-11-055133-4

Tensor Numerical Methods in Quantum Chemistry
Venera Khoromskaia, Boris N. Khoromskij, 2018
ISBN 978-3-11-037015-7, e-ISBN (PDF) 978-3-11-036583-2,
e-ISBN (EPUB) 978-3-11-039137-4

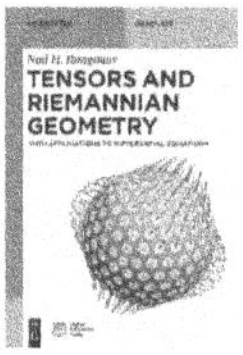

Tensors and Riemannian Geometry.
With Applications to Differential Equations
Nail H. Ibragimov, 2015
ISBN 978-3-11-037949-5, e-ISBN (PDF) 978-3-11-037950-1,
e-ISBN (EPUB) 978-3-11-037964-8

Heinz Schade, Klaus Neemann

Tensor Analysis

Translated by
Andrea Dziubek and Edmond Rusjan

DE GRUYTER

Mathematics Subject Classification 2010
53A45

Authors
Prof. Dr.-Ing. Heinz Schade
App. 430
Erlenweg 72
14532 Kleinmachnow
Germany
heinz.schade@posteo.de

Dr.-Ing. Klaus Neemann
Internationales Studienkolleg
Technische Universität Berlin
Straße des 17. Juni 145
10623 Berlin
Germany
klaus.neemann@tu-berlin.de

Translators
Prof. Dr. Andrea Dziubek
State University of New York
Polytechnic Institute
100 Seymour Road
Utica NY 13502, USA
dziubea@sunyit.edu

Prof. Dr. Edmond Rusjan
State University of New York
Polytechnic Institute
100 Seymour Road
Utica NY 13502, USA
edmond@sunyit.edu

ISBN 978-3-11-040425-8
e-ISBN (PDF) 978-3-11-040426-5
e-ISBN (EPUB) 978-3-11-040549-1

Library of Congress Cataloging-in-Publication Data
Names: Schade, Heinz, author. | Neemann, Klaus, author.
Title: Tensor analysis / Heinz Schade, Klaus Neemann.
Other titles: Tensoranalysis. English
Description: Berlin ; Boston : De Gruyter, [2018] | Series: De Gruyter textbook | Text in English, translated from German. | Includes bibliographical references.
Identifiers: LCCN 2018023807| ISBN 9783110404258 | ISBN 9783110404265 (e-book (pdf) | ISBN 9783110405491 (e-book (epub)
Subjects: LCSH: Calculus of tensors. | Vector analysis.
Classification: LCC QA433 .S2713 2018 | DDC 515/.63--dc23
LC record available at https://lccn.loc.gov/2018023807

Bibliographic information published by the Deutsche Nationalbibliothek
The Deutsche Nationalbibliothek lists this publication in the Deutsche Nationalbibliografie; detailed bibliographic data are available on the Internet at http://dnb.dnb.de.

© 2018 Walter de Gruyter GmbH, Berlin/Boston
Cover image: Klaus Neemann and urfinguss / iStock / Getty Images
Typesetting: VTeX UAB, Lithuania
Printing and binding: CPI books GmbH, Leck

www.degruyter.com

Preface

The first systematic presentation of tensor calculus was published in 1900 by Ricci and Levi-Civita and hence the subject was initially often called Ricci calculus. Gradually, the term tensor calculus became generally accepted. Tensor calculus was essential for the formulation of the theory of relativity, which has in turn stimulated its further development.

Tensor calculus is on the one hand an area of mathematics with applications in differential geometry and on the other hand an essential resource for theoretical physics and, increasingly, also for theoretical engineering. The theory of tensors consists of two parts: tensor algebra and tensor analysis. The term tensor analysis is today often (as in the title of this book) used for both parts. Tensor algebra deals with the general theory of tensors, while tensor analysis is the theory of tensor fields, i. e. tensors, which depend on the points in space, in particular in the three-dimensional Euclidean space of our intuition.

In mathematics, tensor algebra is primarily interesting as a part of multilinear algebra. Consequently, mathematical presentations usually use algebraic methods. Scalars, vectors, and finally tensors are introduced as elements of sets. Then operations between these elements are defined that satisfy certain rules. Finally, the properties of these tensors are investigated. This approach is very abstract for physics students, and even more for engineering students, and leads to a more general concept of vectors and tensors than we need here. [1]

In physics and in engineering we are interested in tensors, because physical quantities can be described as tensors (scalars and vectors are special cases of tensors). Physical equations are equations between tensors and tensor analysis enables us to formulate physical equations and to calculate with them. If we want to express a physical quantity, such as velocity or force, quantitatively, we need a coordinate system in the space of our intuition. The simplest coordinate system is usually a Cartesian coordinate system. We will see that we also need a coordinate system to quantitatively describe all other physical quantities, with the exception of scalar quantities. This means that as long as we only use tensor analysis for computations with physical quantities we can, mathematically speaking, limit ourselves to tensors which can be represented in a three-dimensional Cartesian basis.

This book is about these tensors. It builds tensor analysis in a more intuitive way by defining operations between tensors in terms of their coordinates in a Cartesian

1 The original German edition of this book had a final chapter on this mathematical approach to tensor analysis. We omitted this chapter from the English edition because it appeared that this approach was of less interest to the readers of this book.

https://doi.org/10.1515/9783110404265-201

basis. This naturally leads to the two tensor analysis notations: the so-called symbolic or direct notation and the so-called coordinate or index notation. Both notations have advantages and it is best to learn both notations simultaneously and to be able to translate between them like between languages. Hence the symbolic notation must be able to express, as much as possible, the same information as the coordinate notation with its algorithmic character can express. For example, the order of a tensor is clear in the coordinate notation from the number of the indices and hence it should be also clear in the symbolic notation. We aimed to derive a symbolic notation that does that.

The book has five chapters. In Chapter 1 we collect the tools used later in the book, including the so-called Kronecker symbols and the most important facts about determinants and matrices. Chapter 2 presents tensors in symbolic notation and in Cartesian coordinates while Chapter 4 presents tensors in curvilinear coordinates. Chapter 3 deals with the algebra of second-order tensors, which is closely related to the theory of square order-3 matrices. It also includes some additional theorems about square matrices. Chapter 5 is an introduction to the theory of the representation of tensor functions. We investigate permissible representations of physical quantities under the condition that a physical equation must satisfy tensorial homogeneity. This is similar to dimensional analysis, where an equation must satisfy dimensional homogeneity. (Tensorial homogeneity means that only tensors of the same order can be added or set equal to each other. For example, torque cannot be equal to work, because torque is a vector and work is a scalar.)

Important formulas are indented and labeled by a number. Intermediate results, which are cited later, for example, in a proof or in an auxiliary calculation, are left-aligned and labeled by a letter on the right side. The most important formulas and theorems (which are worth memorizing) are highlighted by a gray background and typeset in a font different from the rest of the text. Italics are used to emphasize text.

The book can be used as a textbook for a one-semester course. It is also well suited for self-study. Both, and especially the latter, require solving exercise problems and hence the text contains assignment problems at appropriate places. They are marked with the symbol of a pen on the left side. Only a limited number of problems is included, so all problems should be solved (unless the corresponding part of the text contains nothing new for the reader), and preferably before continuing reading the text. The relatively small number of problems makes it possible to write in some detail how the problems can be solved (see Appendix A).

The bibliography at the end of the book has been compiled by the translators.

Without the critical questions and suggestions for improvements from many students and assistants, and the generous support of our Institute, we would not have been able to write this book. We want to especially thank (now Professor) Frank Kameier, who professionally and didactically accompanied the course development for many years, Thomas Lauke for careful proofreading and preparing the manuscript for printing, Evelyn Kulzer for drawing the pictures, Andrea Dziubek and Edmond Rusjan for their conscientious translation into English and for initiating this translation, and last but not least Dr. Konrad Kieling from the Walter de Gruyter publishing house for enabling this English edition.

Kleinmachnow and Berlin, Germany — Heinz Schade
March 2018 — Klaus Neemann

Acknowledgments of the Translators

First and foremost, we would like to thank the authors, Professor Heinz Schade and Dr. Klaus Neemann, for writing this beautiful, reader friendly, concise, and self-contained book, for giving us the opportunity to translate it and for their guidance, help, and patience during the translation. We would like to thank the publisher, Walter de Gruyter, and in particular Dr. Konrad Kieling, for making this happen and for their support and patience. We are very grateful to SUNY Poly, Utica, and in particular to our Dean, Professor Andrew Russell, for their support and for making this project possible, and to our alumni Paul Lastowicka and Christopher Indolfi, for their diligent reading of translation drafts and for offering suggestions for improvements.

Utica, NY, USA
March 2018

Andrea Dziubek
Edmond Rusjan

https://doi.org/10.1515/9783110404265-202

Contents

1 Algebraic Tools

In this chapter, we introduce the summation convention, the Kronecker delta symbol δ_{ij}, the generalized Kronecker delta symbol $\delta^{i...j}_{p...q}$, and the Levi-Civita permutation symbol $\varepsilon_{i...j}$. At the same time, we review the most important facts about N-tuples, determinants, and matrices. We will obtain some more, less elementary properties of square matrices later by generalizing tensor algebra results.

1.1 The Summation Convention

1. We distinguish two types of letter subscripts and superscripts:

- Running indices assume values from an index set of natural numbers. They successively assume each value from the given index set. For example, if i takes the values from 1 to 4, then a_i denotes the set (a_1, a_2, a_3, a_4) and we say that the index i runs from 1 to 4. We will use lower-case Latin letters (except x, y, z) for running indices.
- Labels have a fixed meaning. We will use other letters for labels. For example, we write the (physical) cylindrical coordinates of a vector $\underline{a}$ as (a_R, a_ϕ, a_z).

2. In some sections, we need to distinguish between lower and upper indices. Where upper indices can be confused with exponents, we will set the exponents in brackets.

3. For the calculation with matrices and tensors, an agreement proposed by Einstein which is called summation convention has been adopted for running indices. It exists in different versions; we introduce it in the following form:

If a running index appears twice in a term, summation over the values of the index is implied, without explicitly writing the summation sign.

A term is a mathematical expression separated by a plus sign, a minus sign, an equality sign, or an inequality sign.

For example, if i assumes the values 1, 2, 3, then $a_i\, b_i$ is, according to the summation convention, $\sum_{i=1}^{3} a_i\, b_i$, which is $a_1\, b_1 + a_2\, b_2 + a_3\, b_3$. This is independent of the meaning of a_i and b_i; for example, (a_1, a_2, a_3) and (b_1, b_2, b_3) do not have to be vectors.

We often abbreviate $a_i\, a_i$ as a_i^2 or $a_i^{(2)}$. However, $a_{ij}\, a_{ij}$, which is, according to the summation convention, the double sum $\sum_i \sum_j a_{ij}\, a_{ij}$, cannot be written as a square because $a_{ij}^{(2)}$ is defined differently later.

https://doi.org/10.1515/9783110404265-001

4. The summation convention has the following consequences for running indices:

– *A running index can only appear once or twice in a term.*

If a running index appears once, it is called a free index. A free index successively takes each value of its index set. If a running index appears twice, it is called a summation index. A summation index indicates summation over the values of its index set.

For example, if an index i has the index set 1, 2 and a second index j has the index set 1, 2, 3, then the equation $a_{ij}\, b_j = c_i$ abbreviates the two equations

$$a_{11}\, b_1 + a_{12}\, b_2 + a_{13}\, b_3 = c_1,$$
$$a_{21}\, b_1 + a_{22}\, b_2 + a_{23}\, b_3 = c_2.$$

These two equations can also be written as $a_{ik}\, b_k = c_i$ or $a_{mn}\, b_n = c_m$, but not as $a_{ii}\, b_i = c_i$, because in the last equation i appears three times in a term and this is not defined.

It is often necessary to rename indices. For example, if we substitute $A_i = \alpha_{ij} B_j$ into $a_j = A_i\, C_{ij}$ we get formally $a_j = \alpha_{ij}\, B_j\, C_{ij}$, but then j appears three times in a term. To avoid this, we can write the first equation as $A_i = \alpha_{ik}\, B_k$, and now we get an acceptable equation, $a_j = \alpha_{ik}\, B_k\, C_{ij}$.

– *Running indices must have the same index set in all terms of an equation.*

For example, in $a_{ij}\, b_j = c_i$, if i runs from 1 to 3 in a_{ij}, it cannot run from 1 to 2 in c_i, and the same is true for j.

– *A free index must match in all terms of an equation. The order of indices in an equation does not matter.*

For example, the equation $a_{ij}\, b_j = c_k$ is not permissible, while $a_i\, b_j = c_{ij}$ and $b_j\, a_i = c_{ij}$ are permissible and have the same meaning.

The last rule has one exception: we do not add a subscript to the number zero, i. e. the equation $\alpha_i = 0$ means all α_i are zero. We will also need the logical negation of this statement. To express “not all α_i are zero” we introduce our own inequality sign $\mp$ and write $\alpha_i \mp 0$ (read: α_i not all zero). On the other hand, the equation with the common inequality sign $\alpha_i \neq 0$ means all α_i are nonzero. To not deviate more than necessary from common practice, we will use the usual inequality sign if an equation has both meanings. This is the case when the equation does not have free indices, so we write $a \neq 0$ and $a_i\, b_i \neq 0$.

Problem 1.1.

Write out the following expressions:
for $N = 4$, i. e. $i = 1, 2, 3, 4$

A. $a_i\, B_i$, B. A_{ii}, C. $\dfrac{\partial u_i}{\partial x_i}$, D. $\dfrac{\partial^2 \varphi}{\partial x_i^2}$, for $N = 3$

E. $a_{ij}\, b_{ij}$,   F. $a_{ii}\, b_{jj}$,

for $N = 2$

G. $\dfrac{\partial u_i}{\partial x_j} \dfrac{\partial u_i}{\partial x_j}$.

Problem 1.2.

Substitute:

A. $u_i = A_{ik}\, n_k$ in $\varphi = u_k\, v_k$,

B. $u_i = B_{ij}\, v_j$ and $C_{ij} = p_i\, q_j$ in $w_i = C_{mi}\, u_m$,

C. $\phi = \tau_{ij}\, d_{ij}$, $\tau_{ij} = \eta \left(\dfrac{\partial v_i}{\partial x_j} + \dfrac{\partial v_j}{\partial x_i} \right)$, $d_{ij} = \dfrac{1}{2} \left(\dfrac{\partial v_i}{\partial x_j} + \dfrac{\partial v_j}{\partial x_i} \right)$, and

$q_i = -\kappa \dfrac{\partial T}{\partial x_i}$ (κ constant) in $\rho\, T \dfrac{\mathrm{D} s}{\mathrm{D} t} = \phi - \dfrac{\partial q_i}{\partial x_i}$.

Expand the parentheses.

5. In the rare cases that we do not want to apply the summation convention, we underline the corresponding running index. For example, A_{ij} with $i, j = 1, 2, 3$ represents the matrix

$$\begin{pmatrix} A_{11} & A_{12} & A_{13} \\ A_{21} & A_{22} & A_{23} \\ A_{31} & A_{32} & A_{33} \end{pmatrix},$$

A_{ii} represents the *sum* $A_{11} + A_{22} + A_{33}$ of the diagonal elements, and $A_{i\underline{i}}$ represents the *set* (A_{11}, A_{22}, A_{33}) of the diagonal elements. Underlined indices are neither free nor summation indices. We call these indices bound indices, because if a term has a bound index, then the same index must appear either as a free index or as a summation index in the same term. These indices do not count in the rule that a running index can appear at most twice in a term.

This means $\alpha_{\underline{i}}\, A_{ii}$ is permissible and represents $\alpha_1\, A_{11} + \alpha_2\, A_{22} + \alpha_3\, A_{33}$ and $\lambda_{\underline{i}}\, a_{ij}$ is also permissible and represents the matrix

$$\begin{pmatrix} \lambda_1\, a_{11} & \lambda_1\, a_{12} & \lambda_1\, a_{13} \\ \lambda_2\, a_{21} & \lambda_2\, a_{22} & \lambda_2\, a_{23} \\ \lambda_3\, a_{31} & \lambda_3\, a_{32} & \lambda_3\, a_{33} \end{pmatrix}.$$

A short calculation shows that it is possible to multiply a free index by a bound index, but that it is not possible to multiply a summation index by a bound index: From

$a_{ij}\, b_{jk} = c_{ik}$ it follows that $\lambda_{\underline{i}}\, a_{ij}\, b_{jk} = \lambda_{\underline{i}}\, c_{ik}$, but from $a_i\, b_i = c_i\, d_i$ it does *not* follow that $\lambda_{\underline{i}}\, a_i\, b_i = \lambda_{\underline{i}}\, c_i\, d_i$.

If an index is bound to a free index, we can convert the free index into a summation index by multiplication. For example $\lambda_{\underline{i}}\, a_i$ becomes $\lambda_{\underline{i}}\, a_i\, b_i$. However, such an expression is not associative; it should be read as $(\lambda_{\underline{i}}\, a_i) b_i$. The other association $\lambda_{\underline{i}}(a_i\, b_i)$ is not defined, which becomes clear when we substitute $a_i\, b_i = c$.

Problem 1.3.

A. Are the following equations allowed by the summation convention?

1. $A_i\, B_{jj} = C_i\, D_{kk}$, 2. $A_i\, B_i = C_i\, D_j$, 3. $A_i\, B_i = C_i\, D_{\underline{i}}$,

4. $A_{\underline{i}}\, B_i = C_i$, 5. $A_m\, B_n = C_{mn}$, 6. $A_i\, B_k = C_i\, D_j$,

7. $A_i\, B_k = A_k\, B_i$, 8. $\mu_{mn}\, A_m = c_n\, F_{ii}$, 9. $A = \alpha_{\underline{m}}\, C_{mm}$,

10. $A_{\underline{ii}} = B_{ij}\, C_{jji}$.

B. Write out all permissible equations for $N = 2$.

1.2 N-tuples

1.2.1 Definitions

A sequence of N numbers is called an N-tuple. We write an N-tuple as $(a_1, a_2, \ldots, a_N)$ or short (a_i). The length of the sequence N is called the size of the N-tuple and does not appear in the abbreviation.

$$(a_i) := (a_1, a_2, \ldots, a_N). \tag{1.1}$$

The order of elements matters: changing the order of elements in an N-tuple gives a different N-tuple. Special names are often used for small values of N. We call 2-tuples ordered pairs, 3-tuples ordered triples, 4-tuples ordered quadruples, etc.

For example, the three Cartesian coordinates of a point form an N-tuple, in particular, an ordered triple.

We call an N-tuple of arbitrary size with all elements zero a zero-N-tuple and write (0).

1.2.2 Linear Operations

For N-tuples of equal size we define the following so-called linear operations: equality, addition, subtraction, and multiplication by a number:

- Two N-tuples are equal if all homologous elements, i. e. elements at identical positions in both N-tuples and subscribed by the same index, are equal.

- Addition and subtraction of two N-tuples are defined by adding or subtracting their homologous elements.
- Multiplication of an N-tuple by a number is defined by multiplying every element of the N-tuple by that number.

Equality, addition, and subtraction are defined only for N-tuples of equal size.

1.2.3 Linear Independence

1. A set of P N-tuples of equal size $(\overset{1}{a}_i)$, $(\overset{2}{a}_i)$, $\ldots$, $(\overset{P}{a}_i)$ is called linearly independent if the equation

$$\alpha_1 \, (\overset{1}{a}_i) + \alpha_2 \, (\overset{2}{a}_i) + \cdots + \alpha_P \, (\overset{P}{a}_i) = (0) \tag{1.2}$$

is satisfied only if all α_i are zero. This solution is called the trivial solution. Using the summation convention we abbreviate

$$\begin{aligned} &\alpha_j \, (\overset{j}{a}_i) = (0) \quad \text{only if } \alpha_j = 0, \\ &\quad i = 1, \ldots, N, \; j = 1, \ldots, P. \end{aligned} \tag{1.3}$$

If equation (1.2) has also nontrivial solutions, i. e. if at least one of the α_j is nonzero,

$$\alpha_j \, (\overset{j}{a}_i) = (0), \quad \alpha_j \neq 0, \; i = 1, \ldots, N, \; j = 1, \ldots, P, \tag{1.4}$$

then the N-tuples $(\overset{j}{a}_i)$ are called linearly dependent.

2. Equation (1.2) represents a system of N homogeneous, linear equations to determine the P unknown α_j:

$$\begin{aligned} &\alpha_1 \, \overset{1}{a}_1 + \alpha_2 \, \overset{2}{a}_1 + \cdots + \alpha_P \, \overset{P}{a}_1 = 0, \\ &\alpha_1 \, \overset{1}{a}_2 + \alpha_2 \, \overset{2}{a}_2 + \cdots + \alpha_P \, \overset{P}{a}_2 = 0, \\ &\vdots \\ &\alpha_1 \, \overset{1}{a}_N + \alpha_2 \, \overset{2}{a}_N + \cdots + \alpha_P \, \overset{P}{a}_N = 0. \end{aligned}$$

If the number of unknowns is larger than the number of equations, i. e. if $P > N$, the system always has nontrivial solutions. More than N N-tuples are always linearly dependent, while N or fewer than N N-tuples can be linearly independent.

3. A set of N-tuples where each pair is linearly dependent is called collinear; a set of N-tuples where each triple is linearly dependent is called coplanar.

4. The left side of (1.2) is an N-tuple, which is obtained by multiplying each of the N-tuples (of a given set of N-tuples) by a number and then summing up these products. This is called a linear combination of these N-tuples.

1.3 Determinants

We assume the reader is familiar with the algebra of determinants; however, we summarize here the most important definitions and theorems (without proofs) for later reference.

1.3.1 Definitions

1. We can assign a number to a collection of numbers a_{ij} arranged in a square table. This number is called the determinant of this square table and we write

$$\det \underset{\sim}{a} = \det a_{ij} = \begin{vmatrix} a_{11} & a_{12} & \dots & a_{1N} \\ a_{21} & a_{22} & \dots & a_{2N} \\ \vdots & & & \\ a_{N1} & a_{N2} & \dots & a_{NN} \end{vmatrix} . \tag{1.5}$$

A determinant with N rows and N columns is called a determinant of order N. Clearly, each row or column of a determinant is an N-tuple. Let

$$A_{ij...k} := a_{i1}\, a_{j2} \cdots a_{kN} \tag{1.6}$$

be a product of the elements of a determinant of order N with exactly one element from each row and from each column, i. e. the indices $ij \dots k$ represent an arbitrary permutation of the numbers $1, \dots, N$. Since there are $N!$ permutations of N numbers, this yields $N!$ products.

The value of the determinant is the algebraic sum of these $N!$ products. The product $A_{12...N}$ and all products which are obtained by an even number of index exchanges from the sequence of indices $1, 2, \dots, N$ have a positive sign and all products which are obtained by an odd number of exchanges have a negative sign.

This definition can easily be written as a formula: Let

- $p_m(A_{ij...k})$ with $m = 1, \dots, N!$ be the products $A_{ij...k}$ in any order and
- α_m be the number of index exchanges needed to transform the index sequence $1, 2, \dots, N$ into the index sequence $i, j, \dots, k$ (which is the index sequence of the permutation needed to generate the product $p_m(A_{ij...k})$). Then

$$\det \underset{\sim}{a} := (-1)^{\alpha_m}\, p_m(A_{ij...k}). \tag{1.7}$$

2. A determinant, which is obtained from the original determinant by omitting any number of rows and an equal number of columns, is called a subdeterminant of the original determinant. If the omitted columns have the same indices as the omitted rows, the subdeterminant is called a principal subdeterminant. The subdeterminant Δ_{ij} of order $N - 1$, which is obtained by omitting the i-th row and the j-th column, is called a minor;[1] minors corresponding to the main diagonal are also called principal minors. A minor multiplied by $(-1)^{(i+j)}$ is called the cofactor corresponding to the element a_{ij}, for which we write b_{ij}:

$$b_{ij} := (-1)^{(i+j)}\,\Delta_{ij}. \tag{1.8}$$

Cofactors are also called adjuncts or algebraic complements.

3. A determinant of order N has rank $R \leqq N$, if at least one subdeterminant of order R is nonzero and all higher-order subdeterminants are zero. For "the determinant $\underset{\sim}{a}$ has the rank R" we also write $\operatorname{rank}\det \underset{\sim}{a} = R$.

4. A determinant of order N is called regular if it has the rank N, it is called singular if its rank is smaller than N, and it is called M-times singular if its rank is $N - M$; then we also say the determinant has rank defect M.

5. The summation convention does not apply across the symbol det: $\det a_{ij} = 1$ is a valid equation; however, $\det a_{ij} = \delta_{ij}$ is not permissible.

1.3.2 Computing Determinants

1. For a determinant of order two, according to (1.7), we have

$$\det \underset{\sim}{a} = a_{11}\,a_{22} - a_{21}\,a_{12}. \tag{1.9}$$

To easily remember this formula, we write out the elements of the determinant:

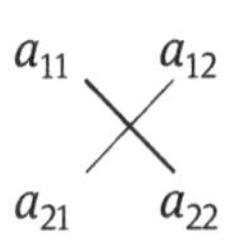

This square table has two diagonals: the diagonal indicated by the bold line is called the main diagonal, and the diagonal indicated by the thin line is called the antidiagonal. To compute the value of a determinant, we subtract from the product of the elements on the main diagonal the product of the elements on the antidiagonal; we shorten this rule to "main diagonal minus antidiagonal".

For a determinant of order three, according to (1.7), we have

$$\begin{aligned}\det \underset{\sim}{a} = {} & a_{11}\,a_{22}\,a_{33} - a_{21}\,a_{12}\,a_{33} - a_{31}\,a_{22}\,a_{13} \\ & - a_{11}\,a_{32}\,a_{23} + a_{21}\,a_{32}\,a_{13} + a_{31}\,a_{12}\,a_{23}.\end{aligned} \tag{1.10}$$

To easily remember this formula we first write out the elements of the determinant and then we repeat the first and the second row below.

1 The term minor is also used with a different meaning in the literature.

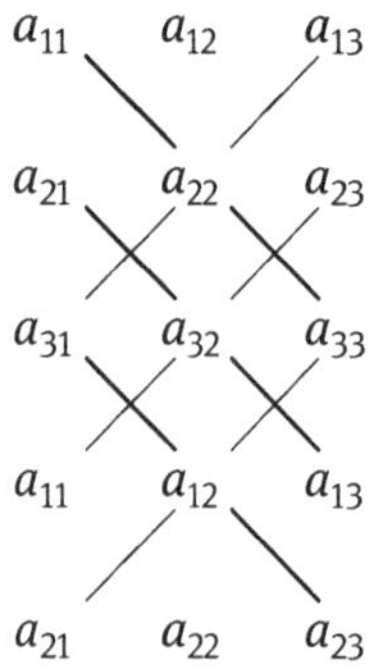
a11 a12 a13
a21 a22 a23
a31 a32 a33
a11 a12 a13
a21 a22 a23

This rectangular table has three main diagonals and three antidiagonals. The value of the determinant is the difference between the sum of the products of the elements on each main diagonal and the sum of the products of the elements on each antidiagonal. Again we shorten this rule to "main diagonals minus antidiagonals" (rule of Sarrus).

Unfortunately, there are no analogous computational patterns or rules for higher-order determinants.

2. We can compute the value of a determinant with the cofactor (or Laplace) expansion formula: we pick any row or column, multiply every element by its cofactor, and add these products. Then we have

$$\det \underset{\sim}{a} = a_{ik}\, b_{i\underline{k}} = a_{ik}\, b_{\underline{i}k}. \tag{1.11}$$

For example, $a_{ik}\, b_{i\underline{k}}$ represents
for $k = 1$: $a_{11}\, b_{11} + a_{21}\, b_{21} + a_{31}\, b_{31}$,
for $k = 2$: $a_{12}\, b_{12} + a_{22}\, b_{22} + a_{32}\, b_{32}$,
for $k = 3$: $a_{13}\, b_{13} + a_{23}\, b_{23} + a_{33}\, b_{33}$.

All three expressions are equal to $\det \underset{\sim}{a}$.

3. In practice, we compute the value of a determinant by first transforming the determinant into triangular form, i. e. a determinant which has only zeros below the main diagonal or only zeros above the main diagonal. To do this, we follow the rules for computations with determinants which we summarize in Section 1.3.3. Once we have transformed a determinant into triangular form, its value is, according to the cofactor expansion, the product of the elements on the main diagonal.

1.3.3 Computations with Determinants

The most important rules for computations with determinants are the following:

I. The value of a determinant does not change,
   a) if we add to a row a linear combination of the remaining rows;
   b) if we add to a column a linear combination of the remaining columns;
   c) if we exchange rows and columns (reflect the determinant at the main diagonal).
II. The sign of a determinant changes if we exchange two rows or columns.
III. A determinant is zero if and only if its rows are linearly dependent. (According to condition Ic, this is equivalent to its columns being linearly dependent.)

IV. Determinants which differ in a single row (or column) are added and subtracted by

$$\begin{vmatrix} a_{11} & a_{12} & \dots & a_{1N} \\ a_{21} & a_{22} & \dots & a_{2N} \\ \vdots & & & \\ a_{N1} & a_{N2} & \dots & a_{NN} \end{vmatrix} \pm \begin{vmatrix} a'_{11} & a'_{12} & \dots & a'_{1N} \\ a_{21} & a_{22} & \dots & a_{2N} \\ \vdots & & & \\ a_{N1} & a_{N2} & \dots & a_{NN} \end{vmatrix} = \begin{vmatrix} a_{11} \pm a'_{11} & a_{12} \pm a'_{12} & \dots & a_{1N} \pm a'_{1N} \\ a_{21} & a_{22} & \dots & a_{2N} \\ \vdots & & & \\ a_{N1} & a_{N2} & \dots & a_{NN} \end{vmatrix}. \tag{1.12}$$

V. To multiply a determinant by a number we multiply any of its rows or columns by this number.

VI. Determinant multiplication is given by

$$(\det a_{im})(\det b_{mj}) = \det(a_{im}\, b_{mj}). \tag{1.13}$$

With condition Ic, we establish that

$$\det(a_{im}\, b_{mj}) = \det(a_{mi}\, b_{mj}) = \det(a_{im}\, b_{jm}) = \det(a_{mi}\, b_{jm}).$$

VII. If b_{ij} is the cofactor of the element a_{ij} of a determinant $\det \underset{\sim}{a}$, then

$$a_{ij}\, b_{ik} = a_{ji}\, b_{ki} = \begin{cases} \det \underset{\sim}{a} & \text{for } j = k, \\ 0 & \text{for } j \neq k. \end{cases} \tag{1.14}$$

Problem 1.4.

Compute the following determinants (consult also Section 1.6.1):

$$\text{A. } \Delta = \begin{vmatrix} 1 & 0 & -1 \\ 3 & 1 & -3 \\ 1 & 2 & -2 \end{vmatrix}, \qquad \text{B. } \Delta = \begin{vmatrix} 1 & 2 & -4 & 4 \\ 3 & 6 & 9 & 0 \\ -2 & 8 & 1 & 9 \\ 4 & 2 & -2 & 0 \end{vmatrix}.$$

1.4 Kronecker Symbols

In this section, we define some quantities which we need later and we explore their most important properties.

1.4.1 δ_{ij}

1. In δ_{ij}, both indices must have the same index set, 1, …, N. We define δ_{ij} as follows.

$$\delta_{ij} := \begin{cases} 1 & \text{for equal indices, i. e. } i = j, \\ 0 & \text{for different indices, i. e. } i \neq j. \end{cases} \tag{1.15}$$

The term δ_{ij} is often called the Kronecker symbol.

Problem 1.5.
Compute for $N = 5$
A. $\delta_{\underline{ii}}$, B. δ_{ii}.

2. The term δ_{ij} is clearly symmetric in its indices, i. e. its value does not change if we change the order of the two indices.

$$\delta_{ij} = \delta_{ji}. \tag{1.16}$$

3. For example, $A_i\,\delta_{ij}$ means for $N = 3$

$$A_i\,\delta_{ij} = A_1\,\delta_{1j} + A_2\,\delta_{2j} + A_3\,\delta_{3j}.$$

Substituting 1, 2, and 3 for j, we get
for $j = 1$: $A_i\,\delta_{i1} = A_1$,
for $j = 2$: $A_i\,\delta_{i2} = A_2$,
for $j = 3$: $A_i\,\delta_{i3} = A_3$.

We can summarize these three equations as $A_i\,\delta_{ij} = A_j$, and we get similar results for $N \neq 3$. If A has additional indices (possibly with a different index set) not appearing in δ_{ij}, these indices are not affected by multiplication with δ_{ij}.

$$A_{i \ldots jmp \ldots q}\,\delta_{mn} = A_{i \ldots jnp \ldots q}. \tag{1.17}$$

Let us write this important formula in words: If we sum over one index of δ_{mn}, we replace this index in the other quantity (here the m in $A_{i \ldots jmp \ldots q}$) by the second index of δ_{mn} (here the n) and we drop the δ_{mn}.

Problem 1.6.
Simplify the following expressions:
A. $\delta_{ij}\,\delta_{jk}$, B. $\delta_{i2}\,\delta_{ik}\,\delta_{3k}$, C. $\delta_{1k}\,A_k$, D. $\delta_{i2}\,\delta_{jk}\,A_{ij}$.

4. Analogously, we prove that

$$\lambda_{\underline{m}}\,A_{i...jmp...q}\,\delta_{mn} = \lambda_{\underline{n}}\,A_{i...jnp...q}. \tag{1.18}$$

1.4.2 $\delta^{i...j}_{p...q}$

1. In $\delta^{i...j}_{p...q}$ the number M of the upper and the lower indices is the same. We also say that $\delta^{i...j}_{p...q}$ has M pairs of indices; in addition, all $2M$ indices are assumed to have the same index set $1, \ldots, N$. Then we define

$$\delta^{ij...k}_{pq...r} := \begin{vmatrix} \delta_{ip} & \delta_{iq} & \ldots & \delta_{ir} \\ \delta_{jp} & \delta_{jq} & \ldots & \delta_{jr} \\ \vdots & & & \\ \delta_{kp} & \delta_{kq} & \ldots & \delta_{kr} \end{vmatrix}. \tag{1.19}$$

We call $\delta^{i...j}_{p...q}$ the generalized Kronecker symbol. Clearly, $\delta^{i}_{j} = \delta_{ij}$, so this name is plausible.

2. Let us derive some rules for the values of the elements of $\delta^{i...j}_{p...q}$. We first consider an element of $\delta^{i...j}_{p...q}$ with at least two equal upper or lower indices. Then the corresponding rows or columns of the determinant are equal, which implies that the determinant is zero, independent of the values of the other indices. This must be the case, for example, if the number M of index pairs is greater than the maximal length N of the index set, and thus such $\delta^{i...j}_{p...q}$ are zero.

Next, we consider an element of $\delta^{i...j}_{p...q}$ with one upper index which does not appear among the lower indices (or an element with a lower index which does not appear among the upper indices). Then the elements of the corresponding row (or column) are all zero and the determinant is again zero, independent of the values of the remaining indices.

So only those elements can be nonzero, whose upper indices are all different and whose lower indices are all different, and whose lower indices differ from their upper indices at most in their order.

We now consider the case where both indices of every index pair are equal, i. e. the order of the upper and lower indices is the same. Then all diagonal elements of the determinant are one and all remaining elements are zero, so the determinant and hence also the corresponding element of $\delta^{i...j}_{p...q}$ is one.

Every change in the order of two lower indices (while keeping the order of the upper indices) exchanges the corresponding columns of the determinant and hence changes the sign of the determinant.

This leads us to the following result.

$$\delta^{i\dots j}_{p\dots q} = \begin{cases} 1 & \text{if all upper indices are distinct and the sequence of the lower indices is an even permutation of the sequence of the upper indices,} \\ -1 & \text{if all upper indices are distinct and the sequence of the lower indices is an odd permutation of the sequence of the upper indices,} \\ 0 & \text{otherwise, i. e. if some of the upper indices are equal or when the lower indices are not a permutation of the upper indices.} \end{cases} \tag{1.20}$$

Clearly,

$$\delta^{i\dots j}_{p\dots q} = \delta^{p\dots q}_{i\dots j}, \tag{1.21}$$

since the value of a determinant does not change if we exchange rows and columns.

Problem 1.7.
A. Write out all nonzero elements of δ^{ij}_{pq} and compute their values
1. for $N = 2$, 2. for $N = 3$.

B. Compute δ^{ij}_{ij} for $N = 3$.

1.4.3 $\varepsilon_{i\dots j}$

1. For some of the more frequently used generalized Kronecker symbols we introduce another notation. We define

$$\varepsilon_{ij\dots k} := \delta^{12\dots M}_{ij\dots k} = \delta^{ij\dots k}_{12\dots M}. \tag{1.22}$$

On the one hand, elements of $\varepsilon_{i\dots j}$ are nonzero only if none of the indices $i\dots j$ is larger than M. On the other hand, generalized Kronecker symbols for $N < M$ do not exist anyway. Therefore, without loss of generality, we have $N = M$, i. e. the length of the index set of all indices equals the number of indices in $\varepsilon_{i\dots j}$.

The term $\varepsilon_{i\dots j}$ is called permutation symbol, Levi-Civita symbol, or alternating symbol.

Problem 1.8.
Compute for $N = 2$ all elements of δ_{ij} and ε_{ij}.

2. According to (1.22), the values of $\varepsilon_{i...j}$ are the following.

$$\varepsilon_{i...j} = \begin{cases} 1 & \text{for } \varepsilon_{12...M} \text{ and for all even permutations of these indices,} \\ -1 & \text{for all odd permutations of these indices,} \\ 0 & \text{otherwise (i. e. if an index appears at least twice).} \end{cases} \tag{1.23}$$

For M indices, $\varepsilon_{i...j}$ has M^M elements (number of combinations of M elements including repetitions in M classes), and $M!$ of these elements (number of permutations of M elements without repetitions) are nonzero, where half of them are $+1$ and the other half are -1.

Clearly, $\varepsilon_{i...j}$ is antimetric with respect to each index pair, i. e. exchanging two indices only changes the sign.

From (1.23), we get a formula that is sometimes useful. Let

- $\varepsilon_{i...j}$ be an arbitrary element with M different indices;
- $p_k(\varepsilon_{i...j})$ with $k = 1, \ldots, M!$ be the elements in any order, which can be formed by permutation of the indices; and
- α_k be the number of index exchanges necessary to change the order of indices $\varepsilon_{i...j}$ into the order of indices $p_k(\varepsilon_{i...j})$.

Then we get

$$\varepsilon_{i...j} = \frac{1}{M!}\,(-1)^{\alpha_k}\,p_k(\varepsilon_{i...j}). \tag{1.24}$$

This formula trivially holds when some of the indices of $\varepsilon_{i...j}$ are repeated, but then each term is zero anyway.

From (1.22) and (1.19), we obtain

$$\varepsilon_{ij...k} = \begin{vmatrix} \delta_{1i} & \delta_{1j} & \ldots & \delta_{1k} \\ \delta_{2i} & \delta_{2j} & \ldots & \delta_{2k} \\ \vdots & & & \\ \delta_{Mi} & \delta_{Mj} & \ldots & \delta_{Mk} \end{vmatrix} = \begin{vmatrix} \delta_{i1} & \delta_{i2} & \ldots & \delta_{iM} \\ \delta_{j1} & \delta_{j2} & \ldots & \delta_{jM} \\ \vdots & & & \\ \delta_{k1} & \delta_{k2} & \ldots & \delta_{kM} \end{vmatrix}. \tag{1.25}$$

3. Let us derive two more useful formulas. From (1.25), it follows that

$$\varepsilon_{ij...k}\,\varepsilon_{pq...r} = \begin{vmatrix} \delta_{i1} & \delta_{i2} & \ldots & \delta_{iM} \\ \delta_{j1} & \delta_{j2} & \ldots & \delta_{jM} \\ \vdots & & & \\ \delta_{k1} & \delta_{k2} & \ldots & \delta_{kM} \end{vmatrix} \begin{vmatrix} \delta_{1p} & \delta_{1q} & \ldots & \delta_{1r} \\ \delta_{2p} & \delta_{2q} & \ldots & \delta_{2r} \\ \vdots & & & \\ \delta_{Mp} & \delta_{Mq} & \ldots & \delta_{Mr} \end{vmatrix}.$$

Using the multiplication theorem for determinants (1.13), we have for the upper left element of this product of two determinants $\delta_{i1}\delta_{1p}+\delta_{i2}\delta_{2p}+\cdots+\delta_{iM}\delta_{Mp} = \delta_{im}\delta_{mp} = \delta_{ip}$.

Analogously, we compute the remaining elements of this product of two determinants and we obtain

$$\varepsilon_{ij\ldots k}\,\varepsilon_{pq\ldots r} = \begin{vmatrix} \delta_{ip} & \delta_{iq} & \ldots & \delta_{ir} \\ \delta_{jp} & \delta_{jq} & \ldots & \delta_{jr} \\ \vdots & & & \\ \delta_{kp} & \delta_{kq} & \ldots & \delta_{kr} \end{vmatrix}$$

or, with (1.19),

$$\varepsilon_{ij\ldots k}\,\varepsilon_{pq\ldots r} = \delta^{ij\ldots k}_{pq\ldots r} = \begin{vmatrix} \delta_{ip} & \delta_{iq} & \ldots & \delta_{ir} \\ \delta_{jp} & \delta_{jq} & \ldots & \delta_{jr} \\ \vdots & & & \\ \delta_{kp} & \delta_{kq} & \ldots & \delta_{kr} \end{vmatrix}. \tag{1.26}$$

The product of $\varepsilon_{ij\ldots k}$ and δ_{pq}, where the number of indices of ε and the length of the index set of all indices is N, is

$$\begin{aligned} \varepsilon_{ij\ldots k}\,\delta_{pq} &= \varepsilon_{pj\ldots k}\,\delta_{iq} + \varepsilon_{ip\ldots k}\,\delta_{jq} + \cdots + \varepsilon_{ij\ldots p}\,\delta_{kq} \\ &= \varepsilon_{qj\ldots k}\,\delta_{ip} + \varepsilon_{iq\ldots k}\,\delta_{jp} + \cdots + \varepsilon_{ij\ldots q}\,\delta_{kp}. \end{aligned} \tag{1.27}$$

This formula is easy to understand. According to (1.20), we have for all indices of the index set of length N $\delta^{pij\ldots kl}_{q12\ldots N-1,N} = 0$, and with (1.19), it follows that

$$\begin{vmatrix} \delta_{pq} & \delta_{p1} & \delta_{p2} & \ldots & \delta_{p,N-1} & \delta_{pN} \\ \delta_{iq} & \delta_{i1} & \delta_{i2} & \ldots & \delta_{i,N-1} & \delta_{iN} \\ \delta_{jq} & \delta_{j1} & \delta_{j2} & \ldots & \delta_{j,N-1} & \delta_{jN} \\ \vdots & & & & & \\ \delta_{kq} & \delta_{k1} & \delta_{k2} & \ldots & \delta_{k,N-1} & \delta_{kN} \\ \delta_{lq} & \delta_{l1} & \delta_{l2} & \ldots & \delta_{l,N-1} & \delta_{lN} \end{vmatrix} = 0.$$

Expanding (using the “expansion” formula) according to the first column and using (1.26) gives

$$\begin{aligned} &\delta_{pq}\,\varepsilon_{ij\ldots kl}\,\varepsilon_{12\ldots N} - \delta_{iq}\,\varepsilon_{pj\ldots kl}\,\varepsilon_{12\ldots N} + \delta_{jq}\,\varepsilon_{pi\ldots kl}\,\varepsilon_{12\ldots N} - \cdots \\ &\qquad + (-1)^{N-1}\delta_{kq}\,\varepsilon_{pij\ldots l}\,\varepsilon_{12\ldots N} + (-1)^{N}\delta_{lq}\,\varepsilon_{pij\ldots k}\,\varepsilon_{12\ldots N} = 0. \end{aligned}$$

The factor $\varepsilon_{12\ldots N}$ is one and can be omitted. Now we bring all terms except the first to the right side and perform as many index exchanges in the epsilons as we need to move the index p to the M-th position in the M-th term on the right side, while keeping the order of the remaining indices unchanged. Clearly, $M-1$ exchanges are necessary, and we get

$$\varepsilon_{ij\ldots kl}\,\delta_{pq} = \varepsilon_{pj\ldots kl}\,\delta_{iq} + \varepsilon_{ip\ldots kl}\,\delta_{jq} + \cdots + \varepsilon_{ij\ldots pl}\,\delta_{kq} + \varepsilon_{ij\ldots kp}\,\delta_{lq}.$$

Or, if we change the notation of the indices slightly, we get the first equation (1.27). The second equation we get from the first equation, if we exchange p and q and use the symmetry of δ_{pq}.

1.4.4 Representation of a Determinant by $\varepsilon_{i...j}$

1. The system of $\varepsilon_{i...j}$ has properties which are similar to the properties of determinants; for example, a determinant is zero if two rows (or columns) are equal and it changes sign if we exchange two rows (or columns). The same is true for the elements of $\varepsilon_{i...j}$ with respect to the indices, so it is plausible to write $\varepsilon_{i...j}$ according to (1.25) as a determinant. Conversely, we can assume that a determinant can also be written using $\varepsilon_{i...j}$. To see this, we multiply (1.25) by $a_{i1}\, a_{j2} \cdots a_{kM}$. Then we have

$$\varepsilon_{ij...k}\, a_{i1}\, a_{j2} \cdots a_{kM} = \begin{vmatrix} \delta_{1i} & \delta_{1j} & \dots & \delta_{1k} \\ \delta_{2i} & \delta_{2j} & \dots & \delta_{2k} \\ \vdots & & & \\ \delta_{Mi} & \delta_{Mj} & \dots & \delta_{Mk} \end{vmatrix} a_{i1}\, a_{j2} \cdots a_{kM}.$$

If, on the right side, we multiply a_{i1} into the first column of the determinant, a_{j2} into the second column, and so on and take (1.17) into account, we get

$$\varepsilon_{ij...k}\, a_{i1}\, a_{j2} \cdots a_{kM} = \begin{vmatrix} a_{11} & a_{12} & \dots & a_{1M} \\ a_{21} & a_{22} & \dots & a_{2M} \\ \vdots & & & \\ a_{M1} & a_{M2} & \dots & a_{MM} \end{vmatrix}.$$

We get the same result if we start with

$$\varepsilon_{ij...k}\, a_{1i}\, a_{2j} \cdots a_{Mk} = \begin{vmatrix} \delta_{i1} & \delta_{i2} & \dots & \delta_{iM} \\ \delta_{j1} & \delta_{j2} & \dots & \delta_{jM} \\ \vdots & & & \\ \delta_{k1} & \delta_{k2} & \dots & \delta_{kM} \end{vmatrix} a_{1i}\, a_{2j} \cdots a_{Mk}$$

and then multiply the first row of the determinant by a_{1i}, the second row by a_{2j}, etc. This gives

$$\det \underset{\sim}{a} = \varepsilon_{ij...k}\, a_{i1}\, a_{j2} \cdots a_{kM} = \varepsilon_{ij...k}\, a_{1i}\, a_{2j} \cdots a_{Mk}. \tag{1.28}$$

Using the definition (1.22) of $\varepsilon_{i...j}$, we can also write

$$\det \underset{\sim}{a} = \delta^{ij...k}_{12...M}\, a_{i1}\, a_{j2} \cdots a_{kM}. \tag{a}$$

The value of $\delta^{ij\ldots k}_{12\ldots M}$ does not change if we change the order of two index pairs, so we also have

$$\det \underset{\sim}{a} = \delta^{ji\ldots k}_{21\ldots M}\, a_{i1}\, a_{j2} \cdots a_{kM} = \delta^{ji\ldots k}_{21\ldots M}\, a_{j2}\, a_{i1} \cdots a_{kM}.$$

And if we rename i to j and j to i in the last equation, we get

$$\det \underset{\sim}{a} = \delta^{ij\ldots k}_{21\ldots M}\, a_{i2}\, a_{j1} \cdots a_{kM}.$$

So we can also obtain $\det \underset{\sim}{a}$ if we replace in (a) the lower indices by a permutation of the numbers $1, 2, \ldots, M$ and the second indices of the terms in the product $a_{i1}\, a_{j2} \cdots a_{kM}$ by the same permutation. If we replace in (a) the number indices by running indices, i. e. if we write

$$\delta^{ij\ldots k}_{pq\ldots r}\, a_{ip}\, a_{jq} \cdots a_{kr},$$

this represents an M-fold summation over all M values of the indices $p, q, \ldots, r$. However, of the M^M summands, only those $M!$ terms are nonzero, in which the sequence $p, q, \ldots, r$ is a permutation of the sequence $1, 2, \ldots, M$. Since all summands are equal to $\det \underset{\sim}{a}$, from (1.26) it follows that

$$\det \underset{\sim}{a} = \frac{1}{M!}\, \delta^{ij\ldots k}_{pq\ldots r}\, a_{ip}\, a_{jq} \cdots a_{kr} = \frac{1}{M!}\, \varepsilon_{ij\ldots k}\, \varepsilon_{pq\ldots r}\, a_{ip}\, a_{jq} \cdots a_{kr}$$

$$= \frac{1}{M!} \begin{vmatrix} \delta_{ip} & \delta_{iq} & \ldots & \delta_{ir} \\ \delta_{jp} & \delta_{jq} & \ldots & \delta_{jr} \\ \vdots & & & \\ \delta_{kp} & \delta_{kq} & \ldots & \delta_{kr} \end{vmatrix} a_{ip}\, a_{jq} \cdots a_{kr}.$$

Finally, if we multiply the first row of the determinant by a_{ip}, the second row by a_{jq}, etc., we get

$$\det \underset{\sim}{a} = \frac{1}{M!} \begin{vmatrix} a_{pp} & a_{qp} & \ldots & a_{rp} \\ a_{pq} & a_{qq} & \ldots & a_{rq} \\ \vdots & & & \\ a_{pr} & a_{qr} & \ldots & a_{rr} \end{vmatrix},$$

$$\det \underset{\sim}{a} = \frac{1}{M!}\, \varepsilon_{ij\ldots k}\, \varepsilon_{pq\ldots r}\, a_{ip}\, a_{jq} \cdots a_{kr} = \frac{1}{M!} \begin{vmatrix} a_{pp} & a_{pq} & \ldots & a_{pr} \\ a_{qp} & a_{qq} & \ldots & a_{qr} \\ \vdots & & & \\ a_{rp} & a_{rq} & \ldots & a_{rr} \end{vmatrix}. \tag{1.29}$$

2. We can also compute the cofactors of the elements of a determinant in terms of $\varepsilon_{i...j}$. We start with $(1.29)_1$ and multiply both sides by δ_{mn}, which gives

$$(\det \underset{\sim}{a})\delta_{mn} = \frac{1}{M!}\,\varepsilon_{ij...k}\,\delta_{mn}\,\varepsilon_{pq...r}\,a_{ip}\,a_{jq}\,\cdots\,a_{kr}. \tag{a}$$

Next, we replace $\varepsilon_{ij...k}\,\delta_{mn}$ using $(1.27)_1$ and we get

$$\begin{aligned}(\det \underset{\sim}{a})\delta_{mn} &= \frac{1}{M!}\,(\varepsilon_{mj...k}\,\delta_{in} + \varepsilon_{im...k}\,\delta_{jn} + \cdots + \varepsilon_{ij...m}\,\delta_{kn})\,\varepsilon_{pq...r}\,a_{ip}\,a_{jq}\,\cdots\,a_{kr}\\ &= \frac{1}{M!}\,(\varepsilon_{mj...k}\,\varepsilon_{pq...r}\,a_{np}\,a_{jq}\,\cdots\,a_{kr} + \varepsilon_{im...k}\,\varepsilon_{pq...r}\,a_{ip}\,a_{nq}\,\cdots\,a_{kr}\\ &\quad + \cdots + \varepsilon_{ij...m}\,\varepsilon_{pq...r}\,a_{ip}\,a_{jq}\,\cdots\,a_{nr}).\end{aligned}$$

To ensure that the factor $a_{np}a_{jq}\cdots a_{kr}$ appears in every term, we rename the summation indices. We have

$$\begin{aligned}(\det \underset{\sim}{a})\delta_{mn} = \frac{1}{M!}\,(&\varepsilon_{mj...k}\,\varepsilon_{pq...r}\,a_{np}\,a_{jq}\,\cdots\,a_{kr}\\ &+ \underset{\substack{\uparrow\\ j}}{\varepsilon_{im...k}}\,\underset{\substack{\uparrow\uparrow\\ qp}}{\varepsilon_{pq...r}}\,\underset{\substack{\uparrow\uparrow\\ jq}}{a_{ip}}\,\underset{\substack{\uparrow\\ p}}{a_{nq}}\cdots a_{kr} + \cdots + \underset{\substack{\uparrow\\ k}}{\varepsilon_{ij...m}}\,\underset{\substack{\uparrow\ \uparrow\\ r\ p}}{\varepsilon_{pq...r}}\,\underset{\substack{\uparrow\uparrow\\ kr}}{a_{ip}}\,a_{jq}\cdots \underset{\substack{\uparrow\\ p}}{a_{nr}}).\end{aligned}$$

If we now exchange an index pair in each term except the first term in both epsilons, which does not change the value of the terms, we see that all M terms are equal. Thus we obtain

$$(\det \underset{\sim}{a})\delta_{mn} = \frac{1}{(M-1)!}\,\varepsilon_{mj...k}\,\varepsilon_{pq...r}\,a_{jq}\,\cdots\,a_{kr}\,a_{np}. \tag{b}$$

Next, we introduce

$$b_{mp} := \frac{1}{(M-1)!}\,\varepsilon_{mj...k}\,\varepsilon_{pq...r}\,a_{jq}\,\cdots\,a_{kr}$$

and we want to show that the so-defined b_{mp} is the cofactor to the element a_{mp}. For $m = p = 1$, we get

$$b_{11} = \frac{1}{(M-1)!}\,\varepsilon_{1j...k}\,\varepsilon_{1q...r}\,a_{jq}\,\cdots\,a_{kr}.$$

Since the number of the indices of the epsilons equals the length of the index set and since all epsilons in which an index appears at least twice are zero, we can substitute $\varepsilon_{j...k}$ for $\varepsilon_{1j...k}$ and $\varepsilon_{q...r}$ for $\varepsilon_{1q...r}$ if the $M-1$ indices of the epsilons range from 2 to M. We have

$$b_{11} = \frac{1}{(M-1)!}\,\varepsilon_{j...k}\,\varepsilon_{q...r}\,a_{jq}\,\cdots\,a_{kr}, \qquad \text{for all indices: } 2, \ldots, M.$$

But this is, according to (1.29), the minor to a_{11}, which is, according to (1.8), equal to the cofactor b_{11}. Analogously, we obtain for $m = 1, p = 2$,

$$b_{12} = \frac{1}{(M-1)!}\,\varepsilon_{1j\ldots k}\,\varepsilon_{2q\ldots r}\,a_{jq}\,\cdots\,a_{kr} = -\frac{1}{(M-1)!}\,\varepsilon_{1j\ldots k}\,\varepsilon_{q2\ldots r}\,a_{jq}\,\cdots\,a_{kr}.$$

We can again substitute $\varepsilon_{j\ldots k}$ for $\varepsilon_{1j\ldots k}$ and $\varepsilon_{q\ldots r}$ for $\varepsilon_{q2\ldots r}$ if the indices $j\ldots k$ can take values from 2 to M and the indices $q\ldots r$ can take values 1 and from 3 to M. Without the minus sign this is the minor to a_{12}; with the minus sign it is the cofactor b_{12}. Analogously, the same can be shown for the remaining elements of b_{mp}, so we get for the cofactor of the element a_{ip} of a determinant

$$b_{ip} = \frac{1}{(M-1)!}\,\varepsilon_{ij\ldots k}\,\varepsilon_{pq\ldots r}\,a_{jq}\,\cdots\,a_{kr}, \tag{1.30}$$

where the epsilons have M indices.

With this, from (b) we obtain

$$(\det \underset{\sim}{a})\delta_{mn} = b_{mp}\,a_{np}.$$

Since δ_{ij} is symmetric we can change the order of m and n on the right side, so we can also write

$$(\det \underset{\sim}{a})\delta_{mn} = a_{mp}\,b_{np}.$$

If we substitute $\varepsilon_{pq\ldots r}\,\delta_{mn}$ in (a) using $(1.27)_2$, we similarly obtain

$$(\det \underset{\sim}{a})\delta_{mn} = b_{im}\,a_{in} = a_{im}\,b_{in}.$$

So we have

$$(\det \underset{\sim}{a})\delta_{ij} = b_{ik}\,a_{jk} = a_{ik}\,b_{jk} = b_{ki}\,a_{kj} = a_{ki}\,b_{kj}. \tag{1.31}$$

By this way, we also proved (1.14).

If we take the determinant of (1.31) on both sides, then with (1.13) it follows that $(\det \underset{\sim}{a})^M = (\det \underset{\sim}{b})(\det \underset{\sim}{a})$ and thus for $\det \underset{\sim}{a} \neq 0$ we obtain

$$\det \underset{\sim}{b} = (\det \underset{\sim}{a})^{M-1}. \tag{1.32}$$

1.4.5 ε_{ijk}

1. We will need the $\varepsilon_{i\ldots j}$ mostly for the case $M = 3$ and we wish to write out some formulas for this case.

$$\varepsilon_{ijk} = \begin{cases} 1 & \text{for } ijk = 123 \text{ and its cyclic permutations 231 and 312,} \\ -1 & \text{for } ijk = 321 \text{ and its cyclic permutations 213 and 132,} \\ 0 & \text{for all remaining index combinations, i. e. if an index} \\ & \text{appears at least twice.} \end{cases} \tag{1.33}$$

We can easily convince ourselves that this definition is consistent with (1.23).

2. Clearly, ε_{ijk} is antimetric with respect to each index pair.

$$\varepsilon_{ijk} = \varepsilon_{jki} = \varepsilon_{kij} = -\varepsilon_{kji} = -\varepsilon_{jik} = -\varepsilon_{ikj}. \tag{1.34}$$

Problem 1.9.
Write out the following expressions for $N = 3$:

A. $v_i = \varepsilon_{ijk}\, \omega_j\, r_k$.
Let v_i, ω_i, and r_i be the Cartesian coordinates of the three vectors $\underline{v}$, $\underline{\omega}$, and $\underline{r}$. Which frequently used vector algebraic relation between $\underline{v}$, $\underline{\omega}$, and $\underline{r}$ is expressed by this equation?

B. $\Omega_j = \varepsilon_{ijk} \dfrac{\partial v_i}{\partial x_k}$.
Let v_i and Ω_i be the Cartesian coordinates of the two vector fields $\underline{v}$ and $\underline{\Omega}$, i. e. v_i and Ω_i are functions of the coordinates $x_i = (x_1, x_2, x_3) = (x, y, z)$. Which frequently used vector analytic relation between $\underline{\Omega}$ and $\underline{v}$ is expressed by this equation?

3. By equating two homologous indices (i. e. indices which are at the same position in both epsilons) we obtain from (1.26), after straightforward computations, the so-called Grassmann identity.

$$\varepsilon_{ijk}\, \varepsilon_{mnk} = \varepsilon_{ikj}\, \varepsilon_{mkn} = \varepsilon_{kij}\, \varepsilon_{kmn} = \delta_{im}\, \delta_{jn} - \delta_{in}\, \delta_{jm}. \tag{1.35}$$

Let us describe this important formula in words. Summing over two homologous indices in two epsilons (for $M = 3$) gives the difference of two terms which are both products of two deltas. In the first term of the difference (the minuend), the indices of both deltas are the homologous indices of the epsilons; in the second term (the subtrahend), the indices are not homologous indices of the epsilons.

Setting in (1.35) $n = j$ gives $\varepsilon_{ijk}\, \varepsilon_{mjk} = \delta_{im}\, \delta_{jj} - \delta_{ij}\, \delta_{jm} = 3\,\delta_{im} - \delta_{im} = 2\,\delta_{im}$,

$$\varepsilon_{ijk}\, \varepsilon_{mjk} = 2\,\delta_{im}. \tag{1.36}$$

If we further substitute $m = i$, we obtain

$$\varepsilon_{ijk}\,\varepsilon_{ijk} = 6. \tag{1.37}$$

4. The product of ε_{ijk} and δ_{pq} is, according to (1.27),

$$\begin{aligned}\varepsilon_{ijk}\,\delta_{pq} &= \varepsilon_{pjk}\,\delta_{iq} + \varepsilon_{ipk}\,\delta_{jq} + \varepsilon_{ijp}\,\delta_{kq}\\ &= \varepsilon_{qjk}\,\delta_{ip} + \varepsilon_{iqk}\,\delta_{jp} + \varepsilon_{ijq}\,\delta_{kp}.\end{aligned} \tag{1.38}$$

1.5 Matrices

1.5.1 Definitions

A rectangular table of numbers is called a matrix. The numbers are the elements of the matrix, with elements aligned horizontally forming the rows and elements aligned vertically forming the columns. A matrix with M rows and N columns is also called an M, N-matrix. A matrix with only one row is called a row matrix, a matrix with only one column is called a column matrix, a matrix with N rows and N columns is called a square matrix of order N. A matrix which can be obtained from another matrix by dropping an arbitrary number of rows and columns is called a submatrix of the other matrix.

We denote a matrix by a letter with an underlying tilde and its elements by the same letter with two indices, where the first index refers to the row and the second index refers to the column. We write

$$\begin{pmatrix} A_{11} & A_{12} & \dots & A_{1N} \\ A_{21} & A_{22} & \dots & A_{2N} \\ \vdots & & & \\ A_{M1} & A_{M2} & \dots & A_{MN} \end{pmatrix}.$$

We call the element A_{ij} at the intersection of the i-th row and the j-th column the i, j-element of the matrix. If we take both indices as running indices and let i range from 1 to M and j from 1 to N, then A_{ij} is a symbol for the entire matrix. So we have three equivalent ways to write this:

$$\begin{pmatrix} A_{11} & A_{12} & \dots & A_{1N} \\ A_{21} & A_{22} & \dots & A_{2N} \\ \vdots & & & \\ A_{M1} & A_{M2} & \dots & A_{MN} \end{pmatrix} = A_{ij} = \underset{\sim}{A}. \tag{1.39}$$

We call a matrix whose elements are all zero a zero matrix $\underset{\sim}{0}$, and we call a square matrix whose only nonzero elements are elements with equal indices (the elements on the main diagonal) a diagonal matrix or a diagonalized matrix. A diagonal matrix in which the elements on the main diagonal are all one is called an identity matrix or a one matrix and is denoted by $\underset{\sim}{I}$. Clearly the elements of the identity matrix are the δ_{ij}, so we will refer to them by this symbol.

The determinant computed from the elements of a square matrix is called the determinant of a matrix. The cofactors corresponding to the elements of this determinant are also called cofactors of the elements of the matrix. A determinant computed from the elements of a submatrix which was obtained from a not necessarily square matrix by deleting an arbitrary number of rows and columns is called a subdeterminant of the matrix.

A matrix has rank R, if it has at least one nonzero subdeterminant of order R and no higher-order nonzero subdeterminant. By Theorem III from Section 1.3.3 on determinants a matrix with rank R has exactly R linearly independent rows and exactly R linearly independent columns. We denote the rank of the matrix $\underset{\sim}{A}$ by rank $\underset{\sim}{A}$.

A square matrix of order N is called regular if it has rank N and it is called singular if it has a lower rank; it is called M-times singular if it has the rank $N - M$, or in other words, if it has the rank defect M.

1.5.2 Arithmetic Operations and First Conclusions

1. A matrix is a special case of an N-tuple; the easiest way to see this is to imagine all rows of a matrix written one after another. Now we can define linear operations and the terms linearly independent and linearly dependent for matrices.

- *Equality*: Two matrices $\underset{\sim}{A}$ and $\underset{\sim}{B}$ are equal if and only if their homologous elements are equal:

$$\underset{\sim}{A} = \underset{\sim}{B} \quad \Longleftrightarrow \quad A_{ij} = B_{ij}. \tag{1.40}$$

- *Addition (subtraction)*: A matrix $\underset{\sim}{C}$ is the sum (difference) of two matrices $\underset{\sim}{A}$ and $\underset{\sim}{B}$ if and only if the same is true for the homologous elements of the three matrices:

$$\underset{\sim}{A} \pm \underset{\sim}{B} = \underset{\sim}{C} \quad \Longleftrightarrow \quad A_{ij} \pm B_{ij} = C_{ij}. \tag{1.41}$$

Clearly equality, addition, and subtraction of matrices are only possible if the matrices have the same number of rows and columns.

- *Multiplication of a matrix by a number*: To multiply a matrix $\underset{\sim}{A}$ by a number α, we multiply every element of the matrix by this number:

$$\alpha \underset{\sim}{A} = \underset{\sim}{B} \quad \Longleftrightarrow \quad \alpha A_{ij} = B_{ij}. \tag{1.42}$$

– *Linear independence*: Q M, N-matrices $\overset{1}{\underset{\sim}{A}}, \overset{2}{\underset{\sim}{A}}, \ldots, \overset{Q}{\underset{\sim}{A}}$ are called linearly independent, if the equation

$$\alpha_i \overset{i}{\underset{\sim}{A}} = \underset{\sim}{0} \quad \Longleftrightarrow \quad \alpha_i \overset{i}{A}_{mn} = 0 \tag{1.43}$$

has only the trivial solution, where all α_i are zero. Otherwise the matrices are called linearly dependent.

Since the equations (1.43) represent MN homogeneous linear equations for the Q unknown α_i, more than MN matrices are always linearly dependent.

2. Now we define *multiplication* of two matrices. A matrix $\underset{\sim}{C}$ is the product of two matrices $\underset{\sim}{A}$ and $\underset{\sim}{B}$ if and only if their elements satisfy the relation $C_{ik} = A_{ij}\, B_{jk}$:

$$\underset{\sim}{A}\,\underset{\sim}{B} = \underset{\sim}{C} \quad \Longleftrightarrow \quad A_{ij}\, B_{jk} = C_{ik}. \tag{1.44}$$

Multiplication of matrices requires that the number of columns of the first factor is equal to the number of rows of the second factor. The so-called Falk scheme is a convenient way to perform the multiplication.

For $A_{ij}\, B_{jk} = C_{ik}$:

$$\begin{array}{cccc|ccc}
 & & & & B_{11} & B_{12} & B_{13} \\
 & & & & B_{21} & B_{22} & B_{23} \\
 & & & & B_{31} & B_{32} & B_{33} \\
 & & & & B_{41} & B_{42} & B_{43} \\
\hline
A_{11} & A_{12} & A_{13} & A_{14} & C_{11} & C_{12} & C_{13} \\
A_{21} & A_{22} & A_{23} & A_{24} & C_{21} & C_{22} & C_{23}
\end{array}$$

For $\underbrace{A_{ij}\, B_{jk}}_{F_{ik}}\, C_{kl} = D_{il}$:

$$\begin{array}{cccc|ccc|cccc}
 & & & & B_{11} & B_{12} & B_{13} & & & & \\
 & & & & B_{21} & B_{22} & B_{23} & C_{11} & C_{12} & C_{13} & C_{14} \\
 & & & & B_{31} & B_{32} & B_{33} & C_{21} & C_{22} & C_{23} & C_{24} \\
 & & & & B_{41} & B_{42} & B_{43} & C_{31} & C_{32} & C_{33} & C_{34} \\
\hline
A_{11} & A_{12} & A_{13} & A_{14} & F_{11} & F_{12} & F_{13} & D_{11} & D_{12} & D_{13} & D_{14} \\
A_{21} & A_{22} & A_{23} & A_{24} & F_{21} & F_{22} & F_{23} & D_{21} & D_{22} & D_{23} & D_{24}
\end{array}$$

The product of two matrices is not commutative in general; in the example given above $\underset{\sim}{B}\,\underset{\sim}{A}$ is not even defined. The product of a matrix and the identity matrix is, however, the original matrix, independent of the order of the factors:

$$\underset{\sim}{A}\,\underset{\sim}{I} = \underset{\sim}{I}\,\underset{\sim}{A} = \underset{\sim}{A} \quad \Longleftrightarrow \quad A_{ij}\, \delta_{jk} = \delta_{ij}\, A_{jk} = A_{ik}. \tag{1.45}$$

When the matrix $\underset{\sim}{A}$ is rectangular, however, the order of the identity matrix differs for the two cases: if we multiply $\underset{\sim}{A}$ by the identity matrix from the right, then the order of the identity matrix equals the number of columns of $\underset{\sim}{A}$; if we multiply $\underset{\sim}{A}$ by the identity matrix from the left, then the order of the identity matrix equals the number of rows of $\underset{\sim}{A}$.

If $\underset{\sim}{A}$ is square and regular, the converse is also true. If the result of the product (1.45) of a regular square matrix $\underset{\sim}{A}$ and an (initially unknown) matrix $\underset{\sim}{I}$ is again the original matrix $\underset{\sim}{A}$, then $\underset{\sim}{I}$ is the identity matrix.

This is easy to see: If $\underset{\sim}{A}$ is square, then according to (1.44), $\underset{\sim}{I}$ also has to be square and has the same order as $\underset{\sim}{A}$, say N. Then (1.45) forms a system of N^2 nonhomogeneous linear equations for the N^2 elements of $\underset{\sim}{I}$. If $\underset{\sim}{A}$ is regular, this equation system clearly always has a unique solution. Since (1.45) is satisfied for every matrix $\underset{\sim}{A}$ by the identity matrix $\underset{\sim}{I}$ and since it has a unique solution if $\underset{\sim}{A}$ is a regular square matrix, the identity matrix is the only solution.

The product of more than two matrices is associative.

3. Next we define the *transpose* of a matrix. To any matrix $\underset{\sim}{A}$ we can associate a matrix $\underset{\sim}{A}^{\mathrm{T}}$, which is obtained from the original matrix by exchanging the rows and columns; the matrix $\underset{\sim}{A}^{\mathrm{T}}$ is called the transpose matrix of $\underset{\sim}{A}$ and we read "A transpose". We denote the i,j-element of the matrix $\underset{\sim}{A}^{\mathrm{T}}$ (not: the transpose of the i,j-element[2] of the matrix $\underset{\sim}{A}$) by $(A_{ij})^{\mathrm{T}}$ or A^{T}_{ij}. Then

$$(A_{ij})^{\mathrm{T}} \equiv A^{\mathrm{T}}_{ij} := A_{ji}, \tag{1.46}$$

the i,j-element of the matrix $\underset{\sim}{A}^{\mathrm{T}}$ is equal to the j,i-element of the matrix $\underset{\sim}{A}$.

Clearly $(A^{\mathrm{T}}_{ij})^{\mathrm{T}} = A^{\mathrm{T}}_{ji} = A_{ij}$, i. e. the transpose of the transpose matrix is the original matrix:

$$(\underset{\sim}{A}^{\mathrm{T}})^{\mathrm{T}} = \underset{\sim}{A} \quad \Longleftrightarrow \quad (A^{\mathrm{T}}_{ij})^{\mathrm{T}} = A_{ij}. \tag{1.47}$$

$A_{im}\,B_{mj}$ is the i,j-element of the matrix $\underset{\sim}{A}\;\underset{\sim}{B}$ and $(A_{im}\,B_{mj})^{\mathrm{T}}$ is the i,j-element of the matrix $(\underset{\sim}{A}\,\underset{\sim}{B})^{\mathrm{T}}$, and so with (1.46) we get

$$(A_{im}\,B_{mj})^{\mathrm{T}} = A_{jm}\,B_{mi} = B_{mi}\,A_{jm} = B^{\mathrm{T}}_{im}\,A^{\mathrm{T}}_{mj}.$$

On the far left is the i,j-element[3] of the matrix $(\underset{\sim}{A}\,\underset{\sim}{B})^{\mathrm{T}}$, next to it is the j,i-element of the matrix $\underset{\sim}{A}\;\underset{\sim}{B}$, the next term cannot be interpreted as an element of a matrix, and

---

2 The transpose of a matrix element is not defined. So we should really read A^{T}_{ij} as "A transpose ij" and not "A ij transpose".

3 We will read this as $(A_{im}\,B_{mj})^{\mathrm{T}}$ "A im B mj (in parentheses) transpose", but let us keep in mind that we really mean the i,j-element of the matrix $(\underset{\sim}{A}\,\underset{\sim}{B})^{\mathrm{T}}$.

the far-right term is the i, j-element of the matrix $\underset{\sim}{B}^{\mathrm{T}}\,\underset{\sim}{A}^{\mathrm{T}}$. Comparing the far-left and the far-right expressions shows that the i, j-elements of the matrices $(\underset{\sim}{A}\,\underset{\sim}{B})^{\mathrm{T}}$ and $\underset{\sim}{B}^{\mathrm{T}}\,\underset{\sim}{A}^{\mathrm{T}}$ are equal, and since this is true for all i and j these two matrices are equal. We have now derived the important formula

$$(\underset{\sim}{A}\,\underset{\sim}{B})^{\mathrm{T}} = \underset{\sim}{B}^{\mathrm{T}}\,\underset{\sim}{A}^{\mathrm{T}} \quad \Longleftrightarrow \quad (A_{im}\,B_{mj})^{\mathrm{T}} = B^{\mathrm{T}}_{im}\,A^{\mathrm{T}}_{mj}. \tag{1.48}$$

4. Another operation defined for square matrices is the *inversion* of a matrix. If we can associate to a square matrix $\underset{\sim}{A}$ a matrix $\underset{\sim}{A}^{-1}$, such that the product of the two matrices is the identity matrix $\underset{\sim}{I}$,

$$\underset{\sim}{A}\,\underset{\sim}{A}^{-1} = \underset{\sim}{I} \quad \Longleftrightarrow \quad A_{ij}\,A^{(-1)}_{jk} = \delta_{ik}, \tag{1.49}$$

then the matrix $\underset{\sim}{A}^{-1}$ is called the inverse matrix[4] of $\underset{\sim}{A}$.

If $\underset{\sim}{A}$ is known, then (1.49) represents a nonhomogeneous system of linear equations for the elements of the matrix $\underset{\sim}{A}^{-1}$. According to the rules for systems of linear equations we can solve the system for $A^{(-1)}_{ij}$ if and only if $\underset{\sim}{A}$ is regular and then the solution is unique. The solution can be derived from (1.31). To see this, we multiply $(\det \underset{\sim}{A})\,\delta_{ij} = B_{ki}\,A_{kj}$ by $A^{(-1)}_{jm}$:

$$(\det \underset{\sim}{A})\,\delta_{ij}\,A^{(-1)}_{jm} = B_{ki}\,A_{kj}\,A^{(-1)}_{jm} = B_{ki}\,\delta_{km},$$
$$(\det \underset{\sim}{A})\,A^{(-1)}_{im} = B_{mi},$$

$$A^{(-1)}_{ij} = \frac{B_{ji}}{\det \underset{\sim}{A}}, \tag{1.50}$$

where B_{ij} is the cofactor of A_{ij}. (Note the different positions of i and j on the two sides of this equation!)

We conclude that a matrix is invertible if and only if it is square and regular. In practice we usually use the Gauss–Jordan algorithm to compute the inverse matrix rather than (1.50).

Finally, from (1.49) it also follows that

$$\underset{\sim}{A}^{-1}\,\underset{\sim}{A} = \underset{\sim}{I} \quad \Longleftrightarrow \quad A^{(-1)}_{ij}\,A_{jk} = \delta_{ik}, \tag{1.51}$$

which means that the inverse of $\underset{\sim}{A}^{-1}$ is again $\underset{\sim}{A}$, i. e.

$$(\underset{\sim}{A}^{-1})^{-1} = \underset{\sim}{A}. \tag{1.52}$$

4 $A^{(-1)}_{ij}$ is the i, j-element of the matrix $\underset{\sim}{A}^{-1}$ and we should read "A inverse ij". Similarly as the transpose matrix, the inverse of a matrix is defined, and the inverse of a matrix element is not defined.

To see this we multiply (1.49) from the right by $\underset{\sim}{A}$:

$$\underset{\sim}{A}\,\underset{\sim}{A}^{-1}\,\underset{\sim}{A} = \underset{\sim}{I}\,\underset{\sim}{A} = \underset{\sim}{A}.$$

If we write for a moment the product $\underset{\sim}{A}^{-1}\,\underset{\sim}{A}$ as $\underset{\sim}{C}$, the last equation becomes

$$\underset{\sim}{A}\,\underset{\sim}{C} = \underset{\sim}{A},$$

and since $\underset{\sim}{A}$ is square and regular, it follows from the converse of (1.45) that $\underset{\sim}{C}$ is the identity matrix, which proves (1.51) and (1.52). Substituting (1.50) into $(1.49)_2$ and $(1.51)_2$ we finally obtain

$$(\det \underset{\sim}{A})\delta_{ik} = A_{ij}\,B_{kj} = B_{ji}\,A_{jk}; \tag{1.53}$$

compare also with (1.31).

Problem 1.10.
Compute the inverse of the following matrices and verify your result, i. e. show that the product of the original matrix and the computed inverse is the identity matrix (compare also with Section 1.6.2):

$$\text{A.}\quad \begin{pmatrix} 1 & 0 & 2 \\ 2 & 1 & 3 \\ 1 & 1 & 2 \end{pmatrix}, \qquad \text{B.}\quad \begin{pmatrix} 0 & 1 & 1 & -1 \\ 1 & 2 & -1 & 1 \\ 1 & 3 & -1 & 0 \\ -1 & -2 & 1 & -2 \end{pmatrix}.$$

Problem 1.11.

A. Show that the operations transpose and inverse of a matrix commute, i. e. show that

$$(\underset{\sim}{A}^{-1})^{\mathrm{T}} = (\underset{\sim}{A}^{\mathrm{T}})^{-1} =: \underset{\sim}{A}^{-T}. \tag{1.54}$$

B. Show that analogously to (1.48) the following holds:

$$(\underset{\sim}{A}\,\underset{\sim}{B})^{-1} = \underset{\sim}{B}^{-1}\,\underset{\sim}{A}^{-1}. \tag{1.55}$$

1.5.3 Equations Involving Matrices and Equations Involving Matrix Elements

We distinguished between matrices (which we denoted by letters with an underlying tilde) and their elements (which we denoted by letters with two indices) and we defined arithmetic operations for matrices (which are not numbers) by writing out the arithmetic operations which have to be carried out between their elements (which are numbers). By this, we found two notations for matrix computations: Every equation for matrices can be written as an equation for the elements of the matrices, or in other words, every equation for matrices can be translated into an equivalent equation for matrix elements. From the definition of matrix equality (1.40) it follows that an equation for matrix elements has the property that the order of free indices coincides in all its terms. Conversely, an equation for matrix elements can only be translated into an equation for matrices, when the order of free indices coincides in all terms. This is not the case in (1.46) and (1.50), for example.

We will meet the same facts again when we compute with tensors.

1.5.4 Elementary Transformations, Normal Form, Equivalent Matrices, Similar Matrices

1. The following transformations are called elementary transformations of an M, N-matrix:
- exchange of two rows or two columns;
- multiplication of a row or a column by a nonzero number;
- addition of a linear combination of the remaining rows (columns) to a row (column).

Theorems II, V, Ia, and Ib for determinants immediately imply that elementary transformations do not change the rank of a matrix.

2. The normal form of any M, N-matrix with rank R is defined as the matrix

$$D = \left(\begin{array}{cccc|ccc} \multicolumn{4}{c}{\overbrace{\hphantom{1 \quad 0 \quad \dots \quad 0}}^{R}} & \multicolumn{3}{c}{\overbrace{\hphantom{0 \quad \dots \quad 0}}^{N-R}} \\ 1 & 0 & \dots & 0 & 0 & \dots & 0 \\ 0 & 1 & \dots & 0 & 0 & \dots & 0 \\ \vdots & & & & \vdots & & \\ 0 & 0 & \dots & 1 & 0 & \dots & 0 \\ \hline 0 & 0 & \dots & 0 & 0 & \dots & 0 \\ \vdots & & & & \vdots & & \\ 0 & 0 & \dots & 0 & 0 & \dots & 0 \end{array}\right) \begin{array}{l} \left.\vphantom{\begin{array}{c}1\\0\\\vdots\\0\end{array}}\right\} R \\ \left.\vphantom{\begin{array}{c}0\\\vdots\\0\end{array}}\right\} M-R \end{array} . \tag{1.56}$$

Obviously, every M, N-matrix of rank R can be transformed into this normal form by elementary transformations and every M, N-matrix, which can be transformed into this normal form by elementary transformations, has rank R, so the name normal form makes sense.

Problem 1.12.
Determine the rank of the matrices (see also Section 1.6.3)

$$\text{A.}\ \begin{pmatrix} 0 & 3 & 6 & 9 & -3 \\ 0 & 0 & 2 & -4 & 8 \\ 1 & 2 & -3 & 4 & 5 \\ 1 & 5 & 5 & 9 & 10 \end{pmatrix}, \qquad \text{B.}\ \begin{pmatrix} 1 & -3 & 2 & 0 \\ 4 & -11 & 10 & -1 \\ -2 & 8 & -5 & 3 \end{pmatrix}.$$

3. The elements of a set are equivalent and form an equivalence class, if there exists a relation $A \sim B$ between two arbitrary elements A and B of the set with the following properties. The relation is
- reflexive, i. e. $A \sim A$ (the relation exists also for $B = A$);
- symmetric, i. e. if $A \sim B$ then $B \sim A$;
- transitive, i. e. if $A \sim B$ and $B \sim C$ then $A \sim C$.

Such a relation is called an equivalence relation on the set.

4. It can be shown that for any pair $\underset{\sim}{A}$ and $\underset{\sim}{B}$ of M, N-matrices of equal rank R, there exists a pair of regular square matrices $\underset{\sim}{S}$ and $\underset{\sim}{T}$, such that

$$\underset{\sim}{A} = \underset{\sim}{S}\,\underset{\sim}{B}\,\underset{\sim}{T} \quad \Longleftrightarrow \quad A_{ij} = S_{im}\, B_{mn}\, T_{nj}, \tag{1.57}$$

where clearly the matrix $\underset{\sim}{S}$ has order M and the matrix $\underset{\sim}{T}$ has order N.

Problem 1.13.
Show that (1.57) is an equivalence relation for the two matrices $\underset{\sim}{A}$ and $\underset{\sim}{B}$.

All matrices with the same number of rows, the same number of columns, and the same rank form an equivalence class. Two matrices with this property (or in other words, two matrices $\underset{\sim}{A}$ and $\underset{\sim}{B}$ satisfying the relation $\underset{\sim}{A} = \underset{\sim}{S}\,\underset{\sim}{B}\,\underset{\sim}{T}$, where $\underset{\sim}{S}$ and $\underset{\sim}{T}$ are regular) are called equivalent.

5. In particular, if two equivalent matrices $\underset{\sim}{A}$ and $\underset{\sim}{B}$ are square, then $\underset{\sim}{S}$ and $\underset{\sim}{T}$ have the same number of rows and columns. If, in addition, $\underset{\sim}{S}$ and $\underset{\sim}{T}$ are inverses of each other, then

$$\underset{\sim}{A} = \underset{\sim}{S}\,\underset{\sim}{B}\,\underset{\sim}{S}^{-1}, \tag{1.58}$$

and we call the matrices $\underset{\sim}{A}$ and $\underset{\sim}{B}$ similar.

1.5.5 Orthogonal Matrices

1. A square matrix $\underset{\sim}{A}$ is called orthogonal, if its inverse equals its transpose. Then $A_{ik}\,A^{\mathrm{T}}_{kj} = A_{ik}\,A_{jk} = \delta_{ij}$, and at the same time $A^{\mathrm{T}}_{ik}\,A_{kj} = A_{ki}\,A_{kj} = \delta_{ij}$. Thus we get the following equivalent notations for orthogonal matrices:

$$\begin{aligned} &\underset{\sim}{A}^{\mathrm{T}} = \underset{\sim}{A}^{-1}, &\qquad &\underset{\sim}{A}\,\underset{\sim}{A}^{\mathrm{T}} = \underset{\sim}{A}^{\mathrm{T}}\,\underset{\sim}{A} = \underset{\sim}{I}, \\ &A_{ik}\,A_{jk} = \delta_{ij}, &\qquad &A_{ki}\,A_{kj} = \delta_{ij}. \end{aligned} \tag{1.59}$$

The last two equations show that the sum of the squares of each row (or column) is one and the sum of the products of two different rows (or two different columns) is zero.

2. Orthogonal matrices also satisfy

$$\det A_{ij} = \pm 1. \tag{1.60}$$

The easiest way to prove this is to expand $(\det A_{ij})^2$ using the rules for multiplication of determinants and then transpose the second determinant. Then we have

$$\begin{vmatrix} A_{11} & A_{12} & \dots & A_{1N} \\ A_{21} & A_{22} & \dots & A_{2N} \\ \vdots & & & \\ A_{N1} & A_{N2} & \dots & A_{NN} \end{vmatrix} \begin{vmatrix} A_{11} & A_{21} & \dots & A_{N1} \\ A_{12} & A_{22} & \dots & A_{N2} \\ \vdots & & & \\ A_{1N} & A_{2N} & \dots & A_{NN} \end{vmatrix}$$

$$= \begin{vmatrix} A_{1i}\,A_{1i} & A_{1i}\,A_{2i} & \dots & A_{1i}\,A_{Ni} \\ A_{2i}\,A_{1i} & A_{2i}\,A_{2i} & \dots & A_{2i}\,A_{Ni} \\ \vdots & & & \\ A_{Ni}\,A_{1i} & A_{Ni}\,A_{2i} & \dots & A_{Ni}\,A_{Ni} \end{vmatrix}.$$

According to (1.59) this is equal to $\det \delta_{ij} = 1$, which completes the proof.

Note that the converse is not true: From $\det A_{ij} = \pm 1$ it does not follow that A_{ij} is orthogonal.

3. If the determinant of an orthogonal matrix has the value +1, we call the matrix proper orthogonal. If the determinant has the value −1, we call the matrix proper orthogonal.[5]

---

5 A more common name for proper orthogonal is special orthogonal, but there is no corresponding commonly accepted name for $\det A_{ij} = -1$ matrices, so we prefer the names proper and improper.

1.6 Algorithms

In this section we summarize some fundamental algorithms from linear algebra without proof.

A determinant or a matrix is transformed column by column from left to right. We describe the steps for an arbitrary column. The row which has the same diagonal element as the column which is being transformed is called the corresponding row. The steps of the algorithms are carried out subsequently, while omitting the steps which do not satisfy the required assumptions.

1.6.1 Computing the Value of a Determinant

The aim is to transform a determinant into a triangular determinant. Then we compute the value of the determinant as the product of the elements on the main diagonal.

I. If the diagonal element and all elements below this diagonal element are zero, the determinant has the value zero. (stop)

II. If the diagonal element is zero, exchange the corresponding row with a row below to make the diagonal element nonzero by taking into account sign changes of the determinant.

III. Make the elements below the diagonal element zero by adding to each row a convenient multiple of the corresponding row.

1.6.2 Solution of Systems of Linear Equations with the Same Regular Coefficient Matrix ("Division by a Regular Matrix", Gauss–Jordan Algorithm)

The aim is to compute the matrix $\underset{\sim}{X}$ in the equation $\underset{\sim}{A}\,\underset{\sim}{X} = \underset{\sim}{B}$, where $\underset{\sim}{A}$ is a regular (and hence a square) matrix. (According to the rules for matrix multiplication, $\underset{\sim}{A}$, $\underset{\sim}{X}$, and $\underset{\sim}{B}$ have the same number of rows, and $\underset{\sim}{X}$ and $\underset{\sim}{B}$ have the same number of columns.)

If $\underset{\sim}{X}$ has only one column (so $\underset{\sim}{B}$ also has only one column), $\underset{\sim}{A}\,\underset{\sim}{X} = \underset{\sim}{B}$ represents a system of linear equations, which has a unique solution. If $\underset{\sim}{X}$ has two columns, (so $\underset{\sim}{B}$ also has two columns), $\underset{\sim}{A}\,\underset{\sim}{X} = \underset{\sim}{B}$ represents two systems of linear equations with the same coefficient matrix and both equation systems have unique solutions, etc. If $\underset{\sim}{B}$ is the identity matrix, then $\underset{\sim}{X}$ is the inverse matrix of $\underset{\sim}{A}$.

$\underset{\sim}{A}$	$\underset{\sim}{B}$
$\vdots$	$\vdots$
$\underset{\sim}{I}$	$\underset{\sim}{X}$

If we introduced the concept of division by a matrix, we could describe the solution to this problem as division by a regular matrix. To solve the problem, we write the two known matrices, $\underset{\sim}{A}$ and $\underset{\sim}{B}$, side by side and transform this double matrix using elementary row transformations until $\underset{\sim}{A}$ becomes the identity matrix. This will transform $\underset{\sim}{B}$ into the desired matrix. The described method is called the Gauss–Jordan algorithm.

I. If the diagonal element and all elements below the diagonal element of the left side of the double matrix are zero, then $\underset{\sim}{A}$ is singular. (stop)
II. If the diagonal element is zero, exchange the corresponding row with a row below to make the diagonal element nonzero.
III. Divide the corresponding row by the value of the diagonal element.
IV. Make the elements above and below the diagonal element zero by adding to each row a convenient multiple of the corresponding row.

1.6.3 Determining the Rank of a Matrix or a Determinant

The aim is the transformation of the matrix or of the elements of the determinant into normal form. Then the rank of the matrix or the determinant is equal to the number of nonzero (diagonal) elements.

I. If the diagonal element and all elements below the diagonal are zero, all elements of this column are zero. Exchange the column with the most-right column whose elements are not all zero.
II. If the diagonal element is zero but not all elements below the diagonal are zero, then exchange the corresponding row with a row below to make the diagonal element nonzero.
III. Make the elements below the diagonal element zero by adding to each row a convenient multiple of the corresponding row.
IV. Replace the elements to the right of the diagonal element by zeros. (Make these elements zero by adding to each column a convenient multiple of the column which is being transformed.)
V. Replace the diagonal element by the number one. (Divide the corresponding row by the value of the diagonal element.)

2 Tensor Analysis in Symbolic Notation and in Cartesian Coordinates

We are now prepared to explore in the rest of the book the analysis of tensors, which can be quantitatively represented in Cartesian coordinate systems in the space we live in, as is the case with tensors which represent physical quantities.

Since we live in a three-dimensional space, we only need the formulas of Chapter 1 for $N = 3$. In Chapter 2, we develop tensor analysis in Cartesian coordinates. In Chapter 3, we focus on special properties of the simplest tensors, besides scalars and vectors, namely second-order tensors. In Chapter 4, we will extend tensor analysis to include curvilinear coordinates. In Chapter 5, we introduce a method called theory of the representation of tensor functions, which, similarly to dimensional analysis, can be used to reduce the number of possible combinations of physical quantities in an equation.

2.1 Cartesian Coordinates, Points, Position Vectors

2.1.1 Position Vectors and Point Coordinates

To describe a point in our three-dimensional world quantitatively, we need a coordinate system. In this chapter we consider only Cartesian coordinate systems. A Cartesian coordinate system consists of an origin and three pairwise perpendicular unit vectors fixed at the origin. We call these unit vectors a (Cartesian) basis and denote them by $(\underline{e}_x, \underline{e}_y, \underline{e}_z)$ or by $(\underline{e}_1, \underline{e}_2, \underline{e}_3)$ or, remembering the summation convention, briefly by $\underline{e}_i$.

We can then identify a point in a given Cartesian coordinate system either by its coordinates $x_i = (x_1, x_2, x_3) = (x, y, z)$ or by the position vector $\underline{x}$ from the origin of the coordinate system to this point, so we have the following alternative notations:

$$\begin{aligned} \underline{x} &= x\,\underline{e}_x + y\,\underline{e}_y + z\,\underline{e}_z \\ &= x_1\,\underline{e}_1 + x_2\,\underline{e}_2 + x_3\,\underline{e}_3 \\ &= x_i\,\underline{e}_i. \end{aligned} \tag{2.1}$$

2.1.2 Transformations of Cartesian Coordinate Systems

Consider two Cartesian coordinate systems, in an arbitrary orientation relative to each other, as shown in the following figure (to keep it simple, the figure is only two-dimensional, but the considerations are also valid in three dimensions). We label one of these coordinate systems by a tilde, to distinguish one from another, and we call it

https://doi.org/10.1515/9783110404265-002

the tilde coordinate system. We call the other coordinate system the no-tilde coordinate system. The relative position of the two coordinate systems is then characterized by

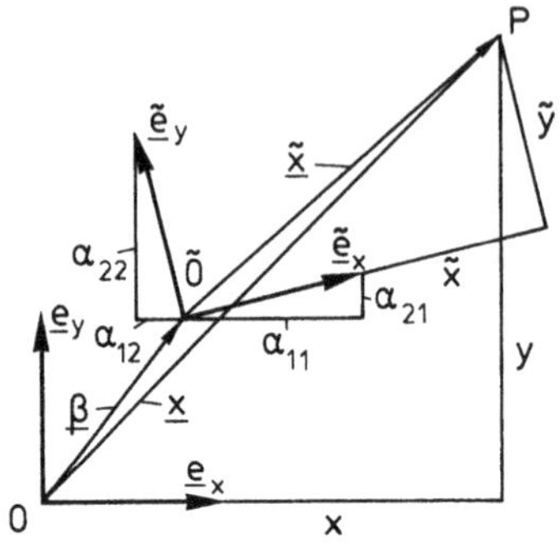

I. The origin $\widetilde{O}$ of the tilde coordinate system has the coordinates $(\beta_x, \beta_y, \beta_z)$ in the no-tilde coordinate system, so its position vector is

$$\underline{\beta} = \beta_x\, \underline{e}_x + \beta_y\, \underline{e}_y + \beta_z\, \underline{e}_z = \beta_i\, \underline{e}_i. \tag{2.2}$$

II. The basis vectors $\widetilde{\underline{e}}_i$ of the tilde coordinate system are represented in the no-tilde system as

$$\begin{aligned} \widetilde{\underline{e}}_x &= \alpha_{11}\, \underline{e}_x + \alpha_{21}\, \underline{e}_y + \alpha_{31}\, \underline{e}_z = \alpha_{i1}\, \underline{e}_i, \\ \widetilde{\underline{e}}_y &= \alpha_{12}\, \underline{e}_x + \alpha_{22}\, \underline{e}_y + \alpha_{32}\, \underline{e}_z = \alpha_{i2}\, \underline{e}_i, \\ \widetilde{\underline{e}}_z &= \alpha_{13}\, \underline{e}_x + \alpha_{23}\, \underline{e}_y + \alpha_{33}\, \underline{e}_z = \alpha_{i3}\, \underline{e}_i, \\ \widetilde{\underline{e}}_j &= \alpha_{ij}\, \underline{e}_i. \end{aligned} \tag{2.3}$$

The transformation between the no-tilde and the tilde coordinate systems is clearly completely described by the β_i and α_{ij}. The α_{ij} are called the transformation coefficients or the transformation matrix.

2.1.3 Properties of the Transformation Coefficients

1. Let $\overset{j}{\widetilde{e}}_i$ be the coordinates of the vector $\widetilde{\underline{e}}_j$ in the basis $\underline{e}_i$, i. e. let $\widetilde{\underline{e}}_j = \overset{j}{\widetilde{e}}_i\, \underline{e}_i$. Comparing this expression with (2.3) yields

$$\alpha_{ij} = \overset{j}{\widetilde{e}}_i, \tag{2.4}$$

i. e. α_{ij} is the i-coordinate of the vector $\widetilde{\underline{e}}_j$ represented in the basis $\underline{e}_i$.

2. Another geometric interpretation of the α_{ij} follows from scalar multiplication of (2.3) by $\underline{e}_k$:[1]

$$\underline{\tilde{e}}_j \cdot \underline{e}_k = \alpha_{ij}\, \underline{e}_i \cdot \underline{e}_k = \alpha_{ij}\, \delta_{ik} = \alpha_{kj}.$$

Since $\underline{\tilde{e}}_j \cdot \underline{e}_k$ is the cosine of the angle between $\underline{\tilde{e}}_j$ and $\underline{e}_k$, we get

$$\alpha_{ij} = \cos(\underline{e}_i, \underline{\tilde{e}}_j), \tag{2.5}$$

where the α_{ij} are the so-called direction cosines. We see immediately that the indices of α_{ij} are not interchangeable; α_{ij} is not the same as α_{ji}. For example, in the figure on the facing page, $\underline{e}_x$ and $\underline{\tilde{e}}_y$ form an obtuse angle, and the corresponding direction cosine is therefore negative; $\underline{e}_y$ and $\underline{\tilde{e}}_x$ form an acute angle, and the corresponding direction cosine is therefore positive.

3. Since both the $\underline{e}_i$ and the $\underline{\tilde{e}}_i$ are pairwise perpendicular unit vectors, they satisfy the relations

$$\underline{e}_i \cdot \underline{e}_j = \delta_{ij}, \quad \underline{\tilde{e}}_i \cdot \underline{\tilde{e}}_j = \delta_{ij}.$$

With (2.3), this gives

$$\delta_{ij} = \underline{\tilde{e}}_i \cdot \underline{\tilde{e}}_j = \alpha_{mi}\, \underline{e}_m \cdot \alpha_{nj}\, \underline{e}_n = \alpha_{mi}\, \alpha_{nj}\, \underline{e}_m \cdot \underline{e}_n = \alpha_{mi}\, \alpha_{nj}\, \delta_{mn} = \alpha_{mi}\, \alpha_{mj},$$

i. e. according to (1.59), the transformation coefficients form an orthogonal matrix, and we have

$$\alpha_{ik}\, \alpha_{jk} = \delta_{ij}, \qquad \alpha_{ki}\, \alpha_{kj} = \delta_{ij}. \tag{2.6}$$

These two equivalent relations are called orthogonality relations.

4. Orthogonal matrices satisfy, according to (1.60),

$$\det \underset{\sim}{\alpha} = \pm 1. \tag{2.7}$$

For a geometric visualization of these two cases, we first assume that both coordinate systems have the same orientation: either both are right-handed systems or both are left-handed systems. We can then convert one system into the other system by a motion.[2]

1 Assuming familiarity with the scalar product of two vectors from elementary vector analysis.

2 By a motion, we mean a composition of a rotation and a translation (but not a reflection). Some authors call this a direct, proper, or rigid motion.

If the coordinate transformation can be interpreted as a motion, we can associate a transformation matrix $\alpha_{ij}(t)$ to every point in time t between the initial time t_0 and the final time t_1 (when both coordinate systems coincide), i. e. for all t with $t_0 \leqq t \leqq t_1$. These matrices, and consequently their determinants, are continuous functions of t. At the final time, t_1, we have $\alpha_{ij}(t_1) = \delta_{ij}$ and hence $\det\underset{\sim}{\alpha}(t_1) = 1$. Since the determinant is continuous, its value cannot jump from 1 to −1 at any point in time, so $\det\underset{\sim}{\alpha} = 1$ at all times and for all transformations and in this case the transformation matrix is proper orthogonal.

The simplest case of a coordinate transformation between two Cartesian coordinate systems with different orientations is a reflection; then two of the three basis vectors do not change, and the third basis vector reverses its direction. In this case, the transformation matrix is a diagonal matrix with two ones and one negative one, for example

$$\begin{pmatrix} 1 & 0 & 0 \\ 0 & 1 & 0 \\ 0 & 0 & -1 \end{pmatrix},$$

and its determinant is −1. An arbitrary transformation between two coordinate systems with different orientations can be composed of a reflection followed by a motion. Since the determinant of a transformation matrix does not change during a motion, it has for every transformation, which changes the orientation of the coordinate system, the value −1. In other words, the transformation is improper orthogonal.

2.1.4 Transformation Law for Basis Vectors

Formula (2.3) can be easily solved for the $\underline{e}_i$ with the help of the orthogonality relation. We formally multiply (2.3) by α_{kj}, which, due to the summation convention, means summation over j. With (2.6), we get

$$\alpha_{kj}\,\underline{\tilde{e}}_j = \alpha_{kj}\,\alpha_{ij}\,\underline{e}_i = \delta_{ki}\,\underline{e}_i = \underline{e}_k .$$

After renaming indices, we obtain the following transformation law for basis vectors:

$$\underline{e}_i = \alpha_{ij}\,\underline{\tilde{e}}_j, \qquad \underline{\tilde{e}}_i = \alpha_{ji}\,\underline{e}_j. \tag{2.8}$$

Since we only used the orthogonality relation, these relations (as well as the transformation laws we will derive later) hold, whether the coordinate system changes its orientation during the transformation or not. In other words, the α_{ij} in (2.8) can represent a proper or an improper orthogonal matrix.

2.1.5 Transformation Law for Point Coordinates

The position vector to a point P can be expressed, as shown in the figure on page 32, in both coordinate systems by

$$\underline{x} = x_i\,\underline{e}_i, \quad \underline{\tilde{x}} = \tilde{x}_i\,\underline{\tilde{e}}_i, \quad \underline{x} = \underline{\beta} + \underline{\tilde{x}}.$$

If all quantities in the last equation are expressed with respect to the no-tilde basis, we obtain

$$x_i\,\underline{e}_i = \beta_i\,\underline{e}_i + \tilde{x}_i\,\alpha_{ki}\,\underline{e}_k.$$

We want to factor out the basis, so we need to rename the indices in the second term. Then we have

$$x_i\,\underline{e}_i = \beta_i\,\underline{e}_i + \tilde{x}_j\,\alpha_{ij}\,\underline{e}_i = (\beta_i + \alpha_{ij}\,\tilde{x}_j)\,\underline{e}_i.$$

Since the representation (2.1) of a position vector with respect to a basis is unique, we have

$$x_i = \beta_i + \alpha_{ij}\,\tilde{x}_j.$$

To solve this equation for $\tilde{x}_j$, we multiply it by α_{ik}:

$$\alpha_{ik}\,x_i = \alpha_{ik}\,\beta_i + \alpha_{ij}\,\alpha_{ik}\,\tilde{x}_j = \alpha_{ik}\,\beta_i + \delta_{jk}\,\tilde{x}_j = \alpha_{ik}\,\beta_i + \tilde{x}_k,$$
$$\tilde{x}_k = \alpha_{ik}\,x_i - \alpha_{ik}\,\beta_i.$$

We have thus derived the transformation law for point coordinates:

$$x_i = \alpha_{ij}\,\tilde{x}_j + \beta_i, \qquad \tilde{x}_i = \alpha_{ji}\,x_j - \alpha_{ji}\,\beta_j. \tag{2.9}$$

2.2 Vectors

2.2.1 Vectors, Vector Components, and Vector Coordinates

1. We first met vectors [3] as arrows in space; the basis vectors of a Cartesian coordinate system are one such example. In order to represent such a vector quantitatively, we

3 Latin: carrier, driver (agent noun to *vehere* [to carry]); short for *radius vector* (traveling ray, i. e. position vector: the ray from a fixed point to a moving point, e. g. in the case of planetary motion from the sun at the focus of the elliptic orbit to the planet). The term vector thus evolved from the term position vector.

need, similarly as for a point, a coordinate system. Analogously to (2.1) we get for a vector $\underline{a}$ in a Cartesian coordinate system with basis $\underline{e}_i$ in different notations

$$\begin{aligned}\underline{a} &= a_x\,\underline{e}_x + a_y\,\underline{e}_y + a_z\,\underline{e}_z \\ &= a_1\,\underline{e}_1 + a_2\,\underline{e}_2 + a_3\,\underline{e}_3 \\ &= a_i\,\underline{e}_i.\end{aligned} \tag{2.10}$$

2. Here, we would like to conceptually distinguish between the components and the coordinates of a vector. We call the quantity $a_x\,\underline{e}_x$ the x-component of the vector $\underline{a}$, and we call the quantity a_x the x-coordinate of the vector $\underline{a}$ (in the given coordinate system or with respect to the given basis). The components of a vector are themselves vectors, but the coordinates are not; the components of a vector are characterized by their magnitude and direction, the coordinates are characterized by their absolute value and sign. Conceptually distinguishing between components and coordinates of a vector is not common practice, but it is meaningful.

2.2.2 The Transformation Law for Vector Coordinates

1. We can obtain the transformation law for the coordinates of an oriented line from the transformation law for point coordinates if we realize that (compare the [again only two-dimensional] figure on this page)

$\underline{a} = \underline{y} - \underline{x} = \underline{\tilde{y}} - \underline{\tilde{x}},$

and so

$$a_i\,\underline{e}_i = (y_i - x_i)\,\underline{e}_i, \quad a_i = y_i - x_i,$$

$$\tilde{a}_i\,\underline{\tilde{e}}_i = (\tilde{y}_i - \tilde{x}_i)\,\underline{\tilde{e}}_i, \quad \tilde{a}_i = \tilde{y}_i - \tilde{x}_i.$$

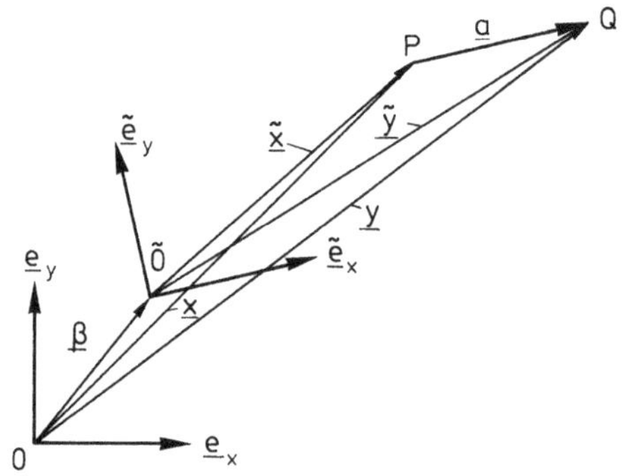

Here we could again cancel out the bases $\underline{e}_i$ and $\underline{\tilde{e}}_i$, because the decomposition of a vector into components is unique, like the decomposition of a position vector. According to (2.9) we have

$$x_i = \alpha_{ij}\,\tilde{x}_j + \beta_i, \quad y_i = \alpha_{ij}\,\tilde{y}_j + \beta_i,$$

which gives

$$a_i = y_i - x_i = \alpha_{ij}(\tilde{y}_j - \tilde{x}_j) = \alpha_{ij}\,\tilde{a}_j.$$

Conversely, according to (2.9) we get

$$\tilde{x}_i = \alpha_{ji}\,x_j - \alpha_{ji}\,\beta_j, \quad \tilde{y}_i = \alpha_{ji}\,y_j - \alpha_{ji}\,\beta_j,$$
$$\tilde{a}_i = \tilde{y}_i - \tilde{x}_i = \alpha_{ji}(y_j - x_j) = \alpha_{ji}\,a_j.$$

Thus we obtained the transformation law for the coordinates of an oriented line

$$a_i = \alpha_{ij}\,\tilde{a}_j, \qquad \tilde{a}_i = \alpha_{ji}\,a_j. \tag{2.11}$$

We arrive at the same result, if in

$$\underline{a} = a_i\,\underline{e}_i = \tilde{a}_i\,\underline{\tilde{e}}_i$$

we substitute $(2.8)_1$ for $\underline{e}_i$:

$$\underline{a} = a_i\,\alpha_{ij}\,\underline{\tilde{e}}_j = \tilde{a}_i\,\underline{\tilde{e}}_i.$$

Clearly, these are two different representations of the vector $\underline{a}$ with respect to the tilde basis. The representation of a vector with respect to a basis is unique, so we can cancel out the basis and obtain an equation for coordinates only. This changes the summation index of the basis into a free index. For all terms in the equation for coordinates to match in this free index, we have to rename the summation indices before canceling out the basis, so that the index of the tilde basis is the same on both sides of the equation. For example, here we can replace on the left side *j* by *i* and *i* by *j*, and we get

$$a_j\,\alpha_{ji}\,\underline{\tilde{e}}_i = \tilde{a}_i\,\underline{\tilde{e}}_i.$$

Then, after canceling out $\underline{\tilde{e}}_i$, we get

$$\tilde{a}_i = \alpha_{ji}\,a_j,$$

which is the transformation law $(2.11)_2$. Conversely we obtain $(2.11)_1$ if we replace in the original equation $\underline{\tilde{e}}_i$ by $(2.8)_2$.

2. Two typical examples of physical quantities described by vectors are velocities and angular velocities. However, there is an important difference between these two types of quantities. Velocity is characterized by its magnitude and direction, and angular velocity is characterized by its magnitude and the direction of rotation. We readily associate a velocity with an oriented line and therefore with a vector; its magnitude is the speed of the velocity and its direction is the direction of the velocity vector. For angular velocity we need a convention, for example the right-hand screw rule, to associate the direction of rotation with the direction of the vector. Let us, for the lack of a better argument, arbitrarily agree to use the right-hand screw rule. Then we associate the angular velocity with a vector whose magnitude is the speed of the angular velocity and whose direction, together with the direction of rotation, forms a right-handed screw. Vectors which describe physical quantities such as velocity are called polar vectors and vectors which describe physical quantities like angular velocity are called axial vectors (or pseudovectors).

We generally assume in physics that a physical process is unaffected by a translation, rotation, or reflection. If we move or reflect the coordinate system in the same way, we consequently expect that the physical quantities have the same coordinates for the different cases. (We can neglect here the violation of reflection invariance in elementary particle physics.)

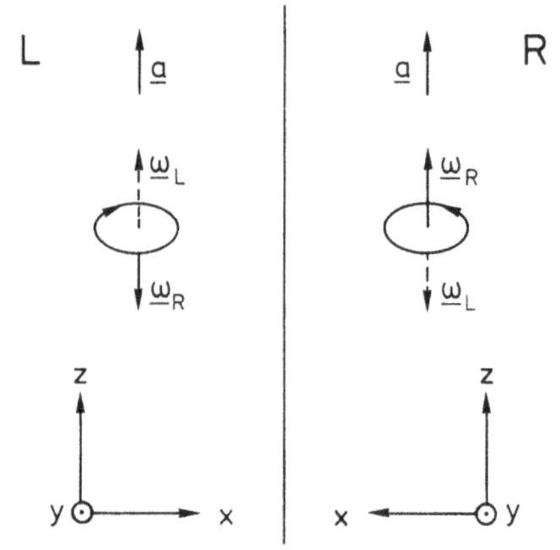

Clearly, polar vectors satisfy these invariances inherently. For a quantity associated with an axial vector, both the direction of rotation and the corresponding vector have to be transformed. In a motion, the correlation between the direction of rotation and the direction of the corresponding vector remains unchanged; a reflection, however, reverses this correlation: a right-hand screw correlation becomes a left-hand screw correlation and vice versa. If we do not adjust the correlation, then the signs of the vector coordinates are reversed in the reflected coordinate system, i. e. the invariance requirement is violated. To preserve reflection invariance, we introduce the following convention.

A physical quantity which is characterized by a magnitude and a direction of rotation is associated with a vector corresponding to the right-hand screw rule in a right-handed system and with a vector corresponding to the left-hand screw rule in a left-handed system.

This is a convention in the sense that the reflection invariance would also be satisfied if we associate with such a quantity in a right-handed system a vector corresponding to a left-hand screw rule and in a left-handed system a vector corresponding to the right-hand screw rule.

In a motion (translation and rotation) where the correlation between the direction of rotation and the corresponding vector is preserved, we do not have this problem.

Certainly this convention has to hold also if the same physical quantity is represented in two different coordinate systems. A polar vector is invariant under an arbitrary change of coordinates, i. e. we have $\underline{a} = \underline{\tilde{a}}$, which explains why the vector commonly is denoted by $\underline{a}$ in both coordinate systems. An axial vector, however, is invariant only under a motion, for a coordinate transformation with a change of orientation we have $\underline{a} = -\underline{\tilde{a}}$. This may seem unfamiliar, as it means that a vector describing an angular velocity, angular momentum, or torque depends on the orientation of the coordinate system in which it is represented. In other words, such a vector is not independent of the coordinate system.

Since we have $\det \underset{\sim}{\alpha} = 1$ for a change of coordinates without change of orientation and $\det \underset{\sim}{\alpha} = -1$ for a change of coordinates together with a change of orientation, we can summarize the transformation law for axial vectors by writing

$$\underline{a} = (\det \underset{\sim}{\alpha})\, \underline{\tilde{a}},$$

and we obtain as transformation law for the coordinates of an axial vector

$$a_i = (\det \underset{\sim}{\alpha})\, \alpha_{ij}\, \tilde{a}_j, \qquad \tilde{a}_i = (\det \underset{\sim}{\alpha})\, \alpha_{ji}\, a_j. \tag{2.12}$$

3. From the decomposition (2.10) and the transformation laws (2.11) and (2.12) we find the following definition of a polar vector and an axial vector.

A vector is a quantity which according to (2.10) can be represented in a Cartesian coordinate system and whose coordinates transform according to (2.11) or (2.12) if the Cartesian coordinate system is changed. If the transformation law (2.11) holds, the vector is called polar, and if the transformation law (2.12) holds, the vector is called axial.

4. Finally we remark that the Cartesian coordinates of a vector depend only on the basis of the chosen coordinate system, but the coordinates of a position vector depend also on its origin. This shows that position vectors are no vectors in the sense of the definition above; however, the difference of two position vectors, which can be interpreted geometrically as a directed line from one point to another point, is a vector in the sense of the definition above; compare the figure on page 36 and the derivation of (2.11).

2.3 Tensors

2.3.1 Second-Order Tensors

1. Let B_i be the coordinates of an arbitrary vector $\underline{B} = B_i\,\underline{e}_i$ with respect to a basis $\underline{e}_i$, and let T_{ij} be a matrix which assigns the coordinates A_i of another vector $\underline{A}$ to the B_i according to the relation

$$A_i = T_{ij}\,B_j. \tag{2.13}$$

Let $\underline{\tilde{e}}_j$ be another arbitrary basis such that between the coordinates $\tilde{A}_i$ and $\tilde{B}_j$ of two vectors $\underline{A}$ and $\underline{B}$ the analogous relation

$$\tilde{A}_i = \tilde{T}_{ij}\,\tilde{B}_j \tag{2.14}$$

holds. Then, what is the relation between the two matrices T_{ij} and $\tilde{T}_{ij}$?

Let us assume initially that the two vectors $\underline{A}$ and $\underline{B}$ are polar. Then according to (2.11) we have

$$A_i = \alpha_{ik}\,\tilde{A}_k, \qquad B_j = \alpha_{jn}\,\tilde{B}_n.$$

Substituting this into (2.13), we get

$$\alpha_{ik}\,\tilde{A}_k = T_{ij}\,\alpha_{jn}\,\tilde{B}_n.$$

Multiplication by α_{im} gives with (2.14)

$$\underbrace{\alpha_{im}\,\alpha_{ik}}_{\delta_{mk}}\,\tilde{A}_k = T_{ij}\,\alpha_{im}\,\alpha_{jn}\,\tilde{B}_n, \quad \tilde{A}_m = \alpha_{im}\,\alpha_{jn}\,T_{ij}\,\tilde{B}_n = \tilde{T}_{mn}\,\tilde{B}_n,$$

$$(\alpha_{im}\,\alpha_{jn}\,T_{ij} - \tilde{T}_{mn})\,\tilde{B}_n = 0.$$

For brevity we temporarily introduce $S_{mn} := \alpha_{im}\,\alpha_{jn}\,T_{ij}$, i. e. we have

$$(S_{mn} - \tilde{T}_{mn})\,\tilde{B}_n = 0. \tag{a}$$

Now we choose three linearly independent polar vectors $\underline{a}$, $\underline{b}$, and $\underline{c}$, so that every polar vector $\underline{B}$ can be written as a linear combination of these three vectors; we have

$$\underline{B} = \alpha\,\underline{a} + \beta\,\underline{b} + \gamma\,\underline{c}, \quad \tilde{B}_n = \alpha\,\tilde{a}_n + \beta\,\tilde{b}_n + \gamma\,\tilde{c}_n.$$

If we substitute the last equation into (a), we see that (a) is satisfied for arbitrary $\underline{B}$, if (a) is satisfied for three linearly independent vectors. If we now substitute in (a) for $\underline{B}$ successively $\underline{a}$, $\underline{b}$, and $\underline{c}$, and write out the equations for $m = 1$, we obtain

$$(S_{11} - \tilde{T}_{11})\,\tilde{a}_1 + (S_{12} - \tilde{T}_{12})\,\tilde{a}_2 + (S_{13} - \tilde{T}_{13})\,\tilde{a}_3 = 0,$$
$$(S_{11} - \tilde{T}_{11})\,\tilde{b}_1 + (S_{12} - \tilde{T}_{12})\,\tilde{b}_2 + (S_{13} - \tilde{T}_{13})\,\tilde{b}_3 = 0,$$
$$(S_{11} - \tilde{T}_{11})\,\tilde{c}_1 + (S_{12} - \tilde{T}_{12})\,\tilde{c}_2 + (S_{13} - \tilde{T}_{13})\,\tilde{c}_3 = 0.$$

This is a homogenous system of linear equations for

$$(S_{11} - \tilde{T}_{11}), \quad (S_{12} - \tilde{T}_{12}), \quad \text{and} \quad (S_{13} - \tilde{T}_{13})$$

with nonzero coefficient determinant; i. e. the system of linear equations has only the trivial solution

$$S_{11} = \tilde{T}_{11}, \quad S_{12} = \tilde{T}_{12}, \quad S_{13} = \tilde{T}_{13}.$$

For $m = 2$ and $m = 3$ we find analogously that also the other elements of the two matrices S_{ij} and $\tilde{T}_{ij}$ must be equal, i. e. we have

$$\tilde{T}_{mn} = \alpha_{im}\,\alpha_{jn}\,T_{ij},$$

and from this we compute the inverse relationship by multiplying by $\alpha_{pm}\,\alpha_{qn}$

$$\alpha_{pm}\,\alpha_{qn}\,\tilde{T}_{mn} = \underbrace{\alpha_{pm}\,\alpha_{im}}_{\delta_{pi}}\,\underbrace{\alpha_{qn}\,\alpha_{jn}}_{\delta_{qj}}\,T_{ij} = T_{pq},$$

$$T_{ij} = \alpha_{im}\,\alpha_{jn}\,\tilde{T}_{mn}, \qquad \tilde{T}_{ij} = \alpha_{mi}\,\alpha_{nj}\,T_{mn}. \tag{2.15}$$

This is clearly a generalization of the transformation law (2.11) for coordinates of polar vectors. Similarly as before, where we interpreted the quantities a_i and $\tilde{a}_i$, related by (2.11), as the coordinates of a (coordinate system-independent) polar vector $\underline{a}$, we interpret the matrices T_{ij} and $\tilde{T}_{ij}$, related by (2.15), as the coordinates of a (coordinate system-independent) second-order polar tensor[4] $\underline{\underline{T}}$ with respect to the bases $\underline{e}_i$ and $\underline{\tilde{e}}_i$.

Thus, if in any coordinate system the coordinates B_i of an arbitrary polar vector $\underline{B}$ (or equivalently, the coordinates of three linearly independent polar vectors) are assigned to the coordinates A_i of a polar vector by always the same relation, either (2.13) or (2.14), then the T_{ij} are the coordinates of a polar tensor of order 2.

Since $T_{ij} = T^{\mathrm{T}}_{ji}$ we arrive at the same transformation law if we start with the equations $A_i = T^{\mathrm{T}}_{ji}\,B_j$ and $\tilde{A}_i = \tilde{T}^{\mathrm{T}}_{ji}\,\tilde{B}_j$, i. e. if we sum over the first index of the matrix.

4 Latin: tensioner (agent noun to *tendere* [to tension]); identified initially only with the quantity that we today call the (Cauchy) stress tensor. (The word stress tensor is etymologically a pleonasm.)

Because the α_{ij} are orthogonal and since the transposed matrix of an orthogonal matrix is equal to its inverse, it follows from (1.58) that all coordinate matrices of a second-order polar tensor are similar; however, not all matrices which are similar to a coordinate matrix of a second-order polar tensor are Cartesian coordinates of this tensor.

Problem 2.1.
Prove the converse: If the relation (2.13) or (2.14) is valid in any Cartesian coordinate system and if the B_i are the coordinates of a polar vector and the T_{ij} are the coordinates of a second-order polar tensor (i. e. they satisfy the transformation equations (2.11) and (2.15)), then the A_i are the coordinates of a polar vector. (Note the difference: For the original theorem the B_i must be the coordinates of an *arbitrary* vector, but the converse theorem only requires that they are the coordinates of *a* vector.)

2. To develop an intuition for second-order polar tensors, let us consider the (Cauchy) stress tensor, from which the tensor originally has its name. The stress tensor describes the relation between the stress vector $\underline{\sigma}$, i. e. the surface force per unit surface element in a point of a surface and the orientation $\underline{n}$ of the surface element through this point. The stress tensor depends, even in the simplest case of a fluid at rest, on the orientation of the area element: it is always perpendicular to the surface element. Let p be the pressure at a given surface element. Then we have

$$\sigma_i = -p\, n_i.$$

For a fluid in motion (or a deformable solid body) it is shown in mechanics that the relations

$$\begin{aligned}
\sigma_x(\underline{x},\, t,\, \underline{n}) &= \pi_{xx}(\underline{x},\, t)\, n_x + \pi_{xy}(\underline{x},\, t)\, n_y + \pi_{xz}(\underline{x},\, t)\, n_z, \\
\sigma_y(\underline{x},\, t,\, \underline{n}) &= \pi_{yx}(\underline{x},\, t)\, n_x + \pi_{yy}(\underline{x},\, t)\, n_y + \pi_{yz}(\underline{x},\, t)\, n_z, \\
\sigma_z(\underline{x},\, t,\, \underline{n}) &= \pi_{zx}(\underline{x},\, t)\, n_x + \pi_{zy}(\underline{x},\, t)\, n_y + \pi_{zz}(\underline{x},\, t)\, n_z
\end{aligned}$$

hold, or, using the summation convention,

$$\sigma_i(\underline{x},\, t,\, \underline{n}) = \pi_{ij}(\underline{x},\, t)\, n_j,$$

i. e. every coordinate of the stress vector $\underline{\sigma}$ is a homogeneous linear function of all coordinates of the normal vector $\underline{n}$. (Setting

$$\pi_{ij} = -p\, \delta_{ij}$$

we obtain the formula for the fluid at rest as a special case of the formula above.) Since both $\underline{n}$ and $\underline{\sigma}$ are polar vectors and since we can substitute for $\underline{n}$ in particular the three linear independent unit vectors $\underline{e}_x$, $\underline{e}_y$, and $\underline{e}_z$, the π_{ij} are, according to our definition (2.13), the coordinates of a polar tensor of order 2, which we call stress tensor $\underline{\underline{\pi}}$.

The stress tensor describes the state of stress at a given point by assigning a stress vector $\underline{\sigma}$ to every normal vector $\underline{n}$; hence, we can also say the tensor $\underline{\underline{\pi}}$ maps the vector $\underline{n}$ to the vector $\underline{\sigma}$.

3. If in (2.13) A_i and B_i are the coordinates of two axial vectors, we again obtain the transformation law (2.15); so also for this case the T_{ij} are the coordinates of a second-order polar tensor. If, however, one of the two vectors $\underline{A}$ and $\underline{B}$ is a polar vector and the other vector is an axial vector, then we obtain the transformation law

$$T_{ij} = (\det \underset{\sim}{\alpha})\, \alpha_{im}\, \alpha_{jn}\, \tilde{T}_{mn}, \qquad \tilde{T}_{ij} = (\det \underset{\sim}{\alpha})\, \alpha_{mi}\, \alpha_{nj}\, T_{mn}, \tag{2.16}$$

and the T_{ij} are called the coordinates of an axial tensor of order 2. Often the word pseudotensor is used instead of axial tensor.

2.3.2 Tensors of Arbitrary Order

1. We can show similarly that if in any Cartesian coordinate system the coordinates B_i of an arbitrary vector are assigned to the coordinates A_{ij} of a tensor of order 2 by the analogous relation

$$A_{ij} = T_{ijk}\, B_k,$$

then the relation between T_{ijk} and $\tilde{T}_{ijk}$ with respect to the bases $\underline{e}_i$ and $\underline{\tilde{e}}_i$, where $\underline{\underline{A}}$ and $\underline{B}$ are both either polar or axial, is given by the transformation law

$$T_{ijk} = \alpha_{im}\, \alpha_{jn}\, \alpha_{kp}\, \tilde{T}_{mnp}, \quad \tilde{T}_{ijk} = \alpha_{mi}\, \alpha_{nj}\, \alpha_{pk}\, T_{mnp}$$

or, if one of the two vectors $\underline{\underline{A}}$ and $\underline{B}$ is polar and the other is axial, by the transformation law

$$T_{ijk} = (\det \underset{\sim}{\alpha})\, \alpha_{im}\, \alpha_{jn}\, \alpha_{kp}\, \tilde{T}_{mnp}, \quad \tilde{T}_{ijk} = (\det \underset{\sim}{\alpha})\, \alpha_{mi}\, \alpha_{nj}\, \alpha_{pk}\, T_{mnp}.$$

In the former case the T_{ijk} are called the coordinates of a polar tensor $\underline{\underline{\underline{T}}}$ of order 3, and in the latter case they are called the coordinates of an axial tensor $\underline{\underline{\underline{T}}}$ of order 3.

We obtain the same transformation laws, if we start with the relations

$$A_{ij} = T^{*}_{ikj}\, B_k \quad \text{or} \quad A_{ij} = T^{**}_{kij}\, B_k.$$

2. A scalar is also called a tensor of order 0 and a vector is also called a tensor of order 1. Thus we have the following rule.

The order of a tensor equals the number of underlines of the tensor and the number of indices of its coordinates.

3. The transformation equations for polar tensors and for axial tensors are different. For polar tensors we have the following laws.

$$\begin{array}{lll} a = \tilde{a}, & & \tilde{a} = a, \\ a_i = \alpha_{im}\, \tilde{a}_m, & & \tilde{a}_i = \alpha_{mi}\, a_m, \\ a_{ij} = \alpha_{im}\, \alpha_{jn}\, \tilde{a}_{mn}, & & \tilde{a}_{ij} = \alpha_{mi}\, \alpha_{nj}\, a_{mn}, \\ \text{etc.} & & \text{etc.} \end{array} \tag{2.17}$$

For axial tensors we have the following laws.

$$\begin{array}{lll} a = (\det \underset{\sim}{\alpha})\, \tilde{a}, & & \tilde{a} = (\det \underset{\sim}{\alpha})\, a, \\ a_i = (\det \underset{\sim}{\alpha})\, \alpha_{im}\, \tilde{a}_m, & & \tilde{a}_i = (\det \underset{\sim}{\alpha})\, \alpha_{mi}\, a_m, \\ a_{ij} = (\det \underset{\sim}{\alpha})\, \alpha_{im}\, \alpha_{jn}\, \tilde{a}_{mn}, & & \tilde{a}_{ij} = (\det \underset{\sim}{\alpha})\, \alpha_{mi}\, \alpha_{nj}\, a_{mn}, \\ \text{etc.} & & \text{etc.} \end{array} \tag{2.18}$$

4. For both polar and axial vectors we have in the two coordinate systems

$$\underline{a} = a_i\, \underline{e}_i, \quad \underline{\tilde{a}} = \tilde{a}_i\, \underline{\tilde{e}}_i.$$

Similarly, tensors of order 2 or higher can be defined from their representation with respect to a Cartesian coordinate system and either the transformation law (2.17) or (2.18), respectively, for the coordinates so obtained. However, we need to postpone this definition, until we know how to represent a tensor in a coordinate system.

5. The tensor whose coordinates are all zero is called the zero tensor. We denote such a tensor by 0, $\underline{0}$, $\underline{\underline{0}}$, etc.

6. Sometimes it is convenient to have a symbol for a tensor of arbitrary order. We denote such a tensor by upper-case large cursive script letters ($\mathscr{A}$, $\mathscr{B}$, etc.) and its coordinates by $a_{i\ldots j}$, $b_{i\ldots j}$, etc.

2.3.3 Symmetries in Physics

General invariances in physics, such as invariance under a translation, a rotation, or a reflection, are also called symmetries in physics. Besides these three symmetries, there are the invariance under repetition, the so-called reproducibility, and the invariance of physical quantities and physical equations under a change of units. All these invariances have consequences for the formulation of physical equations, namely it follows that physical equations do not change under a symmetry operation (e. g. when a physical experiment is repeated). As a result, certain measurable properties of physical processes cannot appear in a physical equation (i. e. strictly speaking, they are no physical quantities), and not all mathematically defined combinations of physical quantities are also physically permissible. We only mention these consequences here; a sound justification is beyond the scope of this text.

From the reproducibility of a physical process it follows that points of time cannot appear explicitly in a physical equation but only time intervals; and from the invariance under translation it follows that points (position coordinates) cannot appear explicitly in a physical equation but only distances between points.

From the invariance of physical properties under a change of units it follows that all terms of a physical equation must be measurable in the same unit (in other words, they must be dimensionally homogeneous). For example, energy and temperature are not measured in the same units, so an energy cannot be equal to a temperature. This symmetry is systematically utilized in dimensional analysis, a method which is known to reduce the number of mathematically possible combinations of physical quantities significantly.

The reflection invariance leads to the distinction between polar and axial tensors; and from the rotational invariance it follows that physical quantities must all be (polar or axial) tensors and that in a physical equation all terms must be tensors of the same order that must all be either polar or axial. (Analogous to the concept of dimensional homogeneity, we could call this property tensorial homogeneity.) For example, a torque (vector!) cannot be equal to a work (scalar!), although both quantities are measured in the unit Joule, i. e. we have dimensional homogeneity; and a velocity (polar vector!) cannot be equal to an angular velocity (axial vector!), even if we eliminated their dimensional inhomogeneity by multiplying the axial vector with a polar scalar. These symmetries are systematically exploited in the theory of representation of tensor functions (see Chapter 5).

There are no practical reasons for the use of left-handed systems in physics and technology. Thus we could drop the distinction between polar tensors and axial tensors if we would limit ourselves to right-handed systems. But then we could not use the reflection invariance to exclude equations which are physically not permissible (i. e. which are irrelevant for the description of physical processes).

2.4 Symbolic Notation, Index Notation, and Matrix Notation

We distinguished between tensors themselves (which we denoted by underlined letters) and their Cartesian coordinates (which we denoted by indexed letters). Likewise we can define mathematical operations for the tensors themselves and for their coordinates. As a result we have two different notations of tensor calculus, which we call symbolic notation and index notation. We require tensor equations to be independent of a coordinate system in symbolic notation (i. e. between tensors) as well as in index notation (i. e. between tensor coordinates).

It is important to master both notations and to be able to translate from one notation into the other. Every equation in symbolic notation can be translated into an equation in index notation; the converse is not always possible, but it is in most practically occurring cases. We think it is best to grow up virtually bilingual from the beginning and will therefore write all operations in both notations. When performing the computations we use whichever notation is more convenient. Usually this means we do the computations in index notation and then translate the results into symbolic notation. Occasionally we will also speak seemingly inaccurately, for example of a vector a_i, since the coordinates of a vector in an unspecified Cartesian coordinate system are after all also a symbol for the vector itself.

Let us finally agree to regard the coordinates of vectors as column matrices (rather than row matrices). Then we can write all tensor equations which contain only scalars, vectors, and second-order tensors also as matrix equations. Writing tensor equations in matrix notation combines the conciseness of symbolic notation with the algorithmic character of index notation. It fails however, for tensors of order higher than 2 (we will see that this is already the case for the vector product of two vectors); and it is not obvious how many columns a matrix $\underset{\sim}{a}$ has (for example, whether it represents the coordinates of a vector or of a second-order tensor). Nevertheless, matrix notation is widely used (and often confused with symbolic notation). For this reason we will write the definitions of tensor algebraic operations not only in symbolic notation and index notation, but whenever possible also in matrix notation; however, we will not use it here.

2.5 Equality, Addition, and Subtraction of Tensors. Multiplication of a Tensor by a Scalar. Linear Independence

In the following sections we define mathematical operations for tensors, i. e. we develop a calculus for tensors. We will define an operation between tensors (i. e. in symbolic notation) by writing out the operation that has to be carried out between the coordinates of the involved tensors (in index notation).[5]

5 Since tensor coordinates are (for us: real) numbers (extension to complex numbers is straightforward), for tensor equations in index notation the rules of arithmetics apply.

We start by defining the linear operations equality, addition, subtraction, and multiplication by a scalar. In the next sections we will then introduce additional algebraic operations, in particular three multiplications as well as differential and integral operations for tensors.

1. Two tensors are equal, if they match (in the same coordinate system) in all homologous coordinates.

$$
\begin{array}{lll}
a = A, & a = A, & a = A, \\
\underline{a} = \underline{B}, & a_i = B_i, & \underset{\sim}{a} = \underset{\sim}{B}, \\
\underline{\underline{a}} = \underline{\underline{C}}, & a_{ij} = C_{ij}, & \underset{\sim}{a} = \underset{\sim}{C}. \\
\text{etc.} & \text{etc.} &
\end{array}
\qquad (2.19)
$$

Conversely, this implies that two tensors are not equal if they differ in at least one pair of homologous coordinates.

Two tensors are added or subtracted by adding or subtracting their homologous coordinates.

$$
\begin{array}{lll}
a \pm b = A, & a \pm b = A, & a \pm b = A, \\
\underline{a} \pm \underline{b} = \underline{B}, & a_i \pm b_i = B_i, & \underset{\sim}{a} \pm \underset{\sim}{b} = \underset{\sim}{B}, \\
\underline{\underline{a}} \pm \underline{\underline{b}} = \underline{\underline{C}}, & a_{ij} \pm b_{ij} = C_{ij}, & \underset{\sim}{a} \pm \underset{\sim}{b} = \underset{\sim}{C}. \\
\text{etc.} & \text{etc.} &
\end{array}
\qquad (2.20)
$$

To multiply a tensor by a scalar we multiply all coordinates by this scalar.

$$
\begin{array}{lll}
\alpha\, a = A, & \alpha\, a = A, & \alpha\, a = A, \\
\alpha\, \underline{a} = \underline{B}, & \alpha\, a_i = B_i, & \alpha\, \underset{\sim}{a} = \underset{\sim}{B}, \\
\alpha\, \underline{\underline{a}} = \underline{\underline{C}}, & \alpha\, a_{ij} = C_{ij}, & \alpha\, \underset{\sim}{a} = \underset{\sim}{C}. \\
\text{etc.} & \text{etc.} &
\end{array}
\qquad (2.21)
$$

2. For completeness, we still need to convince ourselves that the quantities A, B_i, C_{ij} defined in this way are tensor coordinates, i. e. that they satisfy the corresponding transformation law, provided that the quantities on the left side of these equations are tensor coordinates. This is easy to see, e. g. for polar tensors we have

$$C_{ij} = a_{ij} \pm b_{ij} = \alpha_{im}\, \alpha_{jn} (\tilde{a}_{mn} \pm \tilde{b}_{mn}) = \alpha_{im}\, \alpha_{jn}\, \tilde{C}_{mn}.$$

So we can only equate, add, or subtract tensors, which obey the same transformation law, and hence have the same order and are either both polar or both axial.

3. We call Q Tensors $\overset{1}{\mathscr{A}}, \overset{2}{\mathscr{A}}, \ldots, \overset{Q}{\mathscr{A}}$ which have the same order and are either all polar or all axial, linearly independent, if the equation

$$\alpha_i \overset{i}{\mathscr{A}} = 0$$

is only satisfied for $\alpha_i = 0$; otherwise these tensors are called linearly dependent. More than three vectors, more than nine second-order tensors, or in general more than 3^N tensors of order N are clearly always linearly dependent.

2.6 Transposed, Isometric, Symmetric, and Antimetric Tensors

1. Here it is useful to remember that the coordinates of a second-order tensor form a square matrix. So we write

$$a_{ij} := \begin{pmatrix} a_{11} & a_{12} & a_{13} \\ a_{21} & a_{22} & a_{23} \\ a_{31} & a_{32} & a_{33} \end{pmatrix} = \begin{pmatrix} a_{xx} & a_{xy} & a_{xz} \\ a_{yx} & a_{yy} & a_{yz} \\ a_{zx} & a_{zy} & a_{zz} \end{pmatrix}. \tag{2.22}$$

2. Two second-order tensors whose coordinates differ only by the order of the indices are called transposed tensors. Clearly, their coordinate matrices are transposed matrices. Similar to transposed matrices we also denote a transposed tensor by the same letter with a T as superscript.

$$a_{ij}^{\mathrm{T}} = a_{ji}. \tag{2.23}$$

3. Tensors whose coordinates differ only in the order of indices are called isomers. The number of isomers of a tensor increases quickly with its order: A second-order tensor has two isomers, namely the tensor itself and the transposed tensor. A tensor of order 3 has already six isomers, namely A_{ijk}, A_{jki}, A_{kij}, A_{kji}, A_{jik}, and A_{ikj}. Evidently tensors of order n have $n!$ isomers, and we usually do not have a symbolic notation for them.

4. A second-order tensor which is equal to its transposed tensor is called symmetric; for such a tensor we have

$$\underline{\underline{a}} = \underline{\underline{a}}^{\mathrm{T}}, \qquad a_{ij} = a_{ji}, \qquad \underset{\sim}{a} = \underset{\sim}{a}^{\mathrm{T}}. \tag{2.24}$$

5. A second-order tensor which is negative equal to its transposed tensor is called antimetric, antisymmetric, skew symmetric or alternating; for such a tensor we have

$$\underline{\underline{a}} = -\underline{\underline{a}}^{\mathrm{T}}, \qquad a_{ij} = -a_{ji}, \qquad \underset{\sim}{a} = -\underset{\sim}{a}^{\mathrm{T}}. \tag{2.25}$$

6. We can uniquely decompose a second-order tensor using the formula

$$a_{ij} = \underbrace{\frac{1}{2}(a_{ij} + a_{ji})}_{a_{(ij)}} + \underbrace{\frac{1}{2}(a_{ij} - a_{ji})}_{a_{[ij]}} \tag{2.26}$$

into a symmetric part $a_{(ij)}$ and an antimetric part $a_{[ij]}$, which is easy to see if we expand the formula. We analogously say that a higher-order tensor is symmetric or antimetric with respect to an index pair; e. g. we have[6] $a_{[ijk]l} = \frac{1}{2}(a_{ijkl} - a_{kjil})$.

7. Finally, for completeness, we need to show that the transposed coordinate matrices of a tensor in different coordinate systems are again the coordinates of a tensor. This is easy to see using the transformation law.

Let a_{ij} and $\tilde{a}_{ij}$ be the coordinates of a second-order polar tensor in two different coordinate systems, i. e. we have

$$a_{ij} = \alpha_{im}\, \alpha_{jn}\, \tilde{a}_{mn},$$

and let $\tilde{a}^{\mathrm{T}}_{nm} = \tilde{a}_{mn}$ be the transposed coordinate matrix in the tilde system. This gives

$$a^{\mathrm{T}}_{ij} = a_{ji} = \alpha_{jm}\, \alpha_{in}\, \tilde{a}_{mn} = \alpha_{in}\, \alpha_{jm}\, \tilde{a}^{\mathrm{T}}_{nm}, \qquad \text{Q. E. D.}^{7}$$

Thus, symmetry or antimetry of a tensor is a property of the tensor itself and not just of its coordinate matrix in a particular coordinate system.

Problem 2.2.
A tensor (in a given Cartesian coordinate system) has the coordinates

$$a_{ij} = \begin{pmatrix} 1 & 0 & 2 \\ 4 & -3 & 6 \\ 2 & -4 & 5 \end{pmatrix}.$$

A. Write out the coordinates of the transposed tensor.
B. Decompose the tensor into its symmetric and its antimetric part and check your result, i. e. show that the sum of both parts gives the original tensor.

6 Some authors denote by $a_{[ijk]l}$ the average of all possible permutations of the indices i, j, k, where an odd number of index exchanges result in a minus sign, i. e.

$$a_{[ijk]l} = \frac{1}{3!}(a_{ijkl} + a_{jkil} + a_{kijl} - a_{kjil} - a_{jikl} - a_{ikjl}).$$

7 Quod erat demonstrandum, meaning: what was to be demonstrated.

2.7 Tensor Multiplication of Tensors

2.7.1 Definition

We introduce three different products of tensors: the tensor product,[8] the scalar product,[9] and the vector product,[10] and we again define them by specifying the arithmetic operations that need to be performed between the coordinates of the two factors.

The tensor or dyadic product is a generalization of the usual product of two scalars, so we denote it similarly by writing the two factors next to each other without a symbol between them.[11] It is defined for tensors of arbitrary order.

$$
\begin{array}{llll}
a\,b = A, & a\,b = A, & a\,b = A, & \\
a\,\underline{b} = \underline{B}, & a\,b_i = B_i, & a\,\underset{\sim}{b} = \underset{\sim}{B}, & \\
\underline{a}\,b = \underline{C}, & a_i\,b = C_i, & \underset{\sim}{a}\,b = \underset{\sim}{C}, & \\
\underline{a}\,\underline{b} = \underline{\underline{D}}, & a_i\,b_j = D_{ij}, & \underset{\sim}{a}\,\underset{\sim}{b}^{\mathrm{T}} = \underset{\sim}{D}, & \\
a\,\underline{\underline{b}} = \underline{\underline{E}}, & a\,b_{ij} = E_{ij}, & a\,\underset{\sim}{b} = \underset{\sim}{E}, & (2.27)\\
\underline{\underline{a}}\,b = \underline{\underline{F}}, & a_{ij}\,b = F_{ij}, & \underset{\sim}{a}\,b = \underset{\sim}{F}. & \\
\underline{a}\,\underline{\underline{b}} = \underline{\underline{\underline{G}}}, & a_i\,b_{jk} = G_{ijk}, & & \\
\underline{\underline{a}}\,\underline{b} = \underline{\underline{\underline{H}}}, & a_{ij}\,b_k = H_{ijk}, & & \\
\text{etc.} & \text{etc.} & &
\end{array}
$$

The name tensor product indicates that the tensor product of two *vectors* is a second-order *tensor*.

Problem 2.3.
Two vectors have (in a given coordinate system) the coordinates $a_i = (2, 3, 4)$, $b_i = (-2, 1, 5)$. Compute the coordinates of the tensor products $\underline{a}\,\underline{b}$ and $\underline{b}\,\underline{a}$.

8 Frequently also called the dyadic product.

9 Frequently also called the dot product.

10 Frequently also called the cross product.

11 Some authors use ⊗ to denote the tensor product; we introduce ⊗ in Section 2.10.1 with another meaning.

2.7.2 Properties

1. First, we need to convince ourselves that the result of the so-defined product between tensor coordinates consists again of tensor coordinates. We do this by means of the equation $a_i\, b_j = D_{ij}$. Writing this equation in two different Cartesian coordinate systems, we have

$$a_i\, b_j = D_{ij} \quad \text{and} \quad \tilde{a}_i\, \tilde{b}_j = \tilde{D}_{ij},$$

and if, for example, a_i and b_j are the coordinates of two polar vectors, we have

$$a_i = \alpha_{im}\, \tilde{a}_m \quad \text{and} \quad b_j = \alpha_{jn}\, \tilde{b}_n.$$

This gives

$$D_{ij} = a_i\, b_j = \alpha_{im}\, \tilde{a}_m\, \alpha_{jn}\, \tilde{b}_n = \alpha_{im}\, \alpha_{jn}\, \tilde{a}_m\, \tilde{b}_n = \alpha_{im}\, \alpha_{jn}\, \tilde{D}_{mn},$$

i. e. the D_{ij} are according to (2.15) the coordinates of a second-order polar tensor.

2. Clearly, the tensor product of two polar tensors or of two axial tensors gives a polar tensor, and the tensor product of two tensors where one is polar and the other is axial gives an axial tensor.

3. If we compare the equations in (2.27) in the two notations, we realize that *for the translation* the order of the indices must be the same on the left side and on the right side of the equation. So sometimes we may have to change the order of the factors or introduce isomers. For example in

$$a_j\, b_i\, c_k + d_k\, e_{ji} = f_{ijk},$$

we must first rewrite the two terms on the left side, i. e.

$$b_i\, a_j\, c_k + e^{\mathrm{T}}_{ij}\, d_k = f_{ijk},$$

and then we obtain the translation

$$\underline{b}\,\underline{a}\,\underline{c} + \underline{\underline{e}}^{\mathrm{T}}\,\underline{d} = \underline{\underline{\underline{f}}}.$$

If we have to generate isomers of tensors with order higher than 2, then we would need additional rules to be able to translate the equation into symbolic notation. An example is the following equation:

$$a_{ik}\, b_j = c_{ijk}.$$

Problem 2.4.

Translate, as far as possible, into the other two notations:

A. $\underline{a}\,\underline{b}\,\underline{c} = \underline{\underline{\underline{d}}}$, B. $a_i\, b_{kl}\, c_j = d_{ijkl}$, C. $a_i\, b_{kj} = A_{ijk}$, D. $a^{\mathrm{T}}_{mn} = b_{nm}$.

4. Tensor coordinates are real numbers; so all operations in index notation must satisfy the rules for calculations with real numbers, e. g. the commutative law and the associative law for multiplication must hold. We need to check for each law if it is also satisfied in symbolic notation.

Clearly addition and tensor multiplication are distributive also in symbolic notation. We have, for example, in index notation $a_i\, b_j + a_i\, c_j = a_i\,(b_j + c_j)$, and translated $\underline{a}\,\underline{b} + \underline{a}\,\underline{c} = \underline{a}(\underline{b} + \underline{c})$.

We can similarly prove the associative law, e. g. we have $(a_i\, b_j)\, c_k = a_i(b_j\, c_k)$, and translated $(\underline{a}\,\underline{b})\,\underline{c} = \underline{a}(\underline{b}\,\underline{c})$.

However, the commutative law does not hold in general. It only holds if one of the factors is a scalar: $a\, b_{ij} = b_{ij}\, a$ gives $a\,\underline{\underline{b}} = \underline{\underline{b}}\, a$, but $a_i\, b_j = b_j\, a_i$ cannot be translated directly because of the different order of the free indices on the two sides of the equation. It is easy to see that in general $\underline{a}\,\underline{b} \neq \underline{b}\,\underline{a}$. We set $\underline{a}\,\underline{b} =: \underline{\underline{A}}$ and $\underline{b}\,\underline{a} =: \underline{\underline{B}}$. Then we have

$$a_i\, b_j = A_{ij} \quad \text{and} \quad b_m\, a_n = B_{mn}.$$

Since the equation on the right side is commutative in index notation, we can also write $a_n\, b_m = B_{mn}$ and then rename the indices for easier comparison with the equation on the left side, i. e. we replace n by i and m by j. This gives $a_i\, b_j = B_{ji}$, so we have $A_{ij} = B_{ji}$, the tensors $\underline{\underline{A}}$ and $\underline{\underline{B}}$ are transposed, i. e. we have $\underline{a}\,\underline{b} = (\underline{b}\,\underline{a})^{\mathrm{T}}$, and thus $\underline{a}\,\underline{b} \neq \underline{b}\,\underline{a}$ in general.

5. A tensor product is zero if and only if at least one factor is zero. Consider, for example, the tensor product $\underline{\underline{\underline{C}}} = \underline{\underline{A}}\,\underline{B}$ and assume that only one coordinate of the two factors is nonzero in the coordinate system under consideration. Then also one coordinate of the product is nonzero. Only if all the coordinates of one factor are zero, then also all the coordinates of the product are zero. And if all the coordinates of the product are zero, then also all the coordinates of at least one factor are zero.

6. We can factor out a factor which is common to several tensor products, if the factor is nonzero. For example, from $\underline{a}\,\underline{\underline{b}} = \underline{a}\,\underline{\underline{c}}$ we first get $\underline{a}(\underline{\underline{b}} - \underline{\underline{c}}) = \underline{\underline{\underline{0}}}$. If $\underline{a} \neq \underline{0}$, it then follows from the last paragraph that $\underline{\underline{b}} - \underline{\underline{c}} = \underline{\underline{0}}$, which gives $\underline{\underline{b}} = \underline{\underline{c}}$.

7. We can now prove the so-called quotient rule: if the tensor product of two factors, where one of the factors is a nonzero tensor of order m, is a tensor of order $(m+n)$, then the other factor is a tensor of order n. It is a polar tensor if the two tensors are either both polar or both axial, and it is axial if of the two given tensors one is polar and the other is axial. Consider, for example, the relation

$$a_i\, b_j = D_{ij} \quad \text{and} \quad \tilde{a}_i\, \tilde{b}_j = \tilde{D}_{ij}$$

in two Cartesian coordinate systems and let a_i be a nonzero polar vector and D_{ij} a second-order polar tensor, i. e.

$$a_i \neq 0, \quad a_i = \alpha_{im}\, \tilde{a}_m, \quad D_{ij} = \alpha_{im}\, \alpha_{jn}\, \tilde{D}_{mn}.$$

Then we should get consequently

$$b_i = \alpha_{im}\, \tilde{b}_m.$$

We prove this as follows:

$$a_i\, b_j = D_{ij} = \alpha_{im}\, \alpha_{jn}\, \tilde{D}_{mn} = \alpha_{im}\, \alpha_{jn}\, \tilde{a}_m\, \tilde{b}_n = a_i\, \alpha_{jn}\, \tilde{b}_n.$$

Since $a_i \neq 0$, we can cancel out a_i, and by this we get $b_j = \alpha_{jn}\, \tilde{b}_n$, which is the required statement, up to the names of the indices.

The quotient rule is obviously the converse of the theorem that the tensor product of a tensor of order m and a tensor of order n is a tensor of order $(m + n)$.

8. Let us summarize the properties of tensor products that we proved above.

- The tensor product of a tensor of order m and a tensor of order n is a tensor of order $(m + n)$.
- If the tensor product of two factors, of which one is a nonzero tensor of order m, is a tensor of order $(m + n)$, then the other factor is a tensor of order n (quotient rule).
- The three tensors of a tensor multiplication (i. e. the two factors and the product) are either all polar, or two of them are axial and the third is polar.
- *For translating into the symbolic notation*, the index order in a tensor product must match on the left and on the right side of an equation.
- Addition and tensor multiplication are distributive also in symbolic notation.
- Tensor multiplication is associative also in symbolic notation.
- Tensor multiplication is in symbolic notation in general only commutative, if one factor is a scalar.
- A tensor product is zero if and only if at least one factor is zero.
- A nonzero common factor in an equation involving tensor products can be canceled out.

2.7.3 Tensors, Tensor Components, and Tensor Coordinates

1. We can now use tensor products to associate second- and higher-order tensors with their coordinates in a Cartesian coordinate system, analogously to (2.10). If we replace the left side of the equation $\underline{a}\,\underline{b} = \underline{\underline{c}}$ using the representation (2.10), we obtain with (2.27)

$$\underline{a}\,\underline{b} = a_i\,\underline{e}_i\,b_j\,\underline{e}_j = a_i\,b_j\,\underline{e}_i\,\underline{e}_j = c_{ij}\,\underline{e}_i\,\underline{e}_j = \underline{\underline{c}},$$

i. e.

$$\underline{\underline{c}} = c_{ij}\,\underline{e}_i\,\underline{e}_j.$$

Here $\underline{e}_i\,\underline{e}_j$ is a tensor product of two basis vectors. Since i and j take the values 1 to 3, we can write $\underline{e}_i\,\underline{e}_j$ as a square matrix with three rows, whose elements are second-order tensors:

$$\underline{e}_i\,\underline{e}_j = \begin{pmatrix} \underline{e}_1\,\underline{e}_1 & \underline{e}_1\,\underline{e}_2 & \underline{e}_1\,\underline{e}_3 \\ \underline{e}_2\,\underline{e}_1 & \underline{e}_2\,\underline{e}_2 & \underline{e}_2\,\underline{e}_3 \\ \underline{e}_3\,\underline{e}_1 & \underline{e}_3\,\underline{e}_2 & \underline{e}_3\,\underline{e}_3 \end{pmatrix}.$$

Generalizing (2.10) gives the following relation between a tensor of arbitrary order and its coordinates with respect to the basis $\underline{e}_i$ of a coordinate system.

$$\begin{aligned} \underline{a} &= a_i\,\underline{e}_i\,, \\ \underline{\underline{a}} &= a_{ij}\,\underline{e}_i\,\underline{e}_j, \\ \underline{\underline{\underline{a}}} &= a_{ijk}\,\underline{e}_i\,\underline{e}_j\,\underline{e}_k, \\ &\text{etc.} \end{aligned} \tag{2.28}$$

2. Just as any vector $\underline{a}$ can be written as a linear combination of the three basis vectors $\underline{e}_i$ with the coordinates a_i as coefficients, we can also represent any second-order tensor $\underline{\underline{a}}$ as a linear combination of the nine basis tensors $\underline{e}_i\,\underline{e}_j$ with the coordinates a_{ij} as coefficients, and the same holds for tensors of higher order. Similarly as for vectors, we also distinguish for tensors of arbitrary order between their components and coordinates; for example, we call the tensor $a_{xy}\,\underline{e}_x\,\underline{e}_y$ the x, y-component of $\underline{\underline{a}}$ in the given coordinate system, and we call the quantity a_{xy} its x, y-coordinate. The components of a tensor are also tensors, whereas the coordinates are not.

3. We can now generalize our definition of a vector from Section 2.2.2 to tensors of arbitrary order.

A tensor is a quantity which, according to (2.28), can be represented in a Cartesian coordinate system and whose coordinates transform, according to (2.17) or (2.18), if the Cartes ian coordinate system is changed. If the transformation law (2.17) applies, the tensor is called polar, and if the transformation law (2.18) applies, the tensor is called axial.

2.7.4 Tensor Equations, Transformation Equations, and Representation Equations

At this point it is useful to distinguish three different types of equations with tensors, namely tensor equations, transformation equations, and representation equations:

- Tensor equations relate different tensors (tensor equations in symbolic notation) or coordinates of different tensors in the same coordinate system (tensor equations in index notation), e. g. $\underline{\underline{a}}\,\underline{b} = \underline{\underline{\underline{c}}}$ or $a_{ij}\, b_k = c_{ijk}$.
- Transformation equations relate the coordinates of the same tensor in different coordinate systems, e. g. $a_i = \alpha_{ij}\, \tilde{a}_j$.
- Representation equations relate a tensor itself and its coordinates in a given coordinate system, e. g. $\underline{\underline{a}} = a_{ij}\, \underline{e}_i\, \underline{e}_j$.

The summation convention applies to all three types of equations.

If tensors describe physical quantities, only tensor equations can be used to represent physical facts, because only tensor equations connect different tensors. In contrast, transformation equations and representation equations are mathematical equations without physical content.

2.8 δ-Tensor, ε-Tensor, Isotropic Tensors

2.8.1 The δ-Tensor

Clearly, $\alpha_{im}\, \alpha_{jn}\, \delta_{mn} = \alpha_{im}\, \alpha_{jm} = \delta_{ij}$, i. e. the δ_{ij} satisfy the transformation law (2.17) for the coordinates of a second-order polar tensor. Thus we can interpret the δ_{ij} as the Cartesian coordinates of a second-order polar tensor

$$\underline{\underline{\delta}} = \delta_{ij}\, \underline{e}_i\, \underline{e}_j \tag{2.29}$$

which is called the δ-tensor, identity tensor, metric tensor, or fundamental tensor. Clearly, it has the property that its coordinates are the same in every Cartesian coordinate system, namely they are equal to the δ_{ij} defined in (1.15).

2.8.2 The ε-Tensor

1. Now we want to show that the ε_{ijk} satisfy the transformation law (2.18) for the coordinates of a third-order axial tensor, i. e. that they are the Cartesian coordinates of an axial tensor

$$\underline{\underline{\underline{\varepsilon}}} = \varepsilon_{ijk}\, \underline{e}_i\, \underline{e}_j\, \underline{e}_k. \tag{2.30}$$

Since the ε_{ijk} are defined by (1.33), independently of a coordinate system, they must satisfy the equation $\varepsilon_{ijk} = (\det \underset{\sim}{\alpha})\, \alpha_{mi}\, \alpha_{nj}\, \alpha_{pk}\, \varepsilon_{mnp}$. Let

$$\tilde{\varepsilon}_{ijk} = (\det \underset{\sim}{\alpha})\, \alpha_{mi}\, \alpha_{nj}\, \alpha_{pk}\, \varepsilon_{mnp},$$

and then we have to show, using the properties of the ε_{ijk}, that $\tilde{\varepsilon}_{ijk} = \varepsilon_{ijk}$.

Only six of the 27 terms in the triple sum on the right side are nonzero, according to (1.33):

$$\begin{aligned}\tilde{\varepsilon}_{ijk} &= (\det \underset{\sim}{\alpha})\, [\alpha_{1i}\, \alpha_{2j}\, \alpha_{3k} + \alpha_{2i}\, \alpha_{3j}\, \alpha_{1k} + \alpha_{3i}\, \alpha_{1j}\, \alpha_{2k} \\ &\qquad - \alpha_{3i}\, \alpha_{2j}\, \alpha_{1k} - \alpha_{1i}\, \alpha_{3j}\, \alpha_{2k} - \alpha_{2i}\, \alpha_{1j}\, \alpha_{3k}].\end{aligned}$$

The bracket can clearly be written as a determinant, i. e.

$$\tilde{\varepsilon}_{ijk} = (\det \underset{\sim}{\alpha}) \begin{vmatrix} \alpha_{1i} & \alpha_{1j} & \alpha_{1k} \\ \alpha_{2i} & \alpha_{2j} & \alpha_{2k} \\ \alpha_{3i} & \alpha_{3j} & \alpha_{3k} \end{vmatrix}.$$

Since changing the order of two indices in $\tilde{\varepsilon}_{ijk}$ exchanges two columns and changes the sign of the determinant, it follows that $\tilde{\varepsilon}_{ijk}$, like ε_{ijk}, is antimetric with respect to all index pairs; i. e. analogously to (1.34) we have

$$\tilde{\varepsilon}_{ijk} = \tilde{\varepsilon}_{jki} = \tilde{\varepsilon}_{kij} = -\tilde{\varepsilon}_{kji} = -\tilde{\varepsilon}_{jik} = -\tilde{\varepsilon}_{ikj}.$$

If two indices have the same value, then two columns of the determinant are equal, i. e. the determinant is zero; so just like ε_{ijk}, $\tilde{\varepsilon}_{ijk}$ is also zero, if at least two indices are equal. Finally, it remains to show that $\tilde{\varepsilon}_{123} = 1$, and indeed for $i = 1,\ j = 2,\ k = 3$ we have

$$\tilde{\varepsilon}_{123} = (\det \underset{\sim}{\alpha})^2 = 1.$$

So we have $\tilde{\varepsilon}_{ijk} = \varepsilon_{ijk}$, which completes the proof.

2. Some authors use, instead of the ε-tensor, a polar tensor whose coordinates coincide in a right-handed system with the coordinates of the ε-tensor, i. e. with the ε_{ijk} from (1.33). To distinguish the two tensors we call this tensor $\underline{\underline{\underline{E}}}$. Then we have

$$\underline{\underline{\underline{E}}} = E_{ijk}\, \underline{e}_i\, \underline{e}_j\, \underline{e}_k, \qquad E_{ijk} = \begin{cases} \varepsilon_{ijk} & \text{in right-handed systems,} \\ -\varepsilon_{ijk} & \text{in left-handed systems.} \end{cases} \tag{2.31}$$

We will not use the tensor $\underline{\underline{\underline{E}}}$.

2.8.3 Isotropic Tensors

Tensors whose coordinates do not change by a transformation to another Cartesian coordinate system, such as the δ-tensor and the ε-tensor, are called isotropic tensors.

Clearly, every polar scalar is isotropic, and there are no isotropic axial scalars, other than zero, nor are there any (polar or axial) isotropic vectors, other than the zero vector. It can be shown that the most general isotropic tensors of order 2 and 3 have the coordinates $A\,\delta_{ij}$ and $A\,\varepsilon_{ijk}$, respectively, where A is an arbitrary polar scalar.

It can further be shown that the most general isotropic tensors of higher order are sums of tensor products of δ-tensors and ε-tensors. Therefore the most general combination of two δ-tensors constitutes the most general isotropic tensor of order 4; we can easily convince ourselves that it has the form

$$a_{ijkl} = A\,\delta_{ij}\,\delta_{kl} + B\,\delta_{ik}\,\delta_{jl} + C\,\delta_{il}\,\delta_{jk}, \tag{2.32}$$

where A, B, and C are arbitrary polar scalars.[12]

2.9 Scalar Multiplication of Tensors

2.9.1 Definition

The scalar product or inner product or dot product of two tensors is denoted by a dot between the two factors. It is obtained from the corresponding tensor product by setting equal the two indices adjacent to the gap between the two factors, i. e. by summing over these indices.

It is therefore defined for two tensors of order at least one.

$$\begin{array}{lll}
\underline{a}\cdot\underline{b} = A, & a_i\,b_i = A, & \underset{\sim}{a}^{\mathrm{T}}\,\underset{\sim}{b} = A, \\
\underline{a}\cdot\underline{\underline{b}} = \underline{B}, & a_i\,b_{ij} = B_j, & \underset{\sim}{a}^{\mathrm{T}}\,\underset{\sim}{b} = \underset{\sim}{B}^{\mathrm{T}}, \\
\underline{\underline{a}}\cdot\underline{b} = \underline{C}, & a_{ij}\,b_j = C_i, & \underset{\sim}{a}\,\underset{\sim}{b} = \underset{\sim}{C}, \\
\underline{\underline{a}}\cdot\underline{\underline{b}} = \underline{\underline{D}}, & a_{ij}\,b_{jk} = D_{ik}, & \underset{\sim}{a}\,\underset{\sim}{b} = \underset{\sim}{D}. \\
\underline{\underline{a}}\cdot\underline{\underline{\underline{b}}} = \underline{\underline{\underline{E}}}, & a_{ij}\,b_{jkl} = E_{ikl}, & \\
\text{etc.} & \text{etc.} &
\end{array} \tag{2.33}$$

Its name stems from the fact that the scalar product of two *vectors* is a *scalar*.

12 Equation (2.32) is another example of an equation that (without additional conventions) cannot be translated into symbolic notation.

2.9.2 Properties

1. First, we need to convince ourselves that the product of tensor coordinates defined in this way are again tensor coordinates. We do this using the example of the equation $a_i b_{ij} = B_j$ for polar tensors. Writing this equation in two different Cartesian coordinate systems, we get

$$a_i\, b_{ij} = B_j \quad \text{and} \quad \tilde{a}_i\, \tilde{b}_{ij} = \tilde{B}_j,$$

and since the a_i are the coordinates of a polar vector and the b_{ij} are the coordinates of a polar tensor, we have

$$a_i = \alpha_{ik}\, \tilde{a}_k \quad \text{and} \quad b_{ij} = \alpha_{im}\, \alpha_{jn}\, \tilde{b}_{mn}.$$

This gives

$$B_j = a_i\, b_{ij} = \alpha_{ik}\, \tilde{a}_k\, \alpha_{im}\, \alpha_{jn}\, \tilde{b}_{mn} = \delta_{km}\, \alpha_{jn}\, \tilde{a}_k\, \tilde{b}_{mn} = \alpha_{jn}\, \tilde{a}_m\, \tilde{b}_{mn} = \alpha_{jn}\, \tilde{B}_n,$$

thus the B_j are the coordinates of a polar vector.

2. Clearly, the scalar product of two polar tensors or two axial tensors gives a polar tensor and the scalar product of a polar tensor and an axial tensor gives an axial tensor.

3. Comparing the equations from (2.33) in the two notations shows that *for the translation* the order of the *free* indices must be equal on both sides of an equation. Furthermore, to be able to write the summation symbolically as a scalar product, the two summation indices must be adjacent to the gap between both tensors.

Problem 2.5.
Translate, as far as possible, into the other two notations:
A. $\underline{\underline{a}} \cdot \underline{\underline{b}} = \underline{\underline{c}}$, B. $\underline{\underline{b}} \cdot \underline{\underline{a}} = \underline{\underline{c}}$, C. $a_{ik}\, b_{ij} = c_{jk}$, D. $a_i\, b_k\, a_i\, c_k = d$,

E. $a_{ik}\, b_{ijk} = c_j$, F. $(\underline{\underline{a}} \cdot \underline{\underline{b}})^{\mathrm{T}} = \underline{\underline{c}}$, G. $(\underline{\underline{a}} \cdot \underline{\underline{b}} \cdot \underline{\underline{c}})^{\mathrm{T}} = \underline{\underline{d}}$.

Hint: Also in F and G (where transposed tensors in symbolic notation appear) only translate but do not transpose, i. e. the ‘T’ for transposition remains in the equation after translation. (In order to translate an equation from index notation we may have to first perform some transpositions; we never have to do this when we translate from symbolic notation.)

4. Addition and scalar multiplication are distributive also in symbolic notation, e. g. $a_i\, b_{ij} + a_i\, c_{ij} = a_i(b_{ij} + c_{ij})$ gives in symbolic notation

$$\underline{a} \cdot \underline{\underline{b}} + \underline{a} \cdot \underline{\underline{c}} = \underline{a} \cdot (\underline{\underline{b}} + \underline{\underline{c}}).$$

5. Clearly $\underline{a}\cdot\underline{b} = \underline{b}\cdot\underline{a}$; if both factors are vectors, the commutative law also holds in symbolic notation. If one of the factors is of second order or higher, the commutative law does not hold in general, e. g. $\underline{a}\cdot\underline{\underline{b}}$ is not equal to $\underline{\underline{b}}\cdot\underline{a}$. To show this, we assume $a_i\, b_{ij} = b_{ij}\, a_i$. To be able to translate this equation, we need to transpose b_{ij} on the right side. Then we have $a_i\, b_{ij} = b^{\mathrm{T}}_{ji}\, a_i$, which gives $\underline{a}\cdot\underline{\underline{b}} = \underline{\underline{b}}^{\mathrm{T}}\cdot\underline{a}$, but in general $\underline{\underline{b}}^{\mathrm{T}}\cdot\underline{a}$ is not equal to $\underline{\underline{b}}\cdot\underline{a}$.

6. The associate law holds in symbolic notation for an arbitrary sequence of tensor and scalar multiplications, if and only if in index notation no pair of summation indices is separated by other indices. The equation $(a_{ij}\, b_j)\, c_k = a_{ij}(b_j\, c_k)$ can be translated into $(\underline{\underline{a}}\cdot\underline{b})\,\underline{c} = \underline{\underline{a}}\cdot(\underline{b}\,\underline{c})$, but the equation $(a_{ij}\, b_j)\, c_i = a_{ij}(b_j\, c_i)$ cannot be translated into $(\underline{\underline{a}}\cdot\underline{b})\cdot\underline{c} = \underline{\underline{a}}\cdot(\underline{b}\cdot\underline{c})$. In symbolic notation the associative law holds, if for each tensor the sum of its adjacent dots is smaller or equal to its order. This is true for $\underline{\underline{a}}\cdot\underline{b}\,\underline{c}$ but not for $\underline{\underline{a}}\cdot\underline{b}\cdot\underline{c}$. (In this form, the statement is also true for the multiple scalar products defined below.)

Problem 2.6.

For which sequences of the factors on the left side can you translate the equation $a_i\, b_k\, a_i\, c_j = d_{jk}$ into symbolic notation? For each case, check if the associative law holds for all five possible associations.

7. Next, we ask ourselves in which cases we can cancel a common factor in scalar products.

We start with the equation $\underline{a}\cdot\underline{A} = \underline{b}\cdot\underline{A}$, which we can also write as $(\underline{a}-\underline{b})\cdot\underline{A} = 0$ or $(a_i - b_i)\, A_i = 0$.

Clearly, we cannot conclude from a zero scalar product of two vectors that one of its factors is zero. Even if $A_i \neq 0$, $a_i - b_i$ does not need to be zero; it can also be perpendicular to A_i. However, if the equation above is satisfied for three linearly independent vectors $\overset{1}{A}_i$, $\overset{2}{A}_i$, and $\overset{3}{A}_i$, then certainly $a_i - b_i = 0$, because no vector can be perpendicular to three linearly independent vectors simultaneously. In particular it follows from $(a_i - b_i)\, \underline{e}_i = \underline{0}$ or $a_i\, \underline{e}_i = b_i\, \underline{e}_i$ that $a_i = b_i$; we have already used this uniqueness of the decomposition of a vector with respect to a basis several times. Thus, since every vector can be represented as a linear combination of three Cartesian basis vectors, we can alternatively require that the equation above be satisfied for any vector $\underline{A}$.

To generalize this proof to the equation $a_{i\ldots jk}\, A_k = b_{i\ldots jk}\, A_k$, we start by writing the proof for $a_i\, A_i = b_i\, A_i$ more formally. The three equations

$$
\begin{aligned}
(a_i - b_i)\, A_i = 0 \quad &\text{or} \quad (a_1 - b_1)\, A_1 + (a_2 - b_2)\, A_2 + (a_3 - b_3)\, A_3 = 0,\\
(a_i - b_i)\, B_i = 0 \quad &\text{or} \quad (a_1 - b_1)\, B_1 + (a_2 - b_2)\, B_2 + (a_3 - b_3)\, B_3 = 0,\\
(a_i - b_i)\, C_i = 0 \quad &\text{or} \quad (a_1 - b_1)\, C_1 + (a_2 - b_2)\, C_2 + (a_3 - b_3)\, C_3 = 0,
\end{aligned}
$$

with known A_i, B_i, and C_i, form a system of homogeneous linear equations for the $a_i - b_i$. This system has only the trivial solution $a_i - b_i = 0$, if and only if the three vectors $\underline{A}$, $\underline{B}$, and $\underline{C}$ are linearly independent.

This proof generalizes to the next higher-order tensors in the equation $\underline{\underline{a}} \cdot \underline{A} = \underline{\underline{b}} \cdot \underline{A}$ or $(a_{ij} - b_{ij}) A_j = 0$ by using the same argument. The three equations

$$
\begin{aligned}
(a_{ij} - b_{ij}) A_j &= 0, \\
(a_{ij} - b_{ij}) B_j &= 0, \\
(a_{ij} - b_{ij}) C_j &= 0,
\end{aligned}
$$

with known A_j, B_j, and C_j, form a system of homogeneous linear equations for $a_{i1} - b_{i1}$, $a_{i2} - b_{i2}$, and $a_{i3} - b_{i3}$, for each value of the free index i. These three equations have only the trivial solution, if and only if the three vectors $\underline{A}$, $\underline{B}$, and $\underline{C}$ are linearly independent.

The proof for the equation $a_{i...jk} A_k = b_{i...jk} A_k$ is analogous. A vector which is a common factor in an equation with several scalar products can be cancelled out, if and only if arbitrary values can be substituted for this vector (e. g. three linearly independent vectors).

The equation $\underline{a} \cdot \underline{\underline{A}} = \underline{b} \cdot \underline{\underline{A}}$ or $(a_i - b_i) A_{ij} = 0$ with known $\underline{\underline{A}}$ forms a system of homogeneous linear equations for $\underline{a} - \underline{b}$. This system has only the trivial solution, if the determinant of A_{ij} is nonzero. This condition must hold for the equation $\underline{\underline{a}} \cdot \underline{\underline{A}} = \underline{\underline{b}} \cdot \underline{\underline{A}}$ or $(a_{ij} - b_{ij}) A_{jk} = 0$ for each value of the free index i separately. We conclude that a second-order tensor can be canceled out of several scalar products, if and only if its determinant does not vanish.

The equation $\underline{a} \cdot \underline{\underline{\underline{A}}} = \underline{b} \cdot \underline{\underline{\underline{A}}}$ or $(a_i - b_i) A_{ijk} = 0$ with known $\underline{\underline{\underline{A}}}$ forms a system of nine homogeneous linear equations for the three coordinates $a_i - b_i$: one equation for each value of the free indices j and k. At most three of these nine equations can be linearly independent. The system of equations has only the trivial solution, if and only if three equations are linearly independent. So we can cancel out the factor $\underline{\underline{\underline{A}}}$, if the matrix

$$
A_{ijk} = \begin{pmatrix}
A_{111} & A_{112} & A_{113} & A_{121} & A_{122} & A_{123} & A_{131} & A_{132} & A_{133} \\
A_{211} & A_{212} & A_{213} & A_{221} & A_{222} & A_{223} & A_{231} & A_{232} & A_{233} \\
A_{311} & A_{312} & A_{313} & A_{321} & A_{322} & A_{323} & A_{331} & A_{332} & A_{333}
\end{pmatrix}
$$

has rank 3. To generalize the proof to equations of the form $a_{i...jk} A_{klm} = b_{i...jk} A_{klm}$ we follow the same approach as above.

In the above matrix formed by the A_{ijk}, we call i a row index, since there corresponds to each value of i one row of the matrix, and we call j and k column indices, since there corresponds to each value pair jk a column of the matrix. With this notion we can now say under which condition a tensor of order 3 or higher can be canceled out. In the equation $a_{i...jk} A_{km...n} = b_{i...jk} A_{km...n}$ we can cancel out $A_{km...n}$, if the matrix formed by the coordinates $A_{km\ldots n}$, with the summation index k as row index and with the free indices $m \ldots n$ as column indices, has rank 3.

Clearly, the much more general condition that the common factor can assume arbitrary values is also sufficient.

8. We showed at the beginning of this section that the scalar product of two tensors results again in a tensor. We can now prove the converse, which is also called the quotient rule and of which we already used a special case in Section 2.3.1, when we introduced tensors: if the scalar product of two factors is a tensor of order $(m + n - 2)$ and one factor is a tensor of order m that can take on arbitrary values, then the other factor is a tensor of order n. We show this for the example $a_{ij} b_{jk} = c_{ik}$ for polar tensors. In two different Cartesian coordinate systems we have

$$a_{ij}\, b_{jk} = c_{ik} \quad \text{and} \quad \tilde{a}_{ij}\, \tilde{b}_{jk} = \tilde{c}_{ik}.$$

Assuming $\underline{\underline{a}}$ and $\underline{\underline{c}}$ are polar tensors, we have

$$\tilde{a}_{mj} = \alpha_{pm}\, \alpha_{qj}\, a_{pq}, \qquad c_{ik} = \alpha_{im}\, \alpha_{kn}\, \tilde{c}_{mn}.$$

This gives

$$\begin{aligned} a_{ij}\, b_{jk} &= c_{ik} = \alpha_{im}\, \alpha_{kn}\, \tilde{c}_{mn} = \alpha_{im}\, \alpha_{kn}\, \tilde{a}_{mj}\, \tilde{b}_{jn} \\ &= \underbrace{\alpha_{im}\, \alpha_{kn}\, \alpha_{pm}}_{\alpha_{kn}\, \delta_{ip}}\, \alpha_{qj}\, a_{pq}\, \tilde{b}_{jn} = \alpha_{kn}\, \alpha_{qj}\, a_{iq}\, \tilde{b}_{jn}, \end{aligned}$$

$$a_{iq}(b_{qk} - \alpha_{kn}\, \alpha_{qj}\, \tilde{b}_{jn}) = 0.$$

If a_{iq} can take any value, we can cancel it out and then we get

$$b_{qk} = \alpha_{qj}\, \alpha_{kn}\, \tilde{b}_{jn},$$

i. e. $\underline{\underline{b}}$ is a polar tensor of order 2.

9. We remember the geometric interpretation of the scalar product of two vectors: It is well known that it is equal to the product of the lengths of the two vectors and the cosine of the enclosed angle, i. e.

$$\underline{a} \cdot \underline{b} = |\underline{a}||\underline{b}| \cos(\underline{a}, \underline{b}).$$

If the two vectors form an acute angle, then the scalar product is positive; if they are perpendicular to each other, it is zero; if they enclose an obtuse angle, it is negative.

10. For the scalar product of a tensor $\mathscr{A}$ of arbitrary order with the δ-tensor it follows as a special case of (1.17) that

$$\mathscr{A} \cdot \underline{\underline{\delta}} = \mathscr{A}; \tag{2.34}$$

scalar multiplication of a tensor with the δ-tensor does not change the tensor. With respect to scalar multiplication, the δ-tensor has the same property as the number one for multiplication in arithmetics. This explains why the δ-tensor is also called identity tensor.

11. Finally we note the important identity

$$(\underline{\underline{a}} \cdot \underline{\underline{b}})^{\mathrm{T}} = \underline{\underline{b}}^{\mathrm{T}} \cdot \underline{\underline{a}}^{\mathrm{T}}, \qquad (a_{ij}\, b_{jk})^{\mathrm{T}} = b_{ij}^{\mathrm{T}}\, a_{jk}^{\mathrm{T}}, \qquad (\underset{\sim}{a}\, \underset{\sim}{b})^{\mathrm{T}} = \underset{\sim}{b}^{\mathrm{T}}\, \underset{\sim}{a}^{\mathrm{T}}, \tag{2.35}$$

which we can easily prove analogously to (1.48) by translating it into index notation.

Problem 2.7.
Prove the following identities by translating into index notation:
A. $(\underline{a}\,\underline{b})^{\mathrm{T}} = \underline{b}\,\underline{a}$, B. $\underline{a} \cdot \underline{\underline{b}} = \underline{\underline{b}}^{\mathrm{T}} \cdot \underline{a}$, C. $(\underline{\underline{a}} \cdot \underline{\underline{b}} \cdot \underline{\underline{c}})^{\mathrm{T}} = \underline{\underline{c}}^{\mathrm{T}} \cdot \underline{\underline{b}}^{\mathrm{T}} \cdot \underline{\underline{a}}^{\mathrm{T}}$.

12. In summary, the scalar product has the following properties.

- The scalar product of a tensor of order m and a tensor of order n is a tensor of order $(m + n - 2)$.
- If the scalar product of two factors is a tensor of order $(m + n - 2)$ and if one factor is a tensor of order m, which can assume arbitrary values, then the other factor is a tensor of order n (quotient rule).
- The three tensors of a scalar multiplication (i. e. the two factors and the product) are either all polar, or two of them are axial and the third is polar.
- *For translation* the order of the *free* indices in a scalar product must be the same on both sides of an equation.
- Addition and scalar multiplication are distributive also in symbolic notation.
- Scalar multiplication is in general commutative only if both factors are vectors.
- Any sequence of tensor and scalar multiplications is associative also in symbolic notation, if for each tensor the sum of its adjacent dots is smaller or equal to its order.
- A scalar product can be zero, even if none of its factors is zero.
- A common factor in an equation with several scalar products that can assume arbitrary values can be canceled out. In particular we have:
  - A vector as a common factor in a scalar product can be cancelled out if we can substitute three linearly independent vectors for it.
  - A second-order tensor as a common factor in a scalar product can be cancelled out if its determinant is nonzero.
  - A higher-order tensor as a common factor in a scalar product can be cancelled out if the matrix of its coordinates, with the summation index as row index and the free indices as column indices, has rank 3.

2.9.3 Contraction and Trace

1. Setting equal an index in both factors of a tensor product of two tensors defines, according to the summation convention, an arithmetic operation. This operation is called contraction of the two tensors on the two indices. A contraction is defined only in index notation. It builds a tensor of order $(m + n - 2)$ from a tensor of order m and a tensor of order n. The scalar product is a special case of a contraction, namely a contraction on two adjacent indices.

2. The operation defined by setting equal two indices of a single tensor is also called contraction. It reduces the order of the original tensor by two. The simplest contraction of a tensor, the contraction of a second-order tensor, has a special name; it is called the trace of a tensor and it is also expressed in symbolic notation.

$$\operatorname{tr} \underline{\underline{a}} = b, \qquad a_{ii} = b. \tag{2.36}$$

So the term contraction is used for two similar but different operations. The contraction of the two factors of a tensor multiplication is simultaneously a contraction of their tensor product.

2.9.4 Multiple Scalar Products

We denote the double scalar product or double inner product or double dot product by two adjacent dots. It is obtained by contracting the two index pairs adjacent to the gap between the two tensors.

$$\begin{aligned} \underline{\underline{a}} \cdot\cdot\, \underline{\underline{b}} &= A, & a_{ij}\, b_{ij} &= A, \\ \underline{\underline{a}} \cdot\cdot\, \underline{\underline{\underline{b}}} &= \underline{B}, & a_{ij}\, b_{ijk} &= B_k, \\ \underline{\underline{\underline{a}}} \cdot\cdot\, \underline{\underline{b}} &= \underline{C}, & a_{ijk}\, b_{jk} &= C_i, \\ &\text{etc.} & &\text{etc.} \end{aligned} \tag{2.37}$$

Multiple scalar products are defined accordingly, e. g.

$$\underline{\underline{\underline{a}}} \cdots \underline{\underline{\underline{b}}} = A, \qquad a_{ijk}\, b_{ijk} = A. \tag{2.38}$$

It is straightforward to generalize the highlighted properties of the single scalar product to multiple scalar products.

Some authors define the double scalar product and similarly multiple scalar products by a mirrored contraction of two neighboring index pairs; we suggest to denote it by two vertical dots to distinguish it from our definition, i. e.

$$
\begin{aligned}
\underline{\underline{a}} : \underline{\underline{b}} &= D, &\qquad a_{ij}\, b_{ji} &= D, \\
\underline{\underline{a}} : \underline{\underline{\underline{b}}} &= \underline{E}, &\qquad a_{ij}\, b_{jik} &= E_k, \\
\underline{\underline{\underline{a}}} \,\vdots\, \underline{\underline{\underline{b}}} &= F, &\qquad a_{ijk}\, b_{kji} &= F.
\end{aligned}
\tag{2.39}
$$

(As a mnemonic hint, the horizontal dots remind of the summation indices on the two sides of the dots and the vertical dots remind of a plane of reflection.)

Problem 2.8.
Translate, using the definitions in (2.37), into symbolic notation:
A. $a_{ij}\, b_{ij} = c$, B. $a_{ij}\, b_{ji} = c$, C. $a_{ij}\, b_{kjl}\, c_{ik} = d_l$, D. $a_{ij}\, b_{ikl}\, c_{klj} = d$.
Are parentheses necessary in symbolic notation in C and D?

Problem 2.9.
Compute
A. $\operatorname{tr} \underline{\underline{\delta}}$, B. $\underline{\underline{\delta}} \cdot \underline{\underline{\delta}}$, C. $\underline{\underline{\delta}} \cdot\cdot\, \underline{\underline{\delta}}$.

Problem 2.10.
A. Under which condition is the double scalar product of two second-order tensors commutative?
B. Under which condition can b_{ij} be cancelled out in $a_{ij}\, b_{ij} = 0$?

Problem 2.11.
A. Prove that the double scalar product of a symmetric tensor and an antimetric tensor is zero.

$$
a_{(ij)}\, b_{[ij]} = 0. \tag{2.40}
$$

B. Prove also the converse: If the double scalar product of two tensors is zero and one of the tensors is an arbitrary symmetric (antimetric) tensor, then the other tensor is antimetric (symmetric).
C. Prove also the following theorem: If the double scalar product of a second-order tensor with the ε-tensor is zero, then this tensor is symmetric (the converse is according to (2.40) true anyway).

$$
\varepsilon_{ijk}\, a_{jk} = 0 \quad \Longleftrightarrow \quad a_{jk} = a_{kj}. \tag{2.41}
$$

D. How many linearly independent symmetric tensors are necessary to write an arbitrary symmetric tensor as their linear combination?

2.10 Vector Multiplication of Tensors

2.10.1 Definition

1. Tensor multiplication assigns a (second-order) tensor to two vectors and scalar multiplication assigns a scalar. A *vector* can be assigned to two *vectors* by contracting twice with the ε-tensor. We define

$$\underline{a} \times \underline{b} := \underline{\underline{\underline{\varepsilon}}} \cdot\cdot\, \underline{a}\, \underline{b} = \underline{A}$$

and call $\underline{A}$ the vector product or outer product or cross product of the vectors $\underline{a}$ and $\underline{b}$. We write in index notation

$$(\underline{a} \times \underline{b})_i := \varepsilon_{ijk}\, a_j\, b_k = A_i.$$

Clearly, we can also write ε_{ijk} between the two vectors or behind them; if the contraction is translatable, we obtain the expressions

$$(\underline{a} \times \underline{b})_i := \varepsilon_{ijk}\, a_j\, b_k = -a_j\, \varepsilon_{jik}\, b_k = a_j\, b_k\, \varepsilon_{jki},$$

i. e. we have three equivalent definitions for the vector product of two vectors:

$$\underline{a} \times \underline{b} := \underline{\underline{\underline{\varepsilon}}} \cdot\cdot\, \underline{a}\, \underline{b} = -\underline{a} \cdot \underline{\underline{\underline{\varepsilon}}} \cdot \underline{b} = \underline{a}\, \underline{b} \cdot\cdot\, \underline{\underline{\underline{\varepsilon}}}.$$

Like the scalar product, the vector product is also a special case of a contraction.

To compute the vector product of two vectors, we remember that the elements of ε_{ijk} are nonzero only if all three indices are different. For example, for $i = 1$ only ε_{123} and ε_{132} are nonzero. Thus we have

$$(\underline{a} \times \underline{b})_1 = \varepsilon_{123}\, a_2\, b_3 + \varepsilon_{132}\, a_3\, b_2 = a_2\, b_3 - a_3\, b_2,$$
$$(\underline{a} \times \underline{b})_2 = \varepsilon_{231}\, a_3\, b_1 + \varepsilon_{213}\, a_1\, b_3 = a_3\, b_1 - a_1\, b_3,$$
$$(\underline{a} \times \underline{b})_3 = \varepsilon_{312}\, a_1\, b_2 + \varepsilon_{321}\, a_2\, b_1 = a_1\, b_2 - a_2\, b_1,$$

i. e. we can write for the components of the vector product

$$\underline{a} \times \underline{b} = (a_2\, b_3 - a_3\, b_2)\, \underline{e}_1 + (a_3\, b_1 - a_1\, b_3)\, \underline{e}_2 + (a_1\, b_2 - a_2\, b_1)\, \underline{e}_3$$

or in the more familiar form

$$\underline{a} \times \underline{b} = \begin{vmatrix} \underline{e}_1 & \underline{e}_2 & \underline{e}_3 \\ a_1 & a_2 & a_3 \\ b_1 & b_2 & b_3 \end{vmatrix}.$$

2. The ε-tensor is an axial tensor, so the vector product of two polar vectors or of two axial vectors gives an axial vector, and the vector product of a polar and an axial vector gives a polar vector. In other words, the three vectors of a vector product are either all axial or two are polar and the third is axial.

3. Some authors use the polar E-tensor from (2.31) instead of the axial ε-tensor to define the vector product. With this definition the result of the vector product of two polar vectors is a polar vector. We denote this vector product by the symbol $\tilde{\times}$ to distinguish it from our definition:

$$(\underline{a} \,\tilde{\times}\, \underline{b}) := \underline{\underline{\underline{E}}} \cdot\cdot\, \underline{a}\, \underline{b}, \qquad (\underline{a} \,\tilde{\times}\, \underline{b})_i = E_{ijk}\, a_j\, b_k. \tag{2.42}$$

Clearly, this definition is necessary, if the introduction of axial tensors is abandoned altogether. For example, the angular velocity vector is then assigned to the direction of rotation by the right-hand screw rule, independent of the coordinate system. The advantage of this definition is that all tensors are independent of the coordinate system, and the disadvantage is that the reflection symmetry of physical equations is violated. We will not use this vector product here.

4. We recall the geometric interpretation of the vector product $\underline{a} \times \underline{b}$. Its magnitude is equal to the area of the parallelogram spanned by the vectors $\underline{a}$ and $\underline{b}$:

$$|\underline{a} \times \underline{b}| = |\underline{a}||\underline{b}| \sin(\underline{a},\, \underline{b}).$$

The angle is the smaller of the two angles enclosed by the vectors $\underline{a}$ and $\underline{b}$, so that it is at most 180° and thus the sine is nonnegative. The vector $\underline{a} \times \underline{b}$ is perpendicular to the vectors $\underline{a}$ and $\underline{b}$ (i. e. to the plane spanned by them) and forms a right-hand screw with the direction of rotation obtained by rotating the vector $\underline{a}$ about the smaller of the two angles enclosed by $\underline{a}$ and $\underline{b}$ into $\underline{b}$ in a right-handed system, and a left-hand screw in a left-handed system.

5. Finally, we generalize also the vector product to higher-order tensors. According to the definition of the vector product, we can write the ε-tensor in front of the two vectors, between the two vectors, or behind them. We generalize to higher-order tensors using the formulation where the ε-tensor is written between the two factors.

$$\begin{aligned}
\underline{a} \times \underline{b} &:= -\underline{a} \cdot \underline{\underline{\underline{\varepsilon}}} \cdot \underline{b} = \underline{A}, \\
\underline{a} \times \underline{\underline{b}} &:= -\underline{a} \cdot \underline{\underline{\underline{\varepsilon}}} \cdot \underline{\underline{b}} = \underline{\underline{B}}, \\
\underline{\underline{a}} \times \underline{b} &:= -\underline{\underline{a}} \cdot \underline{\underline{\underline{\varepsilon}}} \cdot \underline{b} = \underline{\underline{C}}, \\
\underline{\underline{a}} \times \underline{\underline{b}} &:= -\underline{\underline{a}} \cdot \underline{\underline{\underline{\varepsilon}}} \cdot \underline{\underline{b}} = \underline{\underline{\underline{D}}}, \\
&\text{etc.} \\
(\underline{a} \times \underline{b})_j &:= -a_i\, \varepsilon_{ijk}\, b_k = a_i\, \varepsilon_{kji}\, b_k = A_j, \\
(\underline{a} \times \underline{\underline{b}})_{jm} &:= -a_i\, \varepsilon_{ijk}\, b_{km} = a_i\, \varepsilon_{kji}\, b_{km} = B_{jm}, \\
(\underline{\underline{a}} \times \underline{b})_{mj} &:= -a_{mi}\, \varepsilon_{ijk}\, b_k = a_{mi}\, \varepsilon_{kji}\, b_k = C_{mj}, \\
(\underline{\underline{a}} \times \underline{\underline{b}})_{mjn} &:= -a_{mi}\, \varepsilon_{ijk}\, b_{kn} = a_{mi}\, \varepsilon_{kji}\, b_{kn} = D_{mjn}, \\
&\text{etc.}
\end{aligned} \tag{2.43}$$

If the first factor of a vector product is a vector, then we can write the ε-tensor also in front of the two factors, i. e.

$$\begin{aligned}
&\underline{a} \times \underline{\underline{b}} = -\underline{a} \cdot \underline{\underline{\underline{\varepsilon}}} \cdot \underline{\underline{b}} = \underline{\underline{\underline{\varepsilon}}} \cdot\cdot\, \underline{a}\, \underline{\underline{b}}, \\
&(\underline{a} \times \underline{\underline{b}})_{jm} = -a_i\, \varepsilon_{ijk}\, b_{km} = \varepsilon_{jik}\, a_i\, b_{km},
\end{aligned}$$

and if the last factor of a vector product is a vector, then we can write the ε-tensor also behind the two factors, i. e.

$$\begin{aligned}
&\underline{\underline{a}} \times \underline{b} = -\underline{\underline{a}} \cdot \underline{\underline{\underline{\varepsilon}}} \cdot \underline{b} = \underline{\underline{a}}\, \underline{b} \cdot\cdot\, \underline{\underline{\underline{\varepsilon}}}, \\
&(\underline{\underline{a}} \times \underline{b})_{mj} = -a_{mi}\, \varepsilon_{ijk}\, b_k = a_{mi}\, b_k\, \varepsilon_{ikj}.
\end{aligned}$$

The vector product of two vectors is often written as $(\underline{a} \times \underline{b})_i = \varepsilon_{ijk}\, a_j\, b_k$.

Instead of the standard definition we can also introduce its negative and so avoid the minus sign. If we denote this unusual vector product with $\otimes$ (read: circled cross), then we have

$$\begin{aligned}
&\underline{a} \otimes \underline{b} = -\underline{a} \times \underline{b}, \\
&\mathscr{A} \otimes \mathscr{B} = -\mathscr{A} \times \mathscr{B}, \\
&(\underline{a} \otimes \underline{b})_q := a_p\, \varepsilon_{pqr}\, b_r, \\
&(\mathscr{A} \otimes \mathscr{B})_{i\ldots jqm\ldots n} := a_{i\ldots jp}\, \varepsilon_{pqr}\, b_{rm\ldots n}.
\end{aligned} \tag{2.44}$$

Problem 2.12.
Which of the following expressions are ambiguous without parentheses?
A. $\underline{a} \times \underline{b}\, \underline{c}$, B. $\underline{a} \times \underline{b} \cdot \underline{c}$, C. $\underline{\underline{a}} \cdot \underline{b} \times \underline{c}$, D. $\underline{a} \cdot \underline{\underline{b}} \times \underline{c}$.

Hint: First, check if both meanings, in A e. g. $(\underline{a} \times \underline{b})\, \underline{c}$ and $\underline{a} \times (\underline{b}\, \underline{c})$, are defined. If this is true, then check if the associative law holds. An expression is not unique without parentheses only if both meanings are defined and if the associative law does not hold.

Problem 2.13.
Translate the following expressions into the other notation. Then simplify each expression in index notation and translate the result into symbolic notation, thus obtaining a tensor algebraic identity in symbolic notation.
A. $\varepsilon_{ijk}\, a_m\, b_j\, \varepsilon_{min}\, c_k\, d_n$, B. $\varepsilon_{ijk}\, c_m\, b_{km}\, \varepsilon_{qpi}\, a_j\, d_q$, C. $\underline{a} \times [(\underline{\underline{b}} \cdot \underline{c}) \times \underline{d}]$,
D. $\underline{a} \times (\underline{b} \times \underline{c})$. E. Compare the expressions B and C.

Problem 2.14.
Use the Grassmann identity ("backwards") to rewrite the expression $a_i\, b_k - a_k\, b_i$ and translate it into symbolic notation.

2.10.2 Properties

Since the vector products in (2.43) are defined by scalar products, we can include the vector products in the associative law for scalar products.

From $\underline{a} \times \underline{b} := -\underline{a} \cdot \underline{\underline{\underline{\varepsilon}}} \cdot \underline{b}$ it follows that for both factors the cross is equivalent to the dot.

So a sequence of products of tensors, including vector products, is associative if for each tensor the sum of dots and crosses on each side of the tensor is at most equal to the order of the tensor. For example, this is not the case for $\underline{b}$ in $\underline{a} \times \underline{b} \times \underline{c} = \underline{a} \cdot \underline{\underline{\underline{\varepsilon}}} \cdot \underline{b} \cdot \underline{\underline{\underline{\varepsilon}}} \cdot \underline{c}$ and, as we know, $\underline{a} \times (\underline{b} \times \underline{c}) \neq (\underline{a} \times \underline{b}) \times \underline{c}$.

The remaining statements about vector products can be proven analogously to their corresponding statements about scalar products.

- The vector product of a tensor of order m and a tensor of order n gives a tensor of order $(m + n - 1)$.
- The three tensors of a vector multiplication (i. e. the two factors and the product) are either all axial or two of them are polar and the third is axial.
- *For the translation* the order of the *free* indices of a vector product must be the same on both sides of an equation.
- Addition and vector multiplication are distributive also in symbolic notation.

- A vector product is not only zero if one of the factors is zero.
- Vector multiplication is in general not commutative.
- Any sequence of products of tensors is associative also in symbolic form, if and only if the sum of the scalar product dots and the vector product crosses on both sides of each tensor is at most equal to the order of the tensor.
- A common factor that can assume arbitrary values can be canceled out in an equation with several vector products.

2.10.3 Triple Product

1. A vector product followed by a scalar product between three vectors is often called the triple product of these three vectors; we denote the triple product by square brackets. Using (1.28) we have the following formulas.

$$[\underline{a}, \underline{b}, \underline{c}] = \underline{a} \times \underline{b} \cdot \underline{c} = \underline{a} \cdot \underline{b} \times \underline{c} = A,$$
$$\varepsilon_{ijk}\, a_i\, b_j\, c_k = \begin{vmatrix} a_1 & a_2 & a_3 \\ b_1 & b_2 & b_3 \\ c_1 & c_2 & c_3 \end{vmatrix} = A. \tag{2.45}$$

(Here we do not need to set parentheses because only $(\underline{a}\times\underline{b})\cdot\underline{c}$ is defined, but $\underline{a}\times(\underline{b}\cdot\underline{c})$ is not defined.)

If e. g. all three vectors are polar, then the triple product is an axial scalar; instead of axial scalar often the term pseudoscalar is used.

2. We recall the geometric interpretation of the triple product. Its magnitude is equal to the volume of the parallelepiped spanned by the three vectors $\underline{a}$, $\underline{b}$, and $\underline{c}$; it is zero if the three vectors are coplanar. This includes the case that one of the three vectors is zero.

The triple product is positive if the vector product $\underline{a} \times \underline{b}$ forms an acute angle with $\underline{c}$. If this is the case for three polar vectors in a right-handed system, then we say the three vectors $\underline{a}$, $\underline{b}$, and $\underline{c}$ (in this order) form a right-handed system. If this is true for three polar vectors in a left-handed system, then we say they form (in this order) a left-handed system.

A triple product changes its sign if we change the order of two vectors. The sign does not change if we permute the three vectors cyclically.

3. For the product of two triple products we have

$$[\underline{a}, \underline{b}, \underline{c}][\underline{d}, \underline{e}, \underline{f}] \overset{(2.45)}{=} \begin{vmatrix} a_1 & a_2 & a_3 \\ b_1 & b_2 & b_3 \\ c_1 & c_2 & c_3 \end{vmatrix} \begin{vmatrix} d_1 & e_1 & f_1 \\ d_2 & e_2 & f_2 \\ d_3 & e_3 & f_3 \end{vmatrix} \overset{(1.13)}{=} \begin{vmatrix} \underline{a}\cdot\underline{d} & \underline{a}\cdot\underline{e} & \underline{a}\cdot\underline{f} \\ \underline{b}\cdot\underline{d} & \underline{b}\cdot\underline{e} & \underline{b}\cdot\underline{f} \\ \underline{c}\cdot\underline{d} & \underline{c}\cdot\underline{e} & \underline{c}\cdot\underline{f} \end{vmatrix},$$

$$[\underline{a}, \underline{b}, \underline{c}][\underline{d}, \underline{e}, \underline{f}] = \begin{vmatrix} \underline{a}\cdot\underline{d} & \underline{a}\cdot\underline{e} & \underline{a}\cdot\underline{f} \\ \underline{b}\cdot\underline{d} & \underline{b}\cdot\underline{e} & \underline{b}\cdot\underline{f} \\ \underline{c}\cdot\underline{d} & \underline{c}\cdot\underline{e} & \underline{c}\cdot\underline{f} \end{vmatrix}. \tag{2.46}$$

2.11 Summary of Tensor-Algebraic Operations

For convenience we summarize the tensor-algebraic operations in the following table. If an operation is defined for tensors of different orders, we give a typical example for each case.

Name	Index notation	Symbolic notation
For one tensor:		
transpose (order 2 only)	$a^{\mathrm{T}}_{ij} := a_{ji}$	$\underline{\underline{a}}^{\mathrm{T}}$
contraction (order 2 and higher)	$a_{ijik} = b_{jk}$	
also multiple:	$a_{ijij} = b$	
in particular for order 2: computation of the trace	$a_{ii} = b$	$\mathrm{tr}\,\underline{\underline{a}} = b$
For two tensors:		
equality (only for the same order and polarity)	$a_{ij} = b_{ij}$	$\underline{\underline{a}} = \underline{\underline{b}}$
addition (subtraction) (only for the same order and polarity)	$a_{ij} \pm b_{ij} = c_{ij}$	$\underline{\underline{a}} \pm \underline{\underline{b}} = \underline{\underline{c}}$
tensor multiplication	$a_i\, b_{jk} = c_{ijk}$	$\underline{a}\,\underline{\underline{b}} = \underline{\underline{\underline{c}}}$
contraction (order 1 and higher)	$a_{ijk}\, b_{mkn} = c_{ijmn}$	
also multiple:	$a_{ijk}\, b_{mki} = c_{jm}$	
in particular: scalar multiplication	$a_{ijk}\, b_{km} = c_{ijm}$	$\underline{\underline{\underline{a}}} \cdot \underline{\underline{b}} = \underline{\underline{\underline{c}}}$
also multiple:	$a_{ijk}\, b_{jk} = c_i$	$\underline{\underline{\underline{a}}} \cdot\cdot\, \underline{\underline{b}} = \underline{c}$
in particular: vector multiplication	$a_k\, \varepsilon_{ijk}\, b_{im} = c_{jm}$	$\underline{a} \times \underline{\underline{b}} = \underline{\underline{c}}$
For three vectors:		
computation of the triple product	$\varepsilon_{ijk}\, a_i\, b_j\, c_k = d$	$[\underline{a}, \underline{b}, \underline{c}] = d$

2.12 Differential Operations

We now turn to tensor fields, e. g. tensors whose coordinates are functions of the position. Analogously to the three products we define three differential operators for tensor fields: gradient, divergence, and curl.

2.12.1 The Fundamental Theorem of Tensor Analysis

If we differentiate a scalar field $a(x, y, z)$ with respect to the spatial coordinates, we get the coordinates $(\partial a/\partial x, \partial a/\partial y, \partial a/\partial z)$ of a vector field, which is called the gradient of a. We can show in general that the derivatives of the Cartesian coordinates of a tensor field of order n with respect to the spatial coordinates form the Cartesian coordinates of a tensor field of order $(n+1)$. This statement is called the fundamental theorem of tensor analysis.

We prove it here using the example of a second-order polar tensor field $a_{ij}(\underline{x})$, where the argument $\underline{x}$ stands for the Cartesian spatial coordinates (x_1, x_2, x_3). According to (2.17) we have

$$a_{ij}(\underline{x}) = \alpha_{im}\,\alpha_{jn}\,\tilde{a}_{mn}(\underline{x}) = \alpha_{im}\,\alpha_{jn}\,\tilde{a}_{mn}(\underline{x}(\underline{\tilde{x}})) = \alpha_{im}\,\alpha_{jn}\,\tilde{a}_{mn}(\underline{\tilde{x}}),$$

$$\mathrm{d}\,a_{ij} = \alpha_{im}\,\alpha_{jn}\,\frac{\partial \tilde{a}_{mn}}{\partial \tilde{x}_q}\,\mathrm{d}\,\tilde{x}_q,$$

$$\frac{\partial a_{ij}}{\partial x_p} = \alpha_{im}\,\alpha_{jn}\,\frac{\partial \tilde{a}_{mn}}{\partial \tilde{x}_q}\,\frac{\partial \tilde{x}_q}{\partial x_p}.$$

Further from (2.9) it follows that

$$\frac{\partial \tilde{x}_q}{\partial x_p} = \alpha_{pq},$$

and thus we have

$$\frac{\partial a_{ij}}{\partial x_p} = \alpha_{im}\,\alpha_{jn}\,\alpha_{pq}\,\frac{\partial \tilde{a}_{mn}}{\partial \tilde{x}_q},$$

which is the transformation law for the coordinates of a third-order polar tensor.

2.12.2 Gradient

In symbolic notation we start with the tensor field $\underline{\underline{a}}(\underline{x}) = a_{ij}\,\underline{e}_i\,\underline{e}_j$. If we differentiate with respect to x_k, we have, because the bases are independent of the position,

$$\frac{\partial \underline{\underline{a}}}{\partial x_k} = \frac{\partial a_{ij}}{\partial x_k}\,\underline{e}_i\,\underline{e}_j.$$

Since the $\partial a_{ij}/\partial x_k$ are the coordinates of a tensor of order 3, we obtain from $\partial \underline{\underline{a}}/\partial x_k$ a tensor of order 3 by multiplying it by $\underline{e}_k$. This tensor is called the gradient[13] of the original tensor $\underline{\underline{a}}$.

The gradient of a polar tensor field of order n gives a polar tensor field of order $(n+1)$, and the gradient of an axial tensor field of order n gives an axial tensor field of order $(n+1)$.

Since the tensor product of basis vectors is not commutative, we have two different definitions of the gradient, depending on whether we multiply $\partial \underline{\underline{a}}/\partial x_k$ by $\underline{e}_k$ from the left or from the right; they can be distinguished as the left gradient and the right gradient. Both definitions are used in the literature, so one has to check each time, when the gradient of a tensor of at least order 1 is used, how the gradient is defined.

These two different definitions of the gradient depend on which convention is used for the order of the indices in the spatial derivative of tensor coordinates. If the denominator index is put behind the numerator indices, i. e.

$$\frac{\partial a_{ij}}{\partial x_k} = A_{ijk},$$

then contraction with the corresponding basis gives the right gradient

$$\underbrace{\frac{\partial a_{ij}}{\partial x_k}\,\underline{e}_i\,\underline{e}_j\,\underline{e}_k}_{\operatorname{grad}_{\mathrm{R}} \underline{\underline{a}}} = \underbrace{A_{ijk}\,\underline{e}_i\,\underline{e}_j\,\underline{e}_k}_{\underline{\underline{\underline{A}}}}\,;$$

if the denominator index is put in front of the numerator indices, i. e.

$$\frac{\partial a_{ij}}{\partial x_k} = B_{kij},$$

then contraction with the corresponding basis gives the left gradient

$$\underbrace{\frac{\partial a_{ij}}{\partial x_k}\,\underline{e}_k\,\underline{e}_i\,\underline{e}_j}_{\operatorname{grad}_{\mathrm{L}} \underline{\underline{a}}} = \underbrace{B_{kij}\,\underline{e}_k\,\underline{e}_i\,\underline{e}_j}_{\underline{\underline{\underline{B}}}}\,.$$

Introducing the del operator[14]

$$\underline{\nabla} := \frac{\partial}{\partial x_k}\,\underline{e}_k \tag{2.47}$$

we can define the two gradients as tensor products with the del operator

$$\operatorname{grad}_{\mathrm{R}} \mathscr{A} := \mathscr{A}\,\underline{\nabla}, \qquad \operatorname{grad}_{\mathrm{L}} \mathscr{A} := \underline{\nabla}\,\mathscr{A}, \tag{2.48}$$

13 *Gradiens* Latin: (steepest) stepping (increase) (present participle of *gradi* [to step/walk]).

14 Also called Nabla operator.

where in the first case del acts on the tensor from the right and in the second case it acts on the tensor from the left.

Since tensor multiplication of a tensor by a scalar is commutative, the two gradients of a scalar coincide, i. e.

$$\operatorname{grad} a = a \underline{\nabla} = \underline{\nabla}\, a .$$

Thus the left and the right gradient are generalizations of the gradient from vector calculus. However, for tensors of at least order 1 we have to make a decision. We choose here the right gradient, because the order "numerator index in front of denominator index" is how we read a differential quotient $\partial a_{ij}/\partial x_k$. So we agree on the following convention.

The last index of a gradient is the differentiation index.

With this convention we can drop the index R in grad_R from now on.

For example, we write

$$\operatorname{grad} \underline{\underline{a}} = \underline{\underline{a}}\, \underline{\nabla} = a_{ij}\, \underline{e}_i\, \underline{e}_j \frac{\partial}{\partial x_k}\, \underline{e}_k = \frac{\partial a_{ij}}{\partial x_k}\, \underline{e}_i\, \underline{e}_j\, \underline{e}_k .$$

For tensors of various order we have the following definitions.

$$\begin{aligned} \operatorname{grad} a &:= \frac{\partial a}{\partial x_k}\, \underline{e}_k , \\ \operatorname{grad} \underline{a} &:= \frac{\partial a_i}{\partial x_k}\, \underline{e}_i\, \underline{e}_k , \\ \operatorname{grad} \underline{\underline{a}} &:= \frac{\partial a_{ij}}{\partial x_k}\, \underline{e}_i\, \underline{e}_j\, \underline{e}_k , \\ &\text{etc.} \end{aligned} \tag{2.49}$$

In both notations this is written as follows.

$$\begin{aligned} \operatorname{grad} a &= \underline{A}, & \frac{\partial a}{\partial x_k} &= A_k , \\ \operatorname{grad} \underline{a} &= \underline{\underline{A}}, & \frac{\partial a_i}{\partial x_k} &= A_{ik} , \\ \operatorname{grad} \underline{\underline{a}} &= \underline{\underline{\underline{A}}}, & \frac{\partial a_{ij}}{\partial x_k} &= A_{ijk} , \\ &\text{etc.} & &\text{etc.} \end{aligned} \tag{2.50}$$

Some authors use the del operator instead of the grad symbol in symbolic notation and use it in computations like a vector. However, we need additional rules for computations with this "symbolic vector", since it is not a real vector but only behaves like a vector in some ways. For that reason we use del only in definitions and to distinguish the different differential operations in tensor analysis. We will not use it in computations, where we will use index notation instead, like we did in tensor algebra.

2.12.3 (Total) Differential

1. Consider two points with the coordinates x_i and x_i+dx_i, i. e. with the position vectors $\underline{x}$ and $\underline{x} + \mathrm{d}\,\underline{x}$, which do not have to be neighboring points. The difference $\mathrm{d}\,\underline{x}$ between the two position vectors is called the differential of the position vector $\underline{x}$.

Differentials of the position vector in two different coordinate systems are related by

$$\mathrm{d}\,x_i = \frac{\partial x_i}{\partial \tilde{x}_j}\,\mathrm{d}\,\tilde{x}_j,$$

and according to (2.9) we have

$$\mathrm{d}\,x_i = \alpha_{ij}\,\mathrm{d}\,\tilde{x}_j, \tag{2.51}$$

i. e. the differential of the position vector is a polar vector. We write in different notations

$$\mathrm{d}\,\underline{x} = \mathrm{d}\,x_i\,\underline{e}_i = \underbrace{\mathrm{d}\,x\,\underline{e}_x}_{\mathrm{d}\,\underline{x}_1} + \underbrace{\mathrm{d}\,y\,\underline{e}_y}_{\mathrm{d}\,\underline{x}_2} + \underbrace{\mathrm{d}\,z\,\underline{e}_z}_{\mathrm{d}\,\underline{x}_3}, \tag{2.52}$$

where $\mathrm{d}\,\underline{x}_1$, $\mathrm{d}\,\underline{x}_2$, and $\mathrm{d}\,\underline{x}_3$ are the components of $\mathrm{d}\,\underline{x}$ in the coordinate system under consideration.

With our definition of the gradient of a tensor we define the (total) differential of a tensor in the two notations.

$$\begin{aligned}
\mathrm{d}\,a &:= \operatorname{grad} a \cdot \mathrm{d}\,\underline{x}, & \mathrm{d}\,a &:= \frac{\partial a}{\partial x_k}\,\mathrm{d}\,x_k,\\
\mathrm{d}\,\underline{a} &:= (\operatorname{grad} \underline{a}) \cdot \mathrm{d}\,\underline{x}, & \mathrm{d}\,a_i &:= \frac{\partial a_i}{\partial x_k}\,\mathrm{d}\,x_k,\\
\mathrm{d}\,\underline{\underline{a}} &:= (\operatorname{grad} \underline{\underline{a}}) \cdot \mathrm{d}\,\underline{x}, & \mathrm{d}\,a_{ij} &:= \frac{\partial a_{ij}}{\partial x_k}\,\mathrm{d}\,x_k,\\
&\text{etc.} & &\text{etc.}
\end{aligned} \tag{2.53}$$

The differential of a tensor depends clearly on the point $\underline{x}$ at which we compute the gradient and on the differential $\mathrm{d}\,\underline{x}$. Geometrically, it represents the increase of the tensor field on the tangential hyperplane of the tensor field at the point $\underline{x}$.

Sometimes instead of "the differential of a tensor in index notation" we also speak of "the differential of tensor coordinates".

If we use the left gradient instead of the right gradient in the formulas above, then in symbolic notation, the order of the factors changes.

Since the basis vectors are independent of the position, the differential of a tensor and the differential of its coordinates are directly related. For example, we have

$$\mathrm{d}\,\underline{a} = \mathrm{d}\,(a_i\,\underline{e}_i) = \mathrm{d}\,a_i\,\underline{e}_i,$$

and for tensors of various order we write the following equations.

$$\begin{aligned} \mathrm{d}\,a &= \mathrm{d}\,a, \\ \mathrm{d}\,\underline{a} &= \mathrm{d}\,a_i\,\underline{e}_i, \\ \mathrm{d}\,\underline{\underline{a}} &= \mathrm{d}\,a_{ij}\,\underline{e}_i\,\underline{e}_j, \\ &\text{etc.} \end{aligned} \tag{2.54}$$

2. For neighboring points with the position vectors $\underline{x} + \mathrm{d}\,\underline{x}$ and $\underline{x}$, the differential of tensors and the differential of tensor coordinates, respectively, are equal to the differences of the tensors and tensor coordinates at the two points, up to second order in $\mathrm{d}\,\underline{x}$.

$$\begin{aligned} \mathrm{d}\,a &= a(\underline{x} + \mathrm{d}\,\underline{x}) - a(\underline{x}), \\ \mathrm{d}\,\underline{a} &= \underline{a}(\underline{x} + \mathrm{d}\,\underline{x}) - \underline{a}(\underline{x}), \\ \mathrm{d}\,\underline{\underline{a}} &= \underline{\underline{a}}(\underline{x} + \mathrm{d}\,\underline{x}) - \underline{\underline{a}}(\underline{x}), \\ &\text{etc.} \\ \mathrm{d}\,a &= a(x_p + \mathrm{d}\,x_p) - a(x_p), \\ \mathrm{d}\,a_i &= a_i(x_p + \mathrm{d}\,x_p) - a_i(x_p), \\ \mathrm{d}\,a_{ij} &= a_{ij}(x_p + \mathrm{d}\,x_p) - a_{ij}(x_p), \\ &\text{etc.} \end{aligned} \tag{2.55}$$

2.12.4 Divergence

We obtain the divergence[15] from the gradient, if we replace the tensor product of the tensor field and the del by the scalar product, i. e.

$$\operatorname{div}_{\mathrm{R}} \mathscr{A} := \mathscr{A} \cdot \underline{\nabla}, \qquad \operatorname{div}_{\mathrm{L}} \mathscr{A} := \underline{\nabla} \cdot \mathscr{A}. \tag{2.56}$$

So the divergence of a tensor field is defined for tensor fields of at least order 1, and since the scalar product of two vectors is commutative, the right divergence and left divergence of a vector field coincide; thus both divergences are generalizations of the divergence from vector calculus.

Clearly, the divergence of a polar tensor field of order n gives a polar tensor field of order $(n-1)$, and the divergence of an axial tensor field of order n gives a tensor field of order $(n-1)$.

We will use the right divergence from now on and thus drop the index R in $\operatorname{div}_{\mathrm{R}}$. So we agree on the following convention.

The differentiation index in a divergence is equal to the last index of the tensor whose divergence is computed.

For example, we have

$$\operatorname{div} \underline{\underline{a}} = \underline{\underline{a}} \cdot \underline{\nabla} = a_{ij}\, \underline{e}_i\, \underline{e}_j \cdot \frac{\partial}{\partial x_k}\, \underline{e}_k = \frac{\partial a_{ij}}{\partial x_k}\, \underline{e}_i \underbrace{\underline{e}_j \cdot \underline{e}_k}_{\delta_{jk}} = \frac{\partial a_{ij}}{\partial x_j}\, \underline{e}_i .$$

For tensors of various order we have the following definitions.

$$\begin{aligned} \operatorname{div} \underline{a} &:= \frac{\partial a_k}{\partial x_k}, \\ \operatorname{div} \underline{\underline{a}} &:= \frac{\partial a_{ik}}{\partial x_k}\, \underline{e}_i, \\ \operatorname{div} \underline{\underline{\underline{a}}} &:= \frac{\partial a_{ijk}}{\partial x_k}\, \underline{e}_i\, \underline{e}_j, \end{aligned} \tag{2.57}$$

etc.

15 *Divergentia* Latin: divergence, also source strength (verbal noun to modern Latin *divergere* [to diverge], derived from *vergere* [to slope]).

In the two notations this is written as follows.

$$\begin{aligned}
&\operatorname{div} \underline{a} = A, && \frac{\partial a_k}{\partial x_k} = A, \\
&\operatorname{div} \underline{\underline{a}} = \underline{A}, && \frac{\partial a_{ik}}{\partial x_k} = A_i, \\
&\operatorname{div} \underline{\underline{\underline{a}}} = \underline{\underline{A}}, && \frac{\partial a_{ijk}}{\partial x_k} = A_{ij}, \\
&\text{etc.} && \text{etc.}
\end{aligned} \tag{2.58}$$

Problem 2.15.
Translate into the other notation:

A. $\operatorname{div} \operatorname{grad} \underline{\underline{a}} = \underline{\underline{b}}$, B. $\operatorname{grad} \operatorname{div} \underline{\underline{a}} = \underline{\underline{c}}$, C. $\frac{\partial a_i}{\partial x_j} \frac{\partial b_j}{\partial x_k} = c_{ik}$, D. $\frac{\partial a_i}{\partial x_j} \frac{\partial b_i}{\partial x_j} = c$.

2.12.5 Curl

We obtain the curl from the gradient, if we replace the tensor product of the tensor field and the del by the vector product.

So also the curl of a tensor field is defined for tensor fields of at least order 1. Since the sign of a vector product changes if we change the order of the factors, and since we want both the right curl and the left curl to coincide for a vector field with the curl from vector calculus, we define

$$\operatorname{curl}_R \mathscr{A} := \mathscr{A} \otimes \underline{\nabla} = -\mathscr{A} \times \underline{\nabla}, \qquad \operatorname{curl}_L \mathscr{A} := \underline{\nabla} \times \mathscr{A}. \tag{2.59}$$

The curl of a polar tensor field of order n clearly gives an axial tensor field of order n, and the curl of an axial tensor field of order n gives a polar tensor field of order n.

From now on we use the right curl, so we will drop the index R in curl_R. We write e. g.

$$\operatorname{curl} \underline{\underline{a}} = \underline{\underline{a}} \otimes \underline{\nabla}.$$

We translate, starting from the right side, so we obtain

$$a_{mi}\, \varepsilon_{ijk} \frac{\partial}{\partial x_k} = \frac{\partial a_{mi}}{\partial x_k} \varepsilon_{ijk} = (\operatorname{curl} \underline{\underline{a}})_{mj},$$

$$\operatorname{curl} \underline{\underline{a}} = \frac{\partial a_{mi}}{\partial x_k} \varepsilon_{ijk}\, \underline{e}_m\, \underline{e}_j.$$

In contrast to the gradient and the divergence we have to be careful in $a_{mi}\, \varepsilon_{ijk}\, \partial/\partial x_k$ to pull the differentiation in front of the epsilon and not the coordinate behind the epsilon; e. g. in $\varepsilon_{ijk}\,\partial a_{mi}/\partial x_k = (\text{curl}\,\underline{\underline{a}})_{jm}$ the order of the free indices is reversed, so it would be a different tensor.

For tensors of various order we have the following definitions.

$$
\begin{aligned}
\text{curl}\,\underline{a} &:= \frac{\partial a_i}{\partial x_k}\,\varepsilon_{ijk}\,\underline{e}_j,\\
\text{curl}\,\underline{\underline{a}} &:= \frac{\partial a_{mi}}{\partial x_k}\,\varepsilon_{ijk}\,\underline{e}_m\,\underline{e}_j,\\
\text{curl}\,\underline{\underline{\underline{a}}} &:= \frac{\partial a_{mni}}{\partial x_k}\,\varepsilon_{ijk}\,\underline{e}_m\,\underline{e}_n\,\underline{e}_j,\\
&\text{etc.}
\end{aligned}
\qquad (2.60)
$$

In both notations this is written as follows.

$$
\begin{aligned}
\text{curl}\,\underline{a} &= \underline{A}, & \frac{\partial a_i}{\partial x_k}\,\varepsilon_{ijk} &= A_j,\\
\text{curl}\,\underline{\underline{a}} &= \underline{\underline{A}}, & \frac{\partial a_{mi}}{\partial x_k}\,\varepsilon_{ijk} &= A_{mj},\\
\text{curl}\,\underline{\underline{\underline{a}}} &= \underline{\underline{\underline{A}}}, & \frac{\partial a_{mni}}{\partial x_k}\,\varepsilon_{ijk} &= A_{mnj},\\
&\text{etc.} & &\text{etc.}
\end{aligned}
\qquad (2.61)
$$

Problem 2.16.
Compute
A. grad $\underline{x}$, B. div $\underline{x}$, C. curl $\underline{x}$.

Problem 2.17.
For each of the following expressions, derive a tensor analytical identity in index notation and then translate this identity into symbolic notation:

A. div $(\lambda\,\underline{a})$, B. curl $(\lambda\,\underline{a})$, C. curl grad a, D. div curl $\underline{a}$,

E. curl curl $\underline{a}$, F. $\varepsilon_{ijk}\,\varepsilon_{mnk}\,\dfrac{\partial a_{pj}}{\partial x_i}\,b_m$, G. $\varepsilon_{ijk}\,\varepsilon_{mnk}\,\dfrac{\partial a_{pj} b_i}{\partial x_m}$.

Hint: The last two problems are slightly more difficult, because they have two free indices. If this is the case, we have to check, before the translation, if the order of the free indices matches in all terms.

2.12.6 Laplace Operator

The divergence of a gradient is also called the Laplace operator or the Laplacian and is denoted by Δ.

$$\Delta := \operatorname{div}\operatorname{grad}\,. \tag{2.62}$$

The Laplace operator of a polar tensor field of order n gives a polar tensor field of order n, and the Laplace operator of an axial tensor field of order n gives an axial tensor field of order n. For example, $\operatorname{div}\operatorname{grad}\underline{\underline{a}}$ gives translated

$$\frac{\partial}{\partial x_k}\frac{\partial a_{ij}}{\partial x_k} = \frac{\partial^2 a_{ij}}{\partial x_k^2}\,,$$

so for tensors of various order we have the following equations.

$$\begin{aligned}\Delta a &= \frac{\partial^2 a}{\partial x_k^2}\,,\\ \Delta \underline{a} &= \frac{\partial^2 a_i}{\partial x_k^2}\,\underline{e}_i\,,\\ \Delta \underline{\underline{a}} &= \frac{\partial^2 a_{ij}}{\partial x_k^2}\,\underline{e}_i\,\underline{e}_j\,,\\ &\text{etc.}\end{aligned} \tag{2.63}$$

2.13 Rules for Indices and Underlines

We would like to mention that the notation we have introduced leads to rules which must be satisfied by all members of a well-posed equation, in index notation and in symbolic notation. While doing computations, it often turns out worthwhile to check if these rules are satisfied.

An equation in index notation must satisfy the rules from Section 1.1, where we introduced the summation convention: A summation index can appear only once or twice in a term and all terms of an equation must match in their free indices.

For an equation in symbolic notation, the number of underlines must match in all terms of an equation while taking into consideration the effect of the arithmetic symbols. Each dot reduces the order of the tensor by two, each vector product and each divergence reduces the order of the tensor by one, and each gradient increases the order of the tensor by one. The other operations, i. e. the tensor product, the curl, the differential, and the Laplace operator, do not change the order of the tensor.

Finally, for translating between the two notations the order of the free indices in index notation must match in all terms before or after the translation.

2.14 Integrals of Tensor Fields

The coordinate of a tensor field (in the following we assume as the simplest case a polar scalar) is a function of the three spatial coordinates. So we can compute the following spatial integrals:

$$\begin{aligned}&\int a(x,y,z)\,\mathrm{d}x,\\&\int a(x,y,z)\,\mathrm{d}y,\\&\int a(x,y,z)\,\mathrm{d}z,\end{aligned} \tag{2.64}$$

$$\begin{aligned}&\iint a(x,y,z)\,\mathrm{d}y\,\mathrm{d}z,\\&\iint a(x,y,z)\,\mathrm{d}z\,\mathrm{d}x,\\&\iint a(x,y,z)\,\mathrm{d}x\,\mathrm{d}y,\end{aligned} \tag{2.65}$$

$$\iiint a(x,y,z)\,\mathrm{d}x\,\mathrm{d}y\,\mathrm{d}z. \tag{2.66}$$

We will see that the first three integrals can be interpreted as line integrals, the second three as surface integrals, and the last as a volume integral.

2.14.1 Line Integrals of Tensor Coordinates

1. Clearly, the line element from a point $\underline{x}$ on a curve to a neighboring point $\underline{x} + \mathrm{d}\,\underline{x}$ on the curve is the differential $\mathrm{d}\,\underline{x}$ of the position vector at the point $\underline{x}$, so we can combine the three single integrals from (2.64), using (2.52), as the vector integral

$$\int a(\underline{x})\,\mathrm{d}\,\underline{x}, \qquad \int a(\underline{x})\,\mathrm{d}\,x_i, \tag{2.67}$$

where the so-defined vector line element $\mathrm{d}\,\underline{x}$ is, according to (2.51) (and in agreement with our intuition), a polar vector.

(The three spatial coordinates of the argument can be combined as $\underline{x}$, independently of whether the equation is written in symbolic notation or in index notation; i. e. we can do this also in the single integrals of (2.64)). These integrals, where the differential is a vector line element, are called line integrals of the second kind.

If the differential is the magnitude of the vector line element instead of the line element, we have

$$\int a(\underline{x})\,\mathrm{d}\,x. \tag{2.68}$$

Such an integral is called a line integral of the first kind.

2. The domain of integration in a line integral is a segment of a curve in space. If we assign a coordinate u to any point of the curve, we can describe the curve segment by

$$x = x(u), \qquad y = y(u), \qquad u_1 \leqq u \leqq u_2, \qquad z = z(u); \tag{2.69}$$

we can combine these three equations into vector form:

$$\underline{x} = \underline{x}(u), \qquad x_i = x_i(u). \tag{2.70}$$

This gives for a line element

$$\mathrm{d}x_i = \frac{\mathrm{d}x_i}{\mathrm{d}u}\,\mathrm{d}u \tag{2.71}$$

and for its magnitude

$$\mathrm{d}x = \sqrt{(\mathrm{d}x_i)^2} = \sqrt{\left(\frac{\mathrm{d}x_i}{\mathrm{d}u}\right)^2}\,\mathrm{d}u. \tag{2.72}$$

Thus we obtain for the line integrals (2.67) and (2.68) along the curve segment (2.69)

$$\int a(\underline{x})\,\mathrm{d}x_i = \int_{u_1}^{u_2} a(\underline{x}(u))\,\frac{\mathrm{d}x_i}{\mathrm{d}u}\,\mathrm{d}u, \tag{2.73}$$

$$\int a(\underline{x})\,\mathrm{d}x = \int_{u_1}^{u_2} a(\underline{x}(u))\,\sqrt{\left(\frac{\mathrm{d}x_i}{\mathrm{d}u}\right)^2}\,\mathrm{d}u. \tag{2.74}$$

Problem 2.18.
A point mass with mass m is moving along one turn of a helix around the z-axis with radius a and slope h. The forces acting on the point mass are the gravitational force $\underline{F}_1 = -mg\,\underline{e}_z$ and the elastic force $\underline{F}_2 = -\lambda\,\underline{x}$. Compute the work $W = \int \underline{F}\cdot \mathrm{d}\underline{x}$ exerted on the point mass.

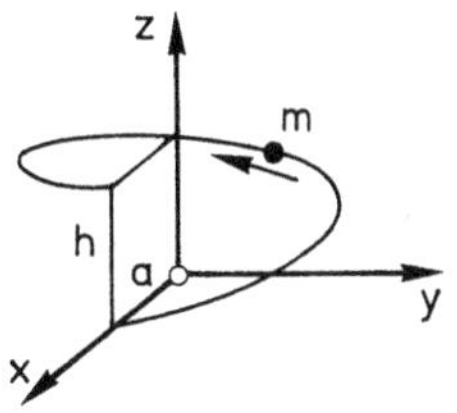

Hint: A parametric representation of a helix is
$x = a\cos\varphi, \quad y = a\sin\varphi, \quad z = h\varphi/(2\pi)$.

2.14.2 Normal Vector and Surface Vector of a Surface Element

1. A surface element is characterized by its magnitude $\mathrm{d}A$ and its normal vector $\underline{n}$ (the unit vector perpendicular to the surface element). The product of the normal vector and the magnitude of the surface element is called the surface vector $\mathrm{d}\underline{A}$.

$$\mathrm{d}\underline{A} = \underline{n}\,\mathrm{d}A. \tag{2.75}$$

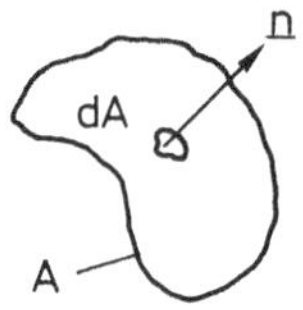

The normal vector $\underline{n}$ and the surface vector $\mathrm{d}\underline{A}$ are not yet uniquely defined, since a surface element has two unit vectors (negatives of each other, in other words, oriented in opposite directions) perpendicular to the surface element. We determine the orientation with a convention, distinguishing between open surfaces (surfaces with a boundary curve) and closed surfaces (surfaces without a boundary curve); in addition we restrict ourselves to two-sided[16] surfaces.

- We can assign a direction of rotation to an open surface (and so also to its surface elements) by the way we travel along the boundary curve. Let us agree to define the normal vector and the surface vector of a surface element by their magnitudes and by the rotational direction of a boundary curve. Then $\underline{n}$ and $\mathrm{d}\underline{A}$ are axial vectors. If an observer travels along the boundary curve and sees the surface on the left side, then $\underline{n}$ (and $\mathrm{d}\underline{A}$) and the direction of rotation are associated according to the right-hand screw rule. For example, if the surface is the ring between two concentric circles, then the boundary has two parts (the outer and the inner circle), and while traveling along each part the surface must be on the same (e. g. left) side of the boundary curve, so the observer travels in different directions along the two parts.
- A closed surface does not have a boundary curve; instead it has an inner side and an outer side. We adopt the following convention.

  The normal vector and the surface vector of the surface element of a closed surface always point outward from the enclosed volume.

  Here $\underline{n}$ and $\mathrm{d}\underline{A}$ are polar vectors.

16 A surface is two-sided, if we cannot go from one side to the other side without passing through the boundary. The most famous example of a one-sided surface is the Möbius strip.

For example, if the enclosed volume is the space between two concentric spheres, then the closed surface has two parts (the surface of the outer sphere and the surface of the inner sphere), and the normal vector points, according to our convention, radially outward from each surface element of the outer sphere and radially inward from each surface element of the inner sphere.

The magnitude of both a polar vector and an axial vector is a polar scalar, i. e. the area of a surface element is a polar scalar for open surfaces and for closed surfaces and thus (in agreement with our intuition) independent of the orientation of the coordinate system.

2. Consider an infinitesimal tetrahedron, with one plane being oriented in arbitrary direction and the other three planes perpendicular to the coordinate axes. The inclined plane has the magnitude $\mathrm{d}A$ and the (outwardly oriented) normal vector $\underline{n}$; the length of the edges along the coordinate axes are $|\mathrm{d}\,a_x|$, $|\mathrm{d}\,a_y|$, and $|\mathrm{d}\,a_z|$. The value of $\mathrm{d}A$ is half of the area of the parallelogram spanned by the edge vectors $\mathrm{d}\,\underline{u}$ and $\mathrm{d}\,\underline{v}$ (see the figure above). According to Section 2.10.1, No. 4 we can write, using the vector product,

$$\underline{n}\,\mathrm{d}A = \frac{1}{2}\,\mathrm{d}\,\underline{u} \times \mathrm{d}\,\underline{v}.$$

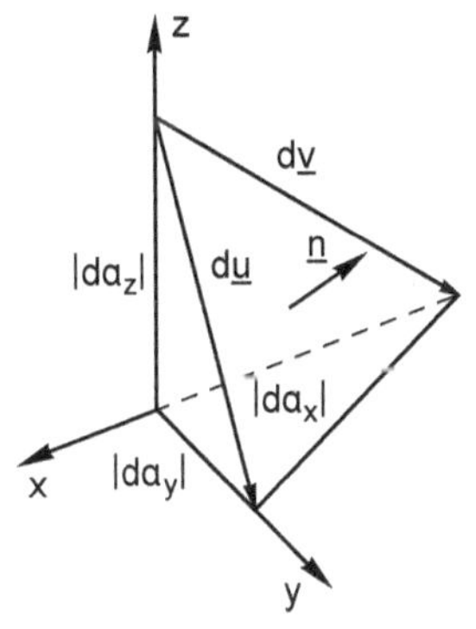

We further see from the figure that

$$\begin{aligned}
\underline{n}\,\mathrm{d}A &= \frac{1}{2}\left(-|\mathrm{d}\,a_z|\,\underline{e}_z + |\mathrm{d}\,a_y|\,\underline{e}_y\right) \times \left(-|\mathrm{d}\,a_z|\,\underline{e}_z - |\mathrm{d}\,a_x|\,\underline{e}_x\right) \\
&= \frac{1}{2}\,(|\mathrm{d}\,a_x|\,|\mathrm{d}\,a_z|\,\underbrace{\underline{e}_z \times \underline{e}_x}_{\underline{e}_y} - |\mathrm{d}\,a_y|\,|\mathrm{d}\,a_z|\,\underbrace{\underline{e}_y \times \underline{e}_z}_{\underline{e}_x} - |\mathrm{d}\,a_x|\,|\mathrm{d}\,a_y|\,\underbrace{\underline{e}_y \times \underline{e}_x}_{-\underline{e}_z}) \\
&= \frac{1}{2}\left(|\mathrm{d}\,a_x|\,|\mathrm{d}\,a_z|\,\underline{e}_y - |\mathrm{d}\,a_y|\,|\mathrm{d}\,a_z|\,\underline{e}_x + |\mathrm{d}\,a_x|\,|\mathrm{d}\,a_y|\,\underline{e}_z\right) \\
&= \underbrace{\frac{1}{2}\,|\mathrm{d}\,a_x|\,|\mathrm{d}\,a_z|}_{|\mathrm{d}\,A_y|}\;\underline{e}_y - \underbrace{\frac{1}{2}\,|\mathrm{d}\,a_y|\,|\mathrm{d}\,a_z|}_{|\mathrm{d}\,A_x|}\;\underline{e}_x + \underbrace{\frac{1}{2}\,|\mathrm{d}\,a_x|\,|\mathrm{d}\,a_y|}_{|\mathrm{d}\,A_z|}\;\underline{e}_z \\
&= -|\mathrm{d}\,A_x|\,\underline{e}_x + |\mathrm{d}\,A_y|\,\underline{e}_y + |\mathrm{d}\,A_z|\,\underline{e}_z. \qquad \text{(a)}
\end{aligned}$$

Here $|\mathrm{d}\,A_x|$, $|\mathrm{d}\,A_y|$, and $|\mathrm{d}\,A_z|$ are the areas of the faces of the tetrahedron perpendicular to the coordinate axes; geometrically speaking they are the projections of the inclined surface onto the coordinate planes. The signs differ because the x-component of $\mathrm{d}\,\underline{A}$ points in the negative x-direction, while the y-component and the z-component point in the direction of the positive y-axis and z-axis, respectively. If we compare the last row of (a) with the general form of the vector $\mathrm{d}\,\underline{A}$ with respect to the basis $\underline{e}_i$,

$$\mathrm{d}\underline{A} = \mathrm{d}\,A_x\,\underline{e}_x + \mathrm{d}\,A_y\,\underline{e}_y + \mathrm{d}\,A_z\,\underline{e}_z,$$

we find the following conclusion.

The absolute values of the coordinates of an arbitrarily oriented surface element are the magnitudes of the projections of this surface element onto the coordinate planes.

3. If we bring in (a) all terms to the left side, we have

$$\underline{n}\,\mathrm{d}A + \underline{e}_x\,|\mathrm{d}\,A_x| - \underline{e}_y\,|\mathrm{d}\,A_y| - \underline{e}_z\,|\mathrm{d}\,A_z| = \underline{0},$$

or with $\underline{n}_x = \underline{e}_x$, $\underline{n}_y = -\underline{e}_y$, $\underline{n}_z = -\underline{e}_z$

$$\underline{n}\,\mathrm{d}A + \underline{n}_x\,|\mathrm{d}\,A_x| + \underline{n}_y\,|\mathrm{d}\,A_y| + \underline{n}_z\,|\mathrm{d}\,A_z| = \underline{0}.$$

The last equation says that the sum of the (outwardly oriented) surface vectors over the total surface of the tetrahedron is equal to the zero vector. Then, for any closed surface, also the integral over the surface vectors of all surface elements gives again the zero vector, since we can split the enclosed volume into infinitesimal tetrahedrons, and the inner surfaces of two adjacent subvolume elements cancel during integration; on the other hand, the integral over all surface area elements gives A.

$$\oint \mathrm{d}\underline{A} = \underline{0}, \qquad \oint \mathrm{d}A = A. \tag{2.76}$$

2.14.3 Surface Integrals of Tensor Coordinates

1. The double integral $\iint a(x,y,z)\,\mathrm{d}y\,\mathrm{d}z$ of (2.65) is an integration of a over the projection of the surface of integration onto the y,z-plane, i. e. we write $\int a(x,y,z)\,\mathrm{d}A_x$. The same applies to the other two double integrals in (2.65), so we have

$$\begin{aligned}\int a(x,y,z)\,\mathrm{d}A_x &:= \pm\iint a(x,y,z)\,\mathrm{d}y\,\mathrm{d}z,\\ \int a(x,y,z)\,\mathrm{d}A_y &:= \pm\iint a(x,y,z)\,\mathrm{d}z\,\mathrm{d}x,\\ \int a(x,y,z)\,\mathrm{d}A_z &:= \pm\iint a(x,y,z)\,\mathrm{d}x\,\mathrm{d}y,\end{aligned} \tag{2.77}$$

where the sign is determined by the orientation of the surface element.

We can combine the three double integrals in vector form as follows:

$$\int a(\underline{x})\,\mathrm{d}\underline{A}, \qquad \int a(\underline{x})\,\mathrm{d}A_i. \tag{2.78}$$

Such an integral, where the differential is a vector surface element, is called a surface integral of the second kind. If the differential is the magnitude of the surface element instead of the vector surface element we have

$$\int a(\underline{x})\,\mathrm{d}A, \tag{2.79}$$

and such an integral is called a surface integral of the first kind.

2. The domain of integration of a surface integral is a segment of a surface in space. If we define two coordinates u and v on the surface and if the surface segment is bounded by two curves $v = v_1(u)$ and $v = v_2(u)$, then the surface segment is described by

$$\begin{aligned} x &= x(u,v), \\ y &= y(u,v), \\ z &= z(u,v), \end{aligned} \qquad \begin{aligned} &u_1 \leqq u \leqq u_2, \\ &v_1(u) \leqq v \leqq v_2(u), \end{aligned} \tag{2.80}$$

which can be combined as

$$\underline{x} = \underline{x}(u,v), \quad x_i = x_i(u,v). \tag{2.81}$$

A surface element of this surface is a parallelogram spanned by two line elements along the coordinate lines. The surface vector of this surface element is then equal to the vector product of these two line elements, where the orientation of the surface vector changes if we change the order of the line elements. From now on we choose the order of the line elements, i. e. the order of the surface coordinates u and v, such that the orientation of the surface vector is in agreement with our convention.

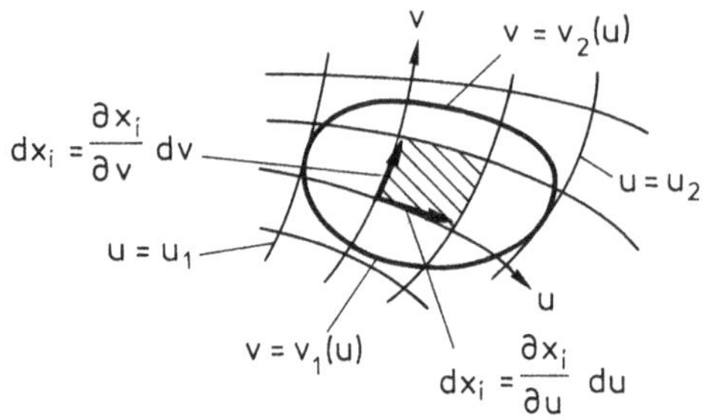

Along a u-line only u changes and v remains constant; i. e. for a line element along a u-line we have $\mathrm{d}x_i = (\partial x_i/\partial u)\,\mathrm{d}u$. Analogously we have for a line element along a v-line $\mathrm{d}x_i = (\partial x_i/\partial v)\,\mathrm{d}v$. This gives for the surface vector of a surface element

$$\mathrm{d}A_i = \varepsilon_{ijk}\frac{\partial x_j}{\partial u}\frac{\partial x_k}{\partial v}\,\mathrm{d}u\,\mathrm{d}v \tag{2.82}$$

and for its coordinates

$$\begin{aligned}\mathrm{d}A_x &= \mathrm{d}y\,\mathrm{d}z = \left(\frac{\partial y}{\partial u}\frac{\partial z}{\partial v} - \frac{\partial z}{\partial u}\frac{\partial y}{\partial v}\right)\mathrm{d}u\,\mathrm{d}v \\ &= \begin{vmatrix}\frac{\partial y}{\partial u} & \frac{\partial y}{\partial v}\\ \frac{\partial z}{\partial u} & \frac{\partial z}{\partial v}\end{vmatrix}\mathrm{d}u\,\mathrm{d}v = \frac{\partial(y,z)}{\partial(u,v)}\,\mathrm{d}u\,\mathrm{d}v,\end{aligned} \tag{2.83}$$

$$\begin{aligned}\mathrm{d}A_y &= \mathrm{d}z\,\mathrm{d}x = \left(\frac{\partial z}{\partial u}\frac{\partial x}{\partial v} - \frac{\partial x}{\partial u}\frac{\partial z}{\partial v}\right)\mathrm{d}u\,\mathrm{d}v \\ &= \begin{vmatrix}\frac{\partial z}{\partial u} & \frac{\partial z}{\partial v}\\ \frac{\partial x}{\partial u} & \frac{\partial x}{\partial v}\end{vmatrix}\mathrm{d}u\,\mathrm{d}v = \frac{\partial(z,x)}{\partial(u,v)}\,\mathrm{d}u\,\mathrm{d}v,\end{aligned} \tag{2.84}$$

$$
\begin{aligned}
\mathrm{d}A_z &= \mathrm{d}x\,\mathrm{d}y = \left(\frac{\partial x}{\partial u}\frac{\partial y}{\partial v} - \frac{\partial y}{\partial u}\frac{\partial x}{\partial v}\right)\mathrm{d}u\,\mathrm{d}v \\
&= \begin{vmatrix} \frac{\partial x}{\partial u} & \frac{\partial x}{\partial v} \\ \frac{\partial y}{\partial u} & \frac{\partial y}{\partial v} \end{vmatrix} \mathrm{d}u\,\mathrm{d}v = \frac{\partial(x,y)}{\partial(u,v)}\,\mathrm{d}u\,\mathrm{d}v.
\end{aligned}
\tag{2.85}
$$

Hence we obtain for the surface integral (2.78)

$$
\int a(\underline{x})\,\mathrm{d}A_i = \int\limits_{u_1}^{u_2}\int\limits_{v_1(u)}^{v_2(u)} a(\underline{x}(u,v))\,\varepsilon_{ijk}\,\frac{\partial x_j}{\partial u}\frac{\partial x_k}{\partial v}\,\mathrm{d}u\,\mathrm{d}v. \tag{2.86}
$$

The magnitude of the surface element is given by

$$
\mathrm{d}A = \sqrt{(\mathrm{d}A_i)^2} = \sqrt{\varepsilon_{ijk}\frac{\partial x_j}{\partial u}\frac{\partial x_k}{\partial v}\,\varepsilon_{imn}\frac{\partial x_m}{\partial u}\frac{\partial x_n}{\partial v}}\,\mathrm{d}u\,\mathrm{d}v.
$$

Applying the Grassmann identity (1.35) gives

$$
\begin{aligned}
\mathrm{d}A &= \sqrt{(\mathrm{d}A_i)^2} = \sqrt{(\delta_{jm}\,\delta_{kn} - \delta_{jn}\,\delta_{km})\frac{\partial x_j}{\partial u}\frac{\partial x_k}{\partial v}\frac{\partial x_m}{\partial u}\frac{\partial x_n}{\partial v}}\,\mathrm{d}u\,\mathrm{d}v \\
&= \sqrt{\left(\frac{\partial x_j}{\partial u}\right)^2\left(\frac{\partial x_k}{\partial v}\right)^2 - \left(\frac{\partial x_j}{\partial u}\frac{\partial x_j}{\partial v}\right)^2}\,\mathrm{d}u\,\mathrm{d}v.
\end{aligned}
$$

With the commonly used notations

$$
\begin{aligned}
E &:= \left(\frac{\partial x_j}{\partial u}\right)^2 = \left(\frac{\partial x}{\partial u}\right)^2 + \left(\frac{\partial y}{\partial u}\right)^2 + \left(\frac{\partial z}{\partial u}\right)^2, \\
F &:= \frac{\partial x_j}{\partial u}\frac{\partial x_j}{\partial v} = \frac{\partial x}{\partial u}\frac{\partial x}{\partial v} + \frac{\partial y}{\partial u}\frac{\partial y}{\partial v} + \frac{\partial z}{\partial u}\frac{\partial z}{\partial v}, \\
G &:= \left(\frac{\partial x_j}{\partial v}\right)^2 = \left(\frac{\partial x}{\partial v}\right)^2 + \left(\frac{\partial y}{\partial v}\right)^2 + \left(\frac{\partial z}{\partial v}\right)^2,
\end{aligned}
\tag{2.87}
$$

we obtain for the surface integral (2.79)

$$
\int a(\underline{x})\,\mathrm{d}A = \int\limits_{u_1}^{u_2}\int\limits_{v_1(u)}^{v_2(u)} a(\underline{x}(u,v))\,\sqrt{EG - F^2}\,\mathrm{d}u\,\mathrm{d}v. \tag{2.88}
$$

Problem 2.19.
Consider a hemispherical shell exposed to the internal pressure $p = \varrho g(H - z)$. Compute the vertical force $F_z = \int p \, \mathrm{d} A_z$ acting on the hemispherical shell.

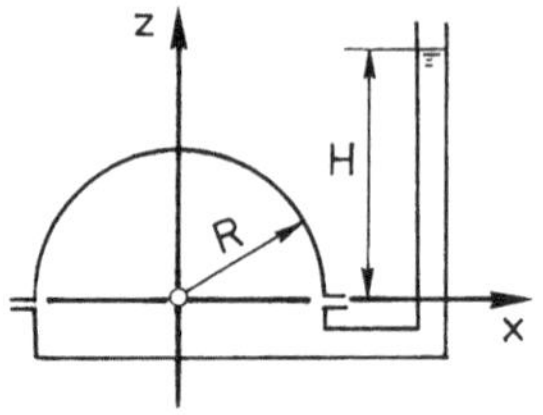

Hint: A parametric representation of a hemispherical shell is
$x = R \sin u \cos v, y = R \sin u \sin v, z = R \cos u$.

2.14.4 Volume Integrals of Tensor Coordinates

1. Consider a volume element spanned by the three line elements $\mathrm{d}\underline{x}_1$, $\mathrm{d}\underline{x}_2$, and $\mathrm{d}\underline{x}_3$ pointing in the positive coordinate directions of a Cartesian coordinate system. For the corresponding triple product we have according to (2.52), independently of the orientation of the coordinate system,

$$\mathrm{d} V = [\mathrm{d}\underline{x}_1, \mathrm{d}\underline{x}_2, \mathrm{d}\underline{x}_3] = \mathrm{d}\underline{x}_1 \times \mathrm{d}\underline{x}_2 \cdot \mathrm{d}\underline{x}_3 = \mathrm{d}x \, \mathrm{d}y \, \mathrm{d}z \, \underbrace{\underline{e}_x \times \underline{e}_y \cdot \underline{e}_z}_{1} = \mathrm{d}x \, \mathrm{d}y \, \mathrm{d}z,$$

i. e. the domain of integration in the triple integral (2.66) is a volume, i. e.

$$\int a(\underline{x}) \, \mathrm{d} V := \iiint a(x, y, z) \, \mathrm{d}x \, \mathrm{d}y \, \mathrm{d}z. \tag{2.89}$$

The volume element d V is a triple product of three polar vectors and so, according to Section 2.10.3, No. 1, an axial scalar, i. e. it changes its sign under a transformation to a coordinate system with a different orientation according to (2.18). However, under the reflection the lower and the upper surface are swapped, so the sign changes again, and thus the volume is (in agreement with our intuition) a polar scalar.

2. If the volume of integration is bounded by two areas $z = z_1(x, y)$ and $z = z_2(x, y)$ which intersect in the two curves $y = y_1(x)$ and $y = y_2(x)$, then we get by taking the integration limits into account

$$\int a(\underline{x}) \, \mathrm{d} V = \int\limits_{x_1}^{x_2} \int\limits_{y_1(x)}^{y_2(x)} \int\limits_{z_1(x, y)}^{z_2(x, y)} a(\underline{x}) \, \mathrm{d}x \, \mathrm{d}y \, \mathrm{d}z. \tag{2.90}$$

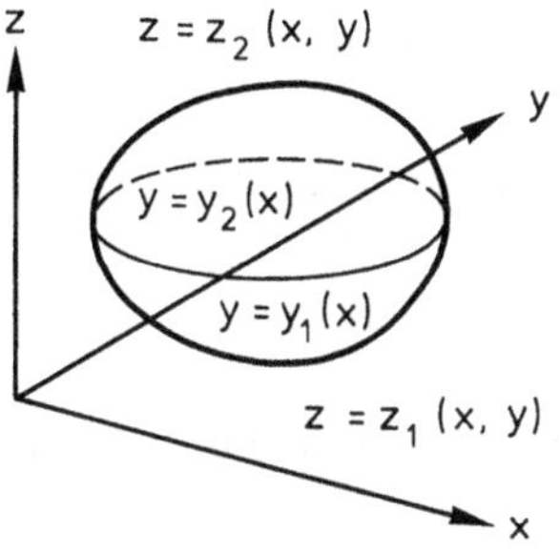

Sometimes the integration limits are conveniently described in a different coordinate system u, v, w. The volume of integration reads in these coordinates

$$
\begin{aligned}
x &= x(u,v,w), & u_1 &\leqq u \leqq u_2, \\
y &= y(u,v,w), & v_1(u) &\leqq v \leqq v_2(u), \\
z &= z(u,v,w), & w_1(u,v) &\leqq w \leqq w_2(u,v).
\end{aligned} \tag{2.91}
$$

Similarly as before, the volume element is spanned in these coordinates by the three line elements pointing in positive direction of the coordinate lines and its value is the absolute value of the corresponding triple product. For a line element in u-direction we have $\mathrm{d}x_i = (\partial x_i/\partial u)\,\mathrm{d}u$, for a line element in v-direction and in w-direction we have accordingly $\mathrm{d}x_i = (\partial x_i/\partial v)\,\mathrm{d}v$ and $\mathrm{d}x_i = (\partial x_i/\partial w)\,\mathrm{d}w$. So we have for the volume element in the coordinates u, v, w

$$
\mathrm{d}V = \left| \varepsilon_{ijk} \frac{\partial x_i}{\partial u} \frac{\partial x_j}{\partial v} \frac{\partial x_k}{\partial w} \right| \mathrm{d}u\,\mathrm{d}v\,\mathrm{d}w. \tag{2.92}
$$

Or, with (2.45),

$$
\mathrm{d}V = \left\| \begin{matrix} \dfrac{\partial x}{\partial u} & \dfrac{\partial x}{\partial v} & \dfrac{\partial x}{\partial w} \\ \dfrac{\partial y}{\partial u} & \dfrac{\partial y}{\partial v} & \dfrac{\partial y}{\partial w} \\ \dfrac{\partial z}{\partial u} & \dfrac{\partial z}{\partial v} & \dfrac{\partial z}{\partial w} \end{matrix} \right\| \mathrm{d}u\,\mathrm{d}v\,\mathrm{d}w = \left| \frac{\partial(x, y, z)}{\partial(u, v, w)} \right| \mathrm{d}u\,\mathrm{d}v\,\mathrm{d}w.
$$

Thus we obtain for the volume integral in the coordinates u, v, w

$$
\int a(\underline{x})\,\mathrm{d}V = \int_{u_1}^{u_2} \int_{v_1(u)}^{v_2(u)} \int_{w_1(u,v)}^{w_2(u,v)} a(\underline{x}(u, v, w)) \left| \frac{\partial(x, y, z)}{\partial(u, v, w)} \right| \mathrm{d}u\,\mathrm{d}v\,\mathrm{d}w. \tag{2.93}
$$

2.14.5 Integrals of Higher-Order Tensor Fields

1. In the formulas above we assumed for simplicity that the integrand is a scalar field. However, we can easily substitute $a_{i...j}(x, y, z)$ for $a(x, y, z)$ in these formulas; also the coordinates of a tensor field of higher order are only functions of the three spatial coordinates.

For example, if the integrand is a second-order tensor we have the following types of integrals.

$$\begin{aligned}
\int \underline{\underline{a}}\,\mathrm{d}x &= \underline{\underline{A}}, & \int a_{ij}\,\mathrm{d}x &= A_{ij},\\
\int \underline{\underline{a}}\,\mathrm{d}\underline{x} &= \underline{\underline{\underline{B}}}, & \int a_{ij}\,\mathrm{d}x_k &= B_{ijk},\\
\int \underline{\underline{a}}\,\mathrm{d}A &= \underline{\underline{C}}, & \int a_{ij}\,\mathrm{d}A &= C_{ij},\\
\int \underline{\underline{a}}\,\mathrm{d}\underline{A} &= \underline{\underline{\underline{D}}}, & \int a_{ij}\,\mathrm{d}A_k &= D_{ijk},\\
\int \underline{\underline{a}}\,\mathrm{d}V &= \underline{\underline{E}}. & \int a_{ij}\,\mathrm{d}V &= E_{ij}.
\end{aligned} \tag{2.94}$$

Here the products of the integrand and the differential are tensor products; we can clearly replace them in integrals of the second kind by scalar products or vector products, and then we get e. g. the formulas

$$\begin{aligned}
\int \underline{\underline{a}} \cdot \mathrm{d}\underline{x} &= \underline{F}, & \int a_{ij}\,\mathrm{d}x_j &= F_i,\\
\int \underline{\underline{a}} \times \mathrm{d}\underline{x} &= \underline{\underline{G}}, & \int a_{ij}\,\varepsilon_{klj}\,\mathrm{d}x_k &= G_{il}.
\end{aligned} \tag{2.95}$$

2. A line integral over a closed curve (the boundary curve of a surface) or a surface integral over a closed surface (the boundary surface of a volume) are also called a closed line integral and a closed surface integral, respectively, and we denote such integrals by $\oint$.

3. If we integrate a tensor over a given spatial domain (e. g. a given curve between two points P and Q), then the result, as for any definite integral, does not depend any more on the integration variables. So integration of a tensor field gives a tensor which is independent of the position.

2.15 Gauss Theorem and Stokes Theorem

2.15.1 Gauss Theorem

1. Consider the Cartesian coordinates $a_{i...j}$ of a tensor field with the following properties:

- the $a_{i...j}$ are continuously differentiable in a given volume,
- the closed surface of this volume is at least piecewise continuously differentiable, and
- if the closed surface (as e. g. for the volume enclosed by the two concentric spheres) has several parts, then the integration is performed over all properly oriented subsurfaces.

Then the Gauss theorem holds.

$$
\begin{aligned}
&\int \operatorname{grad} a \,\mathrm{d}V = \oint a \,\mathrm{d}\underline{A}, && \int \frac{\partial a}{\partial x_k}\,\mathrm{d}V = \oint a \,\mathrm{d}A_k, \\
&\int \operatorname{grad} \underline{a} \,\mathrm{d}V = \oint \underline{a} \,\mathrm{d}\underline{A}, && \int \frac{\partial a_j}{\partial x_k}\,\mathrm{d}V = \oint a_j \,\mathrm{d}A_k, \\
&\int \operatorname{grad} \underline{\underline{a}} \,\mathrm{d}V = \oint \underline{\underline{a}} \,\mathrm{d}\underline{A}, && \int \frac{\partial a_{ij}}{\partial x_k}\,\mathrm{d}V = \oint a_{ij} \,\mathrm{d}A_k, \\
&\text{etc.} && \text{etc.}
\end{aligned}
\tag{2.96}
$$

Here $\mathrm{d}\underline{A}$ is, according to Section 2.14.2, a polar vector.

2. In these equations we can replace the tensor product on the right side with a scalar product, and thus the gradient on the left side with a divergence, if we multiply the equations in index notation (starting with the second one) with δ_{jk}.

$$
\begin{aligned}
&\int \operatorname{div} \underline{a} \,\mathrm{d}V = \oint \underline{a} \cdot \mathrm{d}\underline{A}, && \int \frac{\partial a_k}{\partial x_k}\,\mathrm{d}V = \oint a_k \,\mathrm{d}A_k, \\
&\int \operatorname{div} \underline{\underline{a}} \,\mathrm{d}V = \oint \underline{\underline{a}} \cdot \mathrm{d}\underline{A}, && \int \frac{\partial a_{ik}}{\partial x_k}\,\mathrm{d}V = \oint a_{ik} \,\mathrm{d}A_k, \\
&\text{etc.} && \text{etc.}
\end{aligned}
\tag{2.97}
$$

3. We can also replace the tensor product on the right side with a vector product, and thus the gradient on the left side with a curl, if we multiply (2.96) in index notation (starting with the second one) with ε_{jpk}.

$$\begin{aligned}
&\int \operatorname{curl} \underline{a}\, \mathrm{d}V = \oint \underline{a} \otimes \mathrm{d}\underline{A} = -\oint \underline{a} \times \mathrm{d}\underline{A}, \\
&\int \operatorname{curl} \underline{\underline{a}}\, \mathrm{d}V = \oint \underline{\underline{a}} \otimes \mathrm{d}\underline{A} = -\oint \underline{\underline{a}} \times \mathrm{d}\underline{A}, \\
&\text{etc.} \\
&\int \frac{\partial a_j}{\partial x_k} \varepsilon_{jpk}\, \mathrm{d}V = \oint a_j\, \varepsilon_{jpk}\, \mathrm{d}A_k, \\
&\int \frac{\partial a_{ij}}{\partial x_k} \varepsilon_{jpk}\, \mathrm{d}V = \oint a_{ij}\, \varepsilon_{jpk}\, \mathrm{d}A_k, \\
&\text{etc.}
\end{aligned} \tag{2.98}$$

The equations (2.97) and (2.98) are also versions of the Gauss theorem.

4. When we stated the Gauss theorem we used right derivatives. It is easy to see that if we use left derivatives, then in symbolic notation we need to change the order of the factors and replace $\otimes$ by $\times$. This gives e. g. for $(2.98)_2$

$$\int \mathrm{d}V \operatorname{curl}_{\mathrm{L}} \underline{\underline{a}} = \oint \mathrm{d}\underline{A} \times \underline{\underline{a}}.$$

5. We first prove the basic version of the Gauss theorem, i. e. the equation

$$\int \frac{\partial a}{\partial x_k}\, \mathrm{d}V = \oint a\, \mathrm{d}A_k.$$

Here we assume that $a(x, y, z)$ is a scalar field, which is continuously differentiable with respect to the spatial coordinates.

Let us consider for the moment only the z-coordinate of this equation,

$$\int \frac{\partial a}{\partial z}\, \mathrm{d}V = \oint a\, \mathrm{d}A_z,$$

and let us assume a volume of integration such that the projection of its surface onto the x, y-plane splits into two parts $z = z_1(x, y)$ and $z = z_2(x, y)$, where the two parts intersect in a curve C, and let us further assume that to each value pair (x, y) of the projected surface corresponds exactly one point on z_1 and one point on z_2. In addition we assume that the functions z_1 and z_2 are continuous and at least piecewise

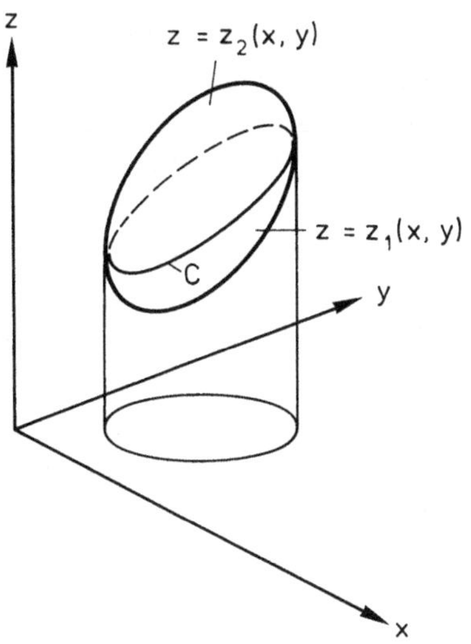

continuously differentiable. Then we have

$$\int \frac{\partial a}{\partial z}\,\mathrm{d}V = \int_{x_1}^{x_2} \mathrm{d}x \int_{y_1(x)}^{y_2(x)} \mathrm{d}y \int_{z_1(x,y)}^{z_2(x,y)} \mathrm{d}z\,\frac{\partial a}{\partial z}$$

$$= \int_{x_1}^{x_2} \mathrm{d}x \int_{y_1(x)}^{y_2(x)} \mathrm{d}y\,[a(x,y,z_2(x,y)) - a(x,y,z_1(x,y))].$$

We can split the last integral into two parts and interpret the first as an integral of a over the surface $z = z_1(x, y)$ and the second as an integral of a over the surface $z = z_2(x, y)$, i. e.

$$\int \frac{\partial a}{\partial z}\,\mathrm{d}V = \int_{z_2(x,y)} a\,\mathrm{d}A_z - \int_{z_1(x,y)} a\,\mathrm{d}A_z,$$

where the two surfaces are oriented so that the z-coordinates of all surface vectors point into the positive z-direction. To have the surface vector oriented outward on the closed surface, as we agreed on, we need to change the sign of the integral over z_1, which gives

$$\int \frac{\partial a}{\partial z}\,\mathrm{d}V = \oint a\,\mathrm{d}A_z,$$

and this is what we wanted to prove.

Now we extend the proof to volumes in which the two surface parts z_1 and z_2 do not intersect along a curve C but are connected by a cylindrical shell whose generating lines are parallel to the z-axis. Such a cylinder does not contribute to the closed surface integral $\oint a\,\mathrm{d}A_z$ because everywhere on the shell we have $\mathrm{d}A_z = 0$; so we conclude that the basic version of the Gauss theorem is also true for these volumes.

We finally extend the proof to volumes with arbitrarily shaped closed surfaces, e. g. multiply connected volumes and volumes with inner closed surfaces, as long as all parts of the closed surfaces are continuous and at least piecewise continuously differentiable. We can always split these volumes into volumes of the type which we already handled earlier, i. e. for which the Gauss theorem holds.

Adding the Gauss theorems for the subvolumes on the left side gives the volume integral over the total volume. When we add the surface integrals on the right side, the parts for the shared faces of adjacent subvolumes cancel out, because the surface vectors of a subvolume are always oriented outward, so for adjacent subvolumes they point in opposite directions; only the surface integral over the outer surface remains, including possibly existing inner parts of the closed surface, where the surface vector is also oriented outward of the volume.

The proof of the basic version of the Gauss theorem for the other two coordinates is similar. To obtain the remaining versions of the Gauss theorem we replace in the basic version the scalar field $a(x, y, z)$ with the coordinates of a tensor field of higher order or perform further algebraic operations.

The only difference between the coordinates of a tensor field of any order and a scalar field is how they transform under a change to another coordinate system. Since we did not make use of the invariance of the scalar tensor field with respect to a coordinate transformation in our proof, this proof also holds for the coordinates of a tensor field of higher order.

2.15.2 Stokes Theorem

1. Consider the Cartesian coordinates $a_{i...j}$ of a tensor field with the following properties:

- the $a_{i...j}$ are continuously differentiable on a given two-sided surface,
- the boundary curve of this surface is at least piecewise continuously differentiable, and
- if the boundary curve (as e. g. for the surface between two concentric circles) has several parts, then the closed line integral is performed over all properly oriented parts.

Then the Stokes theorem holds.

$$\int \operatorname{grad} a \otimes \mathrm{d}\underline{A} = -\int \operatorname{grad} a \times \mathrm{d}\underline{A} = \oint a \, \mathrm{d}\underline{x},$$
$$\int \operatorname{grad} \underline{a} \otimes \mathrm{d}\underline{A} = -\int \operatorname{grad} \underline{a} \times \mathrm{d}\underline{A} = \oint \underline{a} \, \mathrm{d}\underline{x},$$
$$\int \operatorname{grad} \underline{\underline{a}} \otimes \mathrm{d}\underline{A} = -\int \operatorname{grad} \underline{\underline{a}} \times \mathrm{d}\underline{A} = \oint \underline{\underline{a}} \, \mathrm{d}\underline{x},$$
etc.

$$\int \frac{\partial a}{\partial x_i} \varepsilon_{ijk} \, \mathrm{d}A_k = \oint a \, \mathrm{d}x_j,$$
$$\int \frac{\partial a_n}{\partial x_i} \varepsilon_{ijk} \, \mathrm{d}A_k = \oint a_n \, \mathrm{d}x_j, \qquad (2.99)$$
$$\int \frac{\partial a_{mn}}{\partial x_i} \varepsilon_{ijk} \, \mathrm{d}A_k = \oint a_{mn} \, \mathrm{d}x_j,$$
etc.

Here $\mathrm{d}\underline{A}$ is, according to Section 2.14.2, an axial vector.

2. If we replace the tensor product on the right side with a scalar product, by multiplying the equations in index notation (starting with the second one) with δ_{nj}, we obtain the following equations.

$$\int \operatorname{curl} \underline{a} \cdot \mathrm{d}\underline{A} = \oint \underline{a} \cdot \mathrm{d}\underline{x},$$
$$\int \operatorname{curl} \underline{\underline{a}} \cdot \mathrm{d}\underline{A} = \oint \underline{\underline{a}} \cdot \mathrm{d}\underline{x},$$
etc.

$$\int \frac{\partial a_j}{\partial x_i} \varepsilon_{jki} \, \mathrm{d}A_k = \oint a_j \, \mathrm{d}x_j, \qquad (2.100)$$
$$\int \frac{\partial a_{mj}}{\partial x_i} \varepsilon_{jki} \, \mathrm{d}A_k = \oint a_{mj} \, \mathrm{d}x_j,$$
etc.

3. We can also replace in (2.99) the tensor product on the right side with a vector product, if we multiply the equations in index notation (starting with the second one) with $\varepsilon_{npj} = -\varepsilon_{njp}$. This gives e. g. for the second equation

$$\int \frac{\partial a_n}{\partial x_i} \varepsilon_{ijk} \, \mathrm{d}A_k \, (-\varepsilon_{njp}) = \oint a_n \, \varepsilon_{npj} \, \mathrm{d}x_j.$$

However, if we want to avoid the uncommon notation with the double scalar product for the ε-tensor

$$\int (\operatorname{grad} \underline{a} \times \mathrm{d}\underline{A}) \cdot\cdot \underline{\underline{\underline{\varepsilon}}} = -\oint \underline{a} \times \mathrm{d}\underline{x},$$

then the left side cannot be translated into symbolic notation.

4. We stated the Stokes theorem using right derivatives. If we prefer to use left derivatives, we again need to change the order of the factors and replace $\otimes$ by $\times$. Then we have e. g. for $(2.99)_2$

$$\int \mathrm{d}\underline{A} \times \operatorname{grad}_{\mathrm{L}} \underline{a} = \oint \mathrm{d}\underline{x}\, \underline{a}.$$

We see from all these equations that left derivatives lead to symbolic equations with the commonly used vector product $\times$ and right derivatives lead to symbolic equations with the uncommon vector product $\otimes$.

5. We again start with proving the basic version of the Stokes theorem, i. e.

$$\int \frac{\partial a}{\partial x_i}\, \varepsilon_{ijk}\, \mathrm{d}A_k = \oint a\, \mathrm{d}x_j,$$

where, as before, we assume that $a(x, y, z)$ is a continuously differentiable scalar field. We define two coordinates u and v on the surface of integration and then we have, according to (2.82),

$$\begin{aligned}
I_j := \int \frac{\partial a}{\partial x_i}\, \varepsilon_{ijk}\, \mathrm{d}A_k &= \iint \varepsilon_{ijk}\, \varepsilon_{kmn} \frac{\partial a}{\partial x_i} \frac{\partial x_m}{\partial u} \frac{\partial x_n}{\partial v}\, \mathrm{d}u\, \mathrm{d}v \\
&= \iint \varepsilon_{kij}\, \varepsilon_{kmn} \frac{\partial a}{\partial x_i} \frac{\partial x_m}{\partial u} \frac{\partial x_n}{\partial v}\, \mathrm{d}u\, \mathrm{d}v \\
&= \iint (\delta_{im}\, \delta_{jn} - \delta_{in}\, \delta_{jm}) \frac{\partial a}{\partial x_i} \frac{\partial x_m}{\partial u} \frac{\partial x_n}{\partial v}\, \mathrm{d}u\, \mathrm{d}v \\
&= \iint \left(\frac{\partial a}{\partial x_i} \frac{\partial x_i}{\partial u} \frac{\partial x_j}{\partial v} - \frac{\partial a}{\partial x_i} \frac{\partial x_j}{\partial u} \frac{\partial x_i}{\partial v} \right) \mathrm{d}u\, \mathrm{d}v \\
&= \iint \left(\frac{\partial a}{\partial u} \frac{\partial x_j}{\partial v} - \frac{\partial a}{\partial v} \frac{\partial x_j}{\partial u} \right) \mathrm{d}u\, \mathrm{d}v.
\end{aligned}$$

Now we consider the x-coordinate in the basic version of the theorem, i. e. we set the only free index j to 1, and we choose $u = x, \quad v = y$, i. e. the surface of integration is given as $z = z(x, y)$. Let us consider for the moment only surfaces whose projection onto the x, y-plane is a one-to-one image of the surface, whose boundary intersects

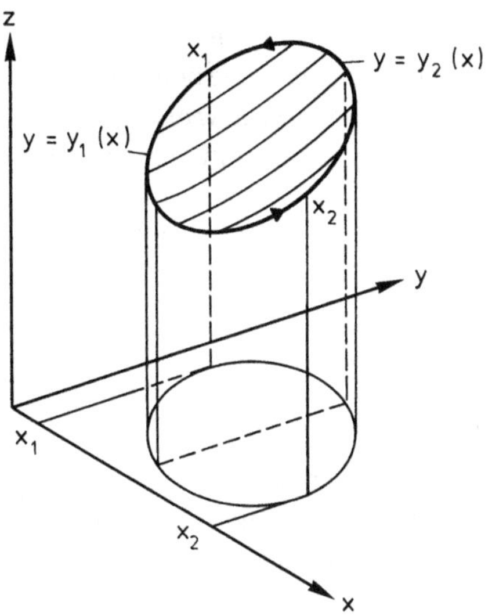

with the coordinate lines $x = \text{const}$ in at most two points, and let the surface be sufficiently smooth so that $z = z(x, y)$ is continuous and at least piecewise continuously differentiable. Then we have with

$$\frac{\partial x_1}{\partial v} = \frac{\partial x}{\partial y} = 0 \quad \text{and} \quad \frac{\partial x_1}{\partial u} = \frac{\partial x}{\partial x} = 1$$

$$I_1 = -\int_{x_1}^{x_2} \mathrm{d}x \int_{y_1(x)}^{y_2(x)} \mathrm{d}y \frac{\partial a}{\partial y} = -\int_{x_1}^{x_2} \mathrm{d}x\, [a(x, y_2(x)) - a(x, y_1(x))].$$

The last integral has two parts. The first can be interpreted as an integral of a along the curve $y = y_1(x)$ and the second as an integral of a along the curve $y = y_2(x)$, so we have

$$I_1 = \int_{y_1(x)} \mathrm{d}x\, a - \int_{y_2(x)} \mathrm{d}x\, a.$$

Here both parts of the curve are oriented so that a traveler on the curves moves towards increasing x. We choose the orientation of the boundary curve as shown in the figure on the last page; then, if we use a right-handed system (as in the figure), the z-component of the surface vector points in the positive z-direction. To make sure that both parts of the curve have this orientation, we need to change the sign of the integral over y_2, which gives

$$I_1 = \int \frac{\partial a}{\partial x_i} \varepsilon_{i1k}\, \mathrm{d}A_k = \oint a\, \mathrm{d}x,$$

which is what we wanted to prove. It is easy to see that this equation also holds, if we change the orientation of the boundary curve or if we use a left-handed system.

Similarly as for the Gauss theorem we now extend the proof to surfaces where the boundary of the projected surface of integration intersects with the coordinate lines $x = \text{const}$ not only in two points x_1 and x_2, but is partially tangent to these coordinate lines. Such a part of the boundary does not contribute to the line integral, since we have $dx = 0$ everywhere, so we conclude that the equation we want to prove also holds for those surfaces.

Finally, we can extend the proof to surfaces of any shape, including multiply connected two-sided surfaces, as long as all parts of the boundary are continuous and at least piecewise continuously differentiable. We can always cut these surfaces into parts of the type of surfaces for which we have already shown that the Stokes theorem holds. When we add the Stokes theorems for these subsurfaces, we obtain on the left side the surface integral over the total surface, and in the closed line integrals on the right side adjacent curves are each traveled through twice in opposite directions, so that the shared parts of the boundary cancel out and only the closed line integral over the outer boundary of the surface remains. For example, in the case of a doubly connected surface with an outer and an inner boundary, both boundary pieces must be oriented so that a traveler along the curve sees the surface in both cases on the same side (e. g. on the left side) of the curve. (It is easy to see that if we split a one-sided surface, like the Möbius strip, the cutting curve is traveled through twice in the same direction.)

Again, since the three Cartesian coordinates are equivalent, the proofs for the other two coordinates of the first version of the Stokes theorem are analogous. Using the same argument as for the Gauss theorem, we can replace the scalar field $a(x, y, z)$ with the coordinates of a higher-order tensor field, and hence extend the proof to the other versions of the Stokes theorem.

3 Algebra of Second-Order Tensors

The coordinates of a second-order tensor clearly form a square matrix, and many of the concepts and results transfer from matrices to second-order tensors. One example is the transpose operation: the definition of a transposed tensor (2.23) corresponds to the definition of a transposed matrix (1.46) and, similarly, the equations (1.48) and (2.35) for the transpose of a product of matrices and of the scalar product of tensors correspond to each other.

3.1 Additive Decomposition of a Tensor

1. We have already seen a (unique) decomposition of a tensor, namely the decomposition (2.26) into its symmetric part $a_{(ij)}$ and its antimetric part $a_{[ij]}$:

$$a_{ij} = \underbrace{\frac{1}{2}(a_{ij} + a_{ji})}_{a_{(ij)}} + \underbrace{\frac{1}{2}(a_{ij} - a_{ji})}_{a_{[ij]}}. \tag{3.1}$$

Sometimes this decomposition is called Cartesian decomposition, because it has the form of a complex number with Cartesian coordinates in the complex plane (or Argand plane) where the transposed tensor corresponds to the complex conjugate number, the symmetric part to the real part, and the antimetric part to the imaginary part.

2. There is another unique additive decomposition of a tensor, namely the decomposition into an isotropic tensor and a deviator; a deviator is a tensor whose trace is zero. We write this decomposition as

$$a_{ij} = \hat{a}\,\delta_{ij} + \mathring{a}_{ij}. \tag{3.2}$$

By deriving unique rules to compute $\hat{a}$ and $\mathring{a}_{ij}$ from the coordinates a_{ij} of an arbitrary tensor, we can show that a unique decomposition is always possible. For this, we compute the trace of (3.2). Since $\mathring{a}_{ii}$ is zero by definition, it follows that $a_{ii} = 3\hat{a}$ or

$$\hat{a} = \frac{1}{3}\,a_{ii}, \tag{3.3}$$

and when we substitute this into (3.2) and solve for $\mathring{a}_{ij}$ we get

$$\mathring{a}_{ij} = a_{ij} - \frac{1}{3}\,a_{kk}\,\delta_{ij}. \tag{3.4}$$

3. These two decompositions are related as follows:

$$a_{ij} = \overbrace{\hat{a}\,\underbrace{\delta_{ij} + (a_{(ij)} - \hat{a}\,\delta_{ij})}_{\mathring{a}_{ij}}}^{a_{(ij)}} + a_{[ij]}. \tag{3.5}$$

https://doi.org/10.1515/9783110404265-003

Using the transformation laws,[1] it is easy to see that all parts of this decomposition, i. e. the isotropic part $\hat{a}\delta_{ij}$, the symmetric part of the deviator $\mathring{a}_{(ij)} := (a_{(ij)} - \hat{a}\delta_{ij})$, the antimetric part $a_{[ij]}$, the symmetric part $a_{(ij)}$, and the deviatoric part $\mathring{a}_{ij}$, transform into a tensor of its original class, under a transformation to another coordinate system.

Problem 3.1.
Consider a tensor (in a given Cartesian coordinate system) with the coordinate matrix

$$a_{ij} = \begin{pmatrix} 4 & 1 & -7 \\ -3 & 5 & 8 \\ 3 & 2 & 9 \end{pmatrix} .$$

Compute (by finding the corresponding coordinate matrices) its isotropic part, its deviatoric part, its symmetric part, its antimetric part, and the symmetric part of its deviator. Check your result by adding the isotropic part, the symmetric deviatoric part, and the antimetric part.

Problem 3.2.
Show that the antimetry of a tensor is invariant under a transformation of coordinates.

3.2 The Determinant of a Tensor

1. For the determinant A of the Cartesian coordinates a_{ij} of a tensor

$$A := \det a_{ij} = \begin{vmatrix} a_{11} & a_{12} & a_{13} \\ a_{21} & a_{22} & a_{23} \\ a_{31} & a_{32} & a_{33} \end{vmatrix} \tag{3.6}$$

we have with (1.29)

$$A := \det a_{ij} = \frac{1}{6}\, \varepsilon_{ijk}\, \varepsilon_{pqr}\; a_{ip}\, a_{jq}\, a_{kr} = \frac{1}{6} \begin{vmatrix} a_{pp} & a_{pq} & a_{pr} \\ a_{qp} & a_{qq} & a_{qr} \\ a_{rp} & a_{rq} & a_{rr} \end{vmatrix} . \tag{3.7}$$

Since the right side of this equation is a scalar, the determinant of the coordinate matrix of a tensor is also a scalar, i. e. it is independent of a chosen coordinate system. Moreover, for a polar tensor it is a polar scalar, and for an axial tensor it is an axial scalar. This justifies to have a symbol for the determinant of a tensor itself and thus we write $A = \det \underline{\underline{a}}$.

1 See Section 2.6, No. 7.

2. A scalar obtained from the coordinates of a tensor is called an invariant of the tensor. So the trace and the determinant of a tensor are invariants of this tensor.

3. The determinant of an antimetric tensor is zero. If we change in the last equation the order of i and p, j and q, and k and r, then we get $A = \frac{1}{6}\,\varepsilon_{pqr}\,\varepsilon_{ijk}\,a_{pi}\,a_{qj}\,a_{rk}$. Comparing this with the last equation gives, for an antimetric tensor, $A = -A$, i. e. $A = 0$.

3.3 The Vector Associated with an Antimetric Tensor

By the relation $A_k = \frac{1}{2}\,\varepsilon_{ijk}\,a_{ij}$ we can associate to any tensor a_{ij} a vector A_k. Conversely, multiplying with ε_{mnk} gives $\varepsilon_{mnk}\,A_k = \frac{1}{2}\,\varepsilon_{ijk}\,\varepsilon_{mnk}\,a_{ij} = \frac{1}{2}\,(\delta_{im}\,\delta_{jn} - \delta_{in}\,\delta_{jm})\,a_{ij} = \frac{1}{2}\,(a_{mn} - a_{nm}) = a_{[mn]}$. If a_{ij} is symmetric, then A_k is zero, i. e. A_k depends only on the antimetric part of a_{ij}. The antimetric part of a_{ij} and the vector A_k are uniquely associated (because in a three-dimensional space an antimetric tensor and a vector are both determined by three parameters). We have

$$A_i = \frac{1}{2}\,\varepsilon_{ijk}\,a_{[jk]}, \qquad a_{[ij]} = \varepsilon_{ijk}\,A_k. \tag{3.8}$$

We call A_k the vector associated with the tensor $a_{[ij]}$,[2] or we say $a_{[ij]}$ is the tensor to A_k. A polar antimetric tensor is associated with an axial vector, and an axial antimetric tensor is associated with a polar vector.

If we introduce a symbol for the antimetric part of a tensor, e. g. $[\underline{\underline{a}}]$, we can write the two equations (3.8) in symbolic notation as follows:[3]

$$\underline{A} = \frac{1}{2}\,\underline{\underline{\underline{\varepsilon}}} \cdot\cdot [\underline{\underline{a}}] = \frac{1}{2}(\underline{\underline{\delta}} \times [\underline{\underline{a}}]) \cdot\cdot \underline{\underline{\delta}}, \quad [\underline{\underline{a}}] = \underline{\underline{\underline{\varepsilon}}} \cdot \underline{A} = -\underline{\underline{\delta}} \times \underline{A}. \tag{3.9}$$

Multiplying $(3.8)_2$ with the coordinates B_m of a vector $\underline{B}$ and contracting the indices m and j gives the often used identity for antimetric tensors

$$[\underline{\underline{a}}] \cdot \underline{B} = \underline{B} \times \underline{A}. \tag{3.10}$$

Problem 3.3.
Consider the antimetric tensor given by

$$a_{[ij]} = \begin{pmatrix} 0 & a_{12} & -a_{31} \\ -a_{12} & 0 & a_{23} \\ a_{31} & -a_{23} & 0 \end{pmatrix}.$$

Compute the coordinates of the associated vector.

---

2 Those authors who consider only polar tensors as tensors and who use only the polar ε-tensor call A_k the axial vector associated with $a_{[ij]}$. This axial vector is then clearly in our sense also a polar vector.

3 It is not very common to use the ε-tensor in symbolic notation; it can be replaced by a suitable vector product with the δ-tensor.

3.4 The Cotensor of a Tensor

1. Let a_{ij} be the coordinate matrix of a second-order tensor. The cofactors b_{ij} of a_{ij} are given, with (1.30), as

$$b_{mn} = \frac{1}{2}\,\varepsilon_{mij}\,\varepsilon_{npq}\,a_{ip}\,a_{jq}. \tag{3.11}$$

This means that the cofactors of a second-order tensor also form the coordinate matrix of a second-order tensor. The so-defined tensor $\underline{\underline{b}}$ is called the cotensor of the tensor $\underline{\underline{a}}$.

Clearly, the cotensor of a second-order tensor is always a polar tensor.

2. In particular, for an antimetric tensor we obtain a simple expression for its cotensor, if we use the vector associated with the tensor according to (3.8). We have

$$\begin{aligned}
b_{ip} &= \frac{1}{2}\,\varepsilon_{ijk}\,\varepsilon_{pqr}\,\varepsilon_{jqm}\,A_m\,\varepsilon_{krn}\,A_n \\
&= \frac{1}{2}\,\varepsilon_{jki}\,\varepsilon_{jqm}\,\varepsilon_{pqr}\,\varepsilon_{nkr}\,A_m\,A_n \\
&= \frac{1}{2}\,(\delta_{kq}\,\delta_{im} - \delta_{km}\,\delta_{iq})(\delta_{pn}\,\delta_{qk} - \delta_{pk}\,\delta_{qn})\,A_m\,A_n \\
&= \frac{1}{2}\,(3\,\delta_{im}\,\delta_{pn} - \delta_{im}\,\delta_{pn} - \delta_{im}\,\delta_{pn} + \delta_{pm}\,\delta_{in})\,A_m\,A_n \\
&= \frac{1}{2}\,(\delta_{im}\,\delta_{pn} + \delta_{pm}\,\delta_{in})\,A_m\,A_n \\
&= \frac{1}{2}\,(A_i\,A_p + A_p\,A_i),
\end{aligned}$$

$$b_{ip} = A_i\,A_p. \tag{3.12}$$

3.5 The Rank of a Tensor

1. The terms rank, regular, and singular also transfer from the matrix of a tensor to the tensor itself.

For the Cartesian coordinates a_{ij} of a tensor $\underline{\underline{a}}$ we have the following table.

Rank	Name	Property
3	regular	$A \neq 0$
2	single singular, rank defect 1	$A = 0,\ b_{ij} \neq 0$
1	double singular, rank defect 2	$b_{ij} = 0,\ a_{ij} \neq 0$
0	triple singular, rank defect 3	$a_{ij} = 0$

Since the property of the coordinate matrices a_{ij} and b_{ij}, to be zero or nonzero, is invariant under coordinate transformations, we conclude that all coordinate matrices of a tensor have the same rank. We call this property the rank of the associated tensor.

2. An antimetric tensor is, as shown in Section 3.2, always singular. If it is not triple singular, then it is, according to (3.12), single singular.

3.6 The Inverse Tensor

Let a_{ij} be a coordinate matrix of a regular tensor $\underline{\underline{a}}$. Then it has an inverse matrix $a_{ij}^{(-1)}$, and (1.49) shows that

$$a_{ij}\, a_{jk}^{(-1)} = \delta_{ik}. \tag{a}$$

The transformation equation between the coordinates a_{ij} of the tensor $\underline{\underline{a}}$ in the original coordinate system and its coordinates $\tilde{a}_{ij}$ in another coordinate system is given by (2.17)

$$a_{ij} = \alpha_{im}\, \alpha_{jn}\, \tilde{a}_{mn}. \tag{b}$$

The matrix $a_{ij}^{(-1)}$ can also be interpreted as the coordinate matrix of a tensor $\underline{\underline{a}}^{-1}$ in the original coordinate system. The transformation equation between the $a_{ij}^{(-1)}$ and the coordinates $\tilde{a}_{ij}^{(-1)}$ of $\underline{\underline{a}}^{-1}$ in the other coordinate system is

$$a_{jk}^{(-1)} = \alpha_{jp}\, \alpha_{kq}\, \tilde{a}_{pq}^{(-1)}. \tag{c}$$

We will show that the coordinate matrices of these two tensors are inverses of each other also in the other coordinate system (and thus in every Cartesian coordinate system).

For this, we substitute (b) and (c) into (a), and we obtain

$$\alpha_{im}\, \underbrace{\alpha_{jn}\, \tilde{a}_{mn}\, \alpha_{jp}}_{\tilde{a}_{mn}\, \delta_{np}}\, \alpha_{kq}\, \tilde{a}_{pq}^{(-1)} = \delta_{ik},$$

$$\alpha_{im}\, \alpha_{kq}\, \tilde{a}_{mn}\, \tilde{a}_{nq}^{(-1)} = \delta_{ik}.$$

Multiplication with $\alpha_{ir}\, \alpha_{ks}$ gives

$$\underbrace{\alpha_{im}\, \alpha_{ir}}_{\delta_{mr}}\, \underbrace{\alpha_{kq}\, \alpha_{ks}}_{\delta_{qs}}\, \tilde{a}_{mn}\, \tilde{a}_{nq}^{(-1)} = \underbrace{\alpha_{kr}\, \alpha_{ks}}_{\delta_{rs}},$$

$$\tilde{a}_{rn}\, \tilde{a}_{ns}^{(-1)} = \delta_{rs}, \quad \text{Q. E. D.}$$

For every regular tensor $\underline{\underline{a}}$ there exists exactly one tensor $\underline{\underline{a}}^{-1}$ with the property that the coordinate matrices of both tensors are inverses of each other in both coordinate systems. We call these two tensors the inverses of each other and then (1.49), (1.51), (1.52), (1.54), (1.50), and (1.53) give

$$
\begin{aligned}
&\underline{\underline{a}} \cdot \underline{\underline{a}}^{-1} = \underline{\underline{\delta}}, && a_{ij}\, a_{jk}^{(-1)} = \delta_{ik},\\
&\underline{\underline{a}}^{-1} \cdot \underline{\underline{a}} = \underline{\underline{\delta}}, && a_{ij}^{(-1)}\, a_{jk} = \delta_{ik},\\
&\left(\underline{\underline{a}}^{-1}\right)^{-1} = \underline{\underline{a}}, && \left(a_{ij}^{(-1)}\right)^{(-1)} = a_{ij},\\
&\left(\underline{\underline{a}}^{-1}\right)^{\mathrm{T}} = \left(\underline{\underline{a}}^{\mathrm{T}}\right)^{-1} =: \underline{\underline{a}}^{-\mathrm{T}}, && \left(a_{ij}^{(-1)}\right)^{\mathrm{T}} = \left(a_{ij}^{\mathrm{T}}\right)^{-1} =: a_{ij}^{(-\mathrm{T})},
\end{aligned}
\tag{3.13}
$$

$$
\underline{\underline{a}}^{-1} = \frac{1}{A}\, \underline{\underline{b}}^{\mathrm{T}}, \qquad a_{ij}^{(-1)} = \frac{1}{A}\, b_{ji}, \tag{3.14}
$$

$$
A\, \underline{\underline{\delta}} = \underline{\underline{b}}^{\mathrm{T}} \cdot \underline{\underline{a}} = \underline{\underline{a}} \cdot \underline{\underline{b}}^{\mathrm{T}}, \qquad A\, \delta_{ik} = b_{ji}\, a_{jk} = a_{ij}\, b_{kj}, \tag{3.15}
$$

where A is the determinant and $\underline{\underline{b}}$ is the cotensor of $\underline{\underline{a}}$.

Problem 3.4.
Consider the tensor with the coordinates

$$
A_{ij} = \begin{pmatrix} 0 & 3 & 2 \\ -3 & 0 & -1 \\ -2 & 1 & 0 \end{pmatrix}.
$$

Find (using all you know):
A. its determinant,
B. the coordinates of its cotensor,
C. if possible, the coordinates of its inverse tensor.

3.7 Orthogonal Tensors

It is easy to show, analogously to how it was derived for the inverse tensor, that the coordinate matrix of a tensor is proper or improper orthogonal in all coordinate systems, if it is proper or improper orthogonal in one coordinate system (compare Problem 3.5). We therefore call such a tensor proper orthogonal or improper orthogonal, respectively, and then, with (1.59) and (1.60), we have

$$
\begin{aligned}
&\underline{\underline{a}}^{\mathrm{T}} = \underline{\underline{a}}^{-1}, && \underline{\underline{a}} \cdot \underline{\underline{a}}^{\mathrm{T}} = \underline{\underline{a}}^{\mathrm{T}} \cdot \underline{\underline{a}} = \underline{\underline{\delta}},\\
&a_{ik}\, a_{jk} = \delta_{ij}, && a_{ki}\, a_{kj} = \delta_{ij},\\
&A = \pm 1.
\end{aligned}
\tag{3.16}
$$

Problem 3.5.
Prove that if the coordinate matrix of a tensor is proper or improper orthogonal in one coordinate system, then it is proper or improper orthogonal in all coordinate systems.

Hint: First show that if the coordinate matrix of a tensor is orthogonal in one coordinate system, then it is orthogonal in all coordinate systems.

Problem 3.6.
Show that a proper orthogonal tensor is equal to its cotensor and that an improper orthogonal tensor is equal to the negative of its cotensor.

Hint: The coordinates A_{ij} of a regular tensor and the coordinates B_{ij} of its cotensor are related, with (1.50), as $B_{ij} = A_{ji}^{(-1)} \det \underline{\underline{A}}$.

3.8 Tensors as Linear Vector Functions

1. The scalar product of a tensor and a vector results in a vector:

$$\underline{U} = \underline{\underline{a}} \cdot \underline{X}, \qquad U_i = a_{ij}\, X_j. \tag{3.17}$$

The tensor $\underline{\underline{a}}$ assigns a vector $\underline{U}$ to each vector $\underline{X}$. We also say $\underline{U}$ is a function of $\underline{X}$; since $\underline{X}$ is a vector, we call it a vector function, and since $\underline{U}$ is a vector, we call it a vector-valued function. We also say the tensor $\underline{\underline{a}}$ maps the vector $\underline{X}$ to the vector $\underline{U}$. The set of all vectors $\underline{X}$ that are mapped by the tensor $\underline{\underline{a}}$ is called the domain and, conversely, the set of all vectors $\underline{U}$ is called the range of the mapping. An element $\underline{U}$ of the range is also referred to as the image of the corresponding $\underline{X}$ in the domain of the mapping and, conversely, $\underline{X}$ is the inverse image of U. Using (3.17), we have $f(\underline{X}+\underline{Y}) = f(\underline{X})+f(\underline{Y})$ and $f(\lambda \underline{X}) = \lambda f(\underline{X})$; we refer to such a function as linear. In summary, we can call a tensor a linear vector-valued vector function and we call the mapping that it generates a linear map.

2. This property of a tensor suggests a geometric interpretation of its coordinates: if we successively substitute for X_i the Cartesian coordinates $\overset{\alpha}{e}_i = \delta_{i\alpha}$ of the unit vectors of the underlying coordinate system, then we get for the image of this Cartesian basis $\overset{\alpha}{U}_i = a_{ij}\, \overset{\alpha}{e}_j = a_{ij}\, \delta_{j\alpha}$,

$$\overset{\alpha}{U}_i = a_{i\alpha}. \tag{3.18}$$

This means that the columns of the coordinate matrix a_{ij} are the coordinates of the images of the unit vectors of the underlying coordinate system.

3. We consider two collinear vectors X_i and $X_i^* = \lambda X_i$. Their images $U_i = a_{ij} X_j$ and $U_i^* = a_{ij}\, X_j^*$ are related as follows:

$$U_i^* = a_{ij}\, X_j^* = a_{ij}\, \lambda X_j = \lambda\, a_{ij}\, X_j = \lambda\, U_i.$$

Two collinear vectors are thus mapped to two also collinear vectors with the same length ratio. Such a map is called affine.

We will see that further properties of a map depend on the rank of the tensor. Hence we consider the values which the rank of a tensor can take separately.

3.8.1 Rank 3

1. If the tensor a_{ij} has rank 3, then the images of the unit vectors are linearly independent, so they are not coplanar, and the inverse tensor $a_{ij}^{(-1)}$ exists and is the inverse map

$$\underline{X} = \underline{\underline{a}}^{-1} \cdot \underline{U}, \qquad X_i = a_{ij}^{(-1)} U_j. \tag{3.19}$$

If we interpret vectors geometrically as position vectors, then to each vector corresponds a point in space, and the tensor a_{ij} assigns to each point in space one-to-one another point in space. We also say the tensor a_{ij} maps the three-dimensional vector space of the X_i one-to-one to the three-dimensional vector space of the U_i.

2. Clearly, the zero vector is mapped to the zero vector; since the map is one-to-one, no vector which is different from zero is mapped to the zero vector.

3. If the tensor is orthogonal, then we get for the square of the image of any vector X_i with (3.17)

$$U_i^2 = a_{ij} X_j a_{ik} X_k = a_{ij} a_{ik} X_j X_k = \delta_{jk} X_j X_k = X_j^2,$$

i. e. each vector is mapped to a vector of the same length. The scalar product of the image vectors $U_i = a_{ij} X_j$ and $V_i = a_{ij} Y_j$ of any two vectors X_i and Y_i satisfies

$$U_i V_i = a_{ij} X_j a_{ik} Y_k = a_{ij} a_{ik} X_j Y_k = \delta_{jk} X_j Y_k = X_j Y_j,$$

i. e. scalar products are preserved under the mapping.

Now, $U_i V_i = U V \cos(U_i, V_i)$ and $X_i Y_i = X Y \cos(X_i, Y_i)$, and since the lengths remain the same, i. e. $U = X$ and $V = Y$, we also have $\cos(U_i, V_i) = \cos(X_i, Y_i)$, i. e. the angles between two vectors also remain the same under the mapping. Such a map, which is length and angle preserving (or isometric and isogonal), is called a congruent map or an orthogonal map.

4. A regular tensor maps a Cartesian basis to three linearly independent (i. e. noncoplanar) vectors, which generally are neither unit vectors nor perpendicular to each other. In particular, an orthogonal tensor maps a Cartesian basis to three mutually perpendicular unit vectors, i. e. again a Cartesian basis. Later we will prove that if the tensor is proper orthogonal, then the orientation of the basis is preserved, i. e.

a right-handed system is mapped to a right-handed system and a left-handed system is mapped to a left-handed system; the map represents a rotation. If the tensor is improper orthogonal, then it changes the orientation: a right-handed system is mapped to a left-handed system and a left-handed system to a right-handed system; the map represents a rotation combined with a reflection.

3.8.2 Rank 2

1. If the tensor a_{ij} has rank 2, then both the rows and the columns of its coordinate matrix are coplanar, but not collinear. If we interpret both the rows and the columns as the coordinates of position vectors or as points in space, then they each define a plane through the origin, which we call the row plane and the column plane, respectively. The images of the unit vectors then lie in the column plane. Since any vector in the domain can be represented as a linear combination of the unit vectors and since this mapping is linear, its image in the image space can also be represented as a linear combination of the $\overset{\alpha}{U}_i$; i. e. geometrically the tensor a_{ij} maps each point of the space to a point in the column plane. We also say the tensor a_{ij} maps the three-dimensional vector space of the X_i onto the two-dimensional vector space of the U_i.

2. Since the matrix of the a_{ij} is singular, the equation $a_{ij}\,X_j = 0$ has nontrivial solutions. This means that there exist vectors X_j whose image is the zero vector, i. e.

$$a_{ij}\,X_j = 0. \tag{3.20}$$

These vectors are called null vectors; they span the null space of the single singular tensor a_{ij}. Geometrically, the null space is a straight line through the origin; the direction of this line is called the null direction. If we interpret again the rows of a_{ij} as position vectors, then this line is, according to (3.20), perpendicular to these position vectors and thus is perpendicular to the row plane. Let $\underline{X}$ and $\underline{Y}$ be any two vectors in the domain, with the images $\underline{U} = \underline{\underline{a}} \cdot \underline{X}$ and $\underline{V} = \underline{\underline{a}} \cdot \underline{Y}$. Then we get by subtracting $\underline{U} - \underline{V} = \underline{\underline{a}} \cdot (\underline{X} - \underline{Y})$. Clearly, the images $\underline{U}$ and $\underline{V}$ are equal if and only if the difference between $\underline{X}$ and $\underline{Y}$ is pointing in the null direction. Thus two vectors whose projections onto the row plane are equal have the same image. In other words, if we decompose a vector $\underline{X}$ into a component $\underline{X}_\mathrm{T}$ in the row plane and a component $\underline{X}_\mathrm{N}$ perpendicular to the row plane, then we get $\underline{\underline{a}} \cdot (\underline{X}_\mathrm{T} + \underline{X}_\mathrm{N}) = \underline{\underline{a}} \cdot \underline{X}_\mathrm{T}$.

Two different vectors in the row plane have different images; geometrically, a rank-2 tensor maps each point in the row plane one-to-one to a point in the column plane, and the corresponding transposed tensor accordingly maps each point in the column plane (of the original tensor) one-to-one to a point in the row plane.

3. For an antimetric tensor, which we know is either single singular or the zero tensor, the vector associated to the tensor points in the null direction. From $(3.8)_2$ it follows by multiplying by A_j that

$$a_{ij}\, A_j = \varepsilon_{ijk}\, A_j\, A_k,$$

and this is zero, according to (2.40).

Problem 3.7.
Consider a tensor (in a given coordinate system) with the coordinate matrix

$$a_{ij} = \begin{pmatrix} 1 & -1 & -1 \\ -2 & 1 & 3 \\ -1 & 0 & 2 \end{pmatrix}.$$

A. Convince yourself that it is single singular.
B. Determine the unit vector which spans the null space.
C. Find a Cartesian basis of the image plane.
Hint: Let $\underline{U}_1$ and $\underline{U}_2$ be two linearly independent vectors in the image plane; first find a vector $\underline{V}_1 = \underline{U}_1 + \alpha\, \underline{U}_2$ which is perpendicular to $\underline{U}_1$.

3.8.3 Rank 1

If the tensor a_{ij} has rank 1, then both the rows and the columns of its coordinate matrix are collinear. Geometrically, both the rows and the columns define a straight line through the origin, which we call the row line and the column line.

The images of all vectors lie on the column line; the tensor maps the three-dimensional vector space of the X_i onto the one-dimensional vector space of the U_i.

There are two linearly independent null directions which are both perpendicular to the row line. These null directions span the null space of the double singular tensor a_{ij}. Geometrically, the null space is a plane through the origin, whose normal vector lies on the row line; we call it the null plane.

A tensor of rank 1 maps each point of the row line one-to-one to a point of the column line; the corresponding transposed tensor accordingly maps each point of the column line (of the original tensor) one-to-one to a point of the row line.

Problem 3.8.
Consider the tensor with the coordinate matrix

$$a_{ij} = \begin{pmatrix} 1 & -1 & 2 \\ -1 & 1 & -2 \\ 3 & -3 & 6 \end{pmatrix}.$$

A. Convince yourself that it is double singular.
B. Determine two linearly independent null vectors.

C. Determine the unit vector normal to the null plane.
D. Find the unit vector in the direction of the image line.

3.8.4 Rank 0

Finally, if the tensor a_{ij} has rank 0, then clearly it transforms any vector X_i into the zero vector: the tensor is the zero tensor.

3.9 Reciprocal Bases

3.9.1 Definition

1. Let three vectors $\underline{g}_1$, $\underline{g}_2$, and $\underline{g}_3$ be noncoplanar, i. e. we have $[\underline{g}_1, \underline{g}_2, \underline{g}_3] \neq 0$. Together they are called a basis or frame and we write $\underline{g}_i$.

2. Now we define the basis $\underline{g}^i$, reciprocal to the basis $\underline{g}_i$, as the three vectors

$$\underline{g}^1 = \frac{\underline{g}_2 \times \underline{g}_3}{[\underline{g}_1, \underline{g}_2, \underline{g}_3]}, \qquad \underline{g}^2 = \frac{\underline{g}_3 \times \underline{g}_1}{[\underline{g}_1, \underline{g}_2, \underline{g}_3]}, \qquad \underline{g}^3 = \frac{\underline{g}_1 \times \underline{g}_2}{[\underline{g}_1, \underline{g}_2, \underline{g}_3]}. \tag{3.21}$$

We also want to write these equations in coordinate notation. If we choose a Cartesian coordinate system and denote e. g. the coordinates of the vector $\underline{g}_1$ in this coordinate system with $\underset{1i}{g}$ and the coordinates of the vector $\underline{g}^1$ with $\overset{1}{g}_i$, then the equations (3.21) become

$$\overset{1}{g}_i = \frac{\varepsilon_{ijk} \underset{2j}{g} \underset{3k}{g}}{\varepsilon_{lmn} \underset{1l}{g} \underset{2m}{g} \underset{3n}{g}}, \qquad \overset{2}{g}_j = \frac{\varepsilon_{ijk} \underset{3k}{g} \underset{1i}{g}}{\varepsilon_{lmn} \underset{1l}{g} \underset{2m}{g} \underset{3n}{g}}, \qquad \overset{3}{g}_k = \frac{\varepsilon_{ijk} \underset{1i}{g} \underset{2j}{g}}{\varepsilon_{lmn} \underset{1l}{g} \underset{2m}{g} \underset{3n}{g}}. \tag{3.22}$$

Since the triple product in the denominator is nonzero, the three vectors $\underline{g}^i$ are uniquely determined, and since three vectors, which are each perpendicular to the other two vectors of a basis, cannot be coplanar, the $\underline{g}^i$ also form a basis.

If the $\underline{g}_i$ are polar vectors, then the $\underline{g}^i$ are also polar vectors.

We can summarize the three equations (3.21) in one equation,

$$\underline{g}^i = \frac{1}{2} \varepsilon_{ijk} \frac{\underline{g}_j \times \underline{g}_k}{[\underline{g}_1, \underline{g}_2, \underline{g}_3]}, \tag{3.23}$$

as we see immediately, if we sequentially substitute 1, 2, and 3 for i.

3.9.2 Orthogonality Relations

1. Clearly, the three vectors $\underline{g}^i$ are, according to (3.21), constructed so that e. g. $\underline{g}^1$ obeys

$$\underline{g}_1 \cdot \underline{g}^1 = 1, \quad \underline{g}_2 \cdot \underline{g}^1 = 0, \quad \underline{g}_3 \cdot \underline{g}^1 = 0.$$

Similar equations hold for $\underline{g}^2$ and $\underline{g}^3$, and we can summarize these nine equations into the so-called orthogonality relations

$$\underline{g}_i \cdot \underline{g}^j = \delta_{ij}, \qquad\qquad \underset{i\,k}{g}\; g_k^{\,j} = \delta_{ij}. \tag{3.24}$$

2. For given $\underline{g}_i$ and unknown $\underline{g}^i$, these orthogonality relations form an inhomogeneous system of nine linear equations for nine unknowns, whose coefficient determinant is nonzero. Since such a system has exactly one solution, $\underline{g}^i$ is, with (3.21), the only basis which satisfies together with $\underline{g}_i$ the orthogonality relations (3.24); the equations (3.21) and (3.24) are therefore equivalent. Conversely, $\underline{g}_i$ is reciprocal to $\underline{g}^i$, because we can change the order of the factors in (3.24).

Writing (3.24) as a matrix equation gives

$$\begin{pmatrix} \underset{11}{g} & \underset{12}{g} & \underset{13}{g} \\ \underset{21}{g} & \underset{22}{g} & \underset{23}{g} \\ \underset{31}{g} & \underset{32}{g} & \underset{33}{g} \end{pmatrix} \begin{pmatrix} g_1^{\,1} & g_1^{\,2} & g_1^{\,3} \\ g_2^{\,1} & g_2^{\,2} & g_2^{\,3} \\ g_3^{\,1} & g_3^{\,2} & g_3^{\,3} \end{pmatrix} = \begin{pmatrix} 1 & 0 & 0 \\ 0 & 1 & 0 \\ 0 & 0 & 1 \end{pmatrix}.$$

The matrix formed by the coordinates of the three vectors $\underline{g}_i$ as rows is the inverse of the matrix formed by the coordinates of the three vectors $\underline{g}^i$ as columns. From this we can compute the basis reciprocal to a given basis with the Gauss–Jordan algorithm.

3. We want to compute $\underline{g}_k\,\underline{g}^k$; clearly this is a second-order tensor, for which we temporarily introduce the unknown $\underline{\underline{X}}$. Then we have in coordinate notation

$$X_{ij} = \underset{k\,i}{g}\; g_j^{\,k}.$$

Multiplying by $\underset{mj}{g}$ gives with (3.24)

$$X_{ij}\,\underset{mj}{g} = \underset{k\,i}{g}\; g_j^{\,k}\,\underset{mj}{g} = \underset{k\,i}{g}\;\delta_{km} = \underset{mi}{g} = \underset{mj}{g}\;\delta_{ij},$$

$$(X_{ij} - \delta_{ij})\,\underset{mj}{g} = 0$$

or, translated back into symbolic notation,

$$(\underline{\underline{X}} - \underline{\underline{\delta}}) \cdot \underline{g}_m = \underline{0}.$$

Thus all three vectors $\underline{g}_m$ are null vectors of the tensor $\underline{\underline{X}} - \underline{\underline{\delta}}$, i. e. the tensor $\underline{\underline{X}} - \underline{\underline{\delta}}$ must be triple singular. In other words, it must be the zero tensor. This yields the following relations (which are also called orthogonality relations):

$$\underline{g}_k\, \underline{g}^k = \underline{\underline{\delta}}, \qquad \underset{k\,i}{g}\; \overset{k}{g}_j = \delta_{ij}. \tag{3.25}$$

3.9.3 Orthogonal Bases and Orthonormal Bases

If the three vectors of a basis are mutually perpendicular, we call the basis orthogonal, and if these vectors are unit vectors, we call it a normalized basis. If the basis (as the basis of a Cartesian coordinate system) is both orthogonal and normalized, it is called orthonormal.

For a pair of reciprocal bases it follows from (3.21) and (3.24) that if the original basis is orthogonal, then its reciprocal basis is also orthogonal, and the homologous vectors of both bases are collinear and their magnitudes are reciprocal. If the original basis is orthonormal, then it is identical to its reciprocal basis.

The orthogonality relations (3.24) and (3.25) for a Cartesian basis $\underline{e}_i$ are, according to (2.4),[4] given by

$$\begin{aligned} &\underline{e}_i \cdot \underline{e}_j = \delta_{ij}, && \overset{i}{e}_k\, \overset{j}{e}_k = \delta_{ij}, && \alpha_{ki}\, \alpha_{kj} = \delta_{ij}, \\ &\underline{e}_k\, \underline{e}_k = \underline{\underline{\delta}}, && \overset{k}{e}_i\, \overset{k}{e}_j = \delta_{ij}, && \alpha_{ik}\, \alpha_{jk} = \delta_{ij}. \end{aligned} \tag{3.26}$$

Problem 3.9.

The following is given with respect to a Cartesian basis $\underline{e}_i$:

A. the (orthogonal) basis $\underline{g}_1 = 3\,\underline{e}_1$, $\underline{g}_2 = 2\,\underline{e}_2$, $\underline{g}_3 = \underline{e}_3$;

B. the (nonorthogonal) basis, whose vectors are given by the three edges of a regular tetrahedron with edge length one, lying in the first octant, with $\underline{g}_1 = \underline{e}_1$ and $\underline{g}_2$ lying in the $\underline{e}_1, \underline{e}_2$-plane.

Compute the reciprocal bases.

3.9.4 Reciprocal Bases in the Plane

1. Let $\underline{g}_1$ and $\underline{g}_2$ be two linearly independent vectors. Then there is, *in the plane spanned by $\underline{g}_1$ and $\underline{g}_2$*, exactly one pair of vectors $\underline{g}^1$ and $\underline{g}^2$ which satisfies the orthogonality re-

4 In (2.4) the general Cartesian basis is denoted by $\underline{\tilde{e}}_i$.

lations (3.24), with i and j running naturally only from 1 to 2. We call this basis the basis reciprocal to $\underline{g}_1$ and $\underline{g}_2$. The easiest way to see why this is plausible is geometrically: for example, $\underline{g}^1$ must be in the plane spanned by $\underline{g}_1$ and $\underline{g}_2$, it must be perpendicular to $\underline{g}_2$, it must enclose an acute angle with $\underline{g}_1$, and its length must be such that $\underline{g}_1 \cdot \underline{g}^1 = 1$.

2. In a Cartesian basis in this plane, all four vectors have only two coordinates, and the two-dimensional Cartesian coordinates of $\underline{g}^1$ and $\underline{g}^2$ are computed, analogously to (3.22), from the equations

$$\overset{1}{g}_i = \frac{\varepsilon_{ij}\, \underset{2j}{g}}{\varepsilon_{mn}\, \underset{1m}{g}\, \underset{2n}{g}}, \qquad \overset{2}{g}_j = \frac{\varepsilon_{ij}\, \underset{1i}{g}}{\varepsilon_{mn}\, \underset{1m}{g}\, \underset{2n}{g}}. \tag{3.27}$$

It is easy to verify that the so-defined two-dimensional vectors satisfy the orthogonality relations.

3.10 Representation of a Tensor by Vectors

In Section 3.8 we analyzed the equation $\underline{U} = \underline{\underline{a}} \cdot \underline{X}$ by successively substituting for $\underline{X}$ the vectors of a *Cartesian* basis. With the concept of a reciprocal basis, we are now in the position to generalize this idea to *any* basis.

Let $\underline{\underline{a}}$ be any tensor and $\underline{g}_i$ be any basis and let the image of $\underline{g}_i$ under the mapping $\underline{\underline{a}}$ be

$$\underline{h}_i = \underline{\underline{a}} \cdot \underline{g}_i. \tag{3.28}$$

Tensor multiplication by the reciprocal basis $\underline{g}^i$ gives with (3.25)

$$\underline{h}_i\, \underline{g}^i = \underline{\underline{a}} \cdot \underline{g}_i\, \underline{g}^i = \underline{\underline{a}} \cdot \underline{\underline{\delta}} = \underline{\underline{a}},$$

$$\underline{\underline{a}} = \underline{h}_i\, \underline{g}^i = \underline{h}_1\, \underline{g}^1 + \underline{h}_2\, \underline{g}^2 + \underline{h}_3\, \underline{g}^3. \tag{3.29}$$

Any tensor $\underline{\underline{a}}$ can therefore be represented by six vectors in this way; here the $\underline{h}_i$ are the images of the basis $\underline{g}_i$ under the mapping by the tensor $\underline{\underline{a}}$, and the basis $\underline{g}^i$ is reciprocal to the basis $\underline{g}_i$. One can start from an arbitrary basis $\underline{g}_i$ and hence the representation (3.29) is possible in infinitely many ways: the basis $\underline{g}^i$ can be freely specified and then all three vectors $\underline{h}_i$ are uniquely determined by $\underline{\underline{a}}$.

The basis $\underline{g}^i$ is always noncoplanar. With respect to the $\underline{h}_i$, we need to distinguish between the cases where $\underline{\underline{a}}$ has rank 3, 2, or 1.

3.10.1 Rank 3

1. If $\underline{\underline{a}}$ has rank 3, then the $\underline{h}_i$ are also noncoplanar. The tensor $\underline{\underline{a}}^{-1}$, the inverse of the tensor $\underline{\underline{a}}$, exists and so does the basis $\underline{h}^i$, reciprocal to $\underline{h}_i$.

2. The converse of (3.29) is also true: if a tensor $\underline{\underline{a}}$ can be written in the form (3.29) and both the $\underline{g}^i$ and the $\underline{h}_i$ are linearly independent, then the tensor $\underline{\underline{a}}$ has rank 3.

Then $\underline{U} = \underline{\underline{a}} \cdot \underline{X}$ gives

$$\underline{U} = \underline{h}_1 \underline{g}^1 \cdot \underline{X} + \underline{h}_2 \underline{g}^2 \cdot \underline{X} + \underline{h}_3 \underline{g}^3 \cdot \underline{X}.$$

We set $\underline{g}^1 \cdot \underline{X} = \alpha^1$, $\underline{g}^2 \cdot \underline{X} = \alpha^2$, and $\underline{g}^3 \cdot \underline{X} = \alpha^3$, and we see that the α^i can take any value, since the $\underline{g}^i$ are assumed linearly independent and $\underline{X}$ can be chosen arbitrarily. So we have

$$\underline{U} = \alpha^1 \underline{h}_1 + \alpha^2 \underline{h}_2 + \alpha^3 \underline{h}_3,$$

i. e. $\underline{U}$ is a linear combination of the $\underline{h}_i$, which are also assumed linearly independent. Therefore we have a mapping from the three-dimensional space of the vectors $\underline{X}$ onto the three-dimensional space of the vectors $\underline{U}$, which means that $\underline{\underline{a}}$ must have rank 3.

3. Scalar multiplication of (3.28) by $\underline{\underline{a}}^{-1}$ from the left gives $\underline{g}_i = \underline{\underline{a}}^{-1} \cdot \underline{h}_i$.

In coordinate notation we see immediately that the transpose of (3.29) is $\underline{\underline{a}}^{\mathrm{T}} = \underline{g}^i \underline{h}_i$. Then scalar multiplication by $\underline{h}^j$ from the right gives $\underline{\underline{a}}^{\mathrm{T}} \cdot \underline{h}^j = \underline{g}^i \delta_{ij}$ or $\underline{g}^j = \underline{\underline{a}}^{\mathrm{T}} \cdot \underline{h}^j$.

Finally, scalar multiplication of the last equation from the left by $\underline{\underline{a}}^{-\mathrm{T}}$ gives $\underline{h}^j = \underline{\underline{a}}^{-\mathrm{T}} \cdot \underline{g}^j$.

Hence there are four equivalent relations,

$$\underline{h}_i = \underline{\underline{a}} \cdot \underline{g}_i, \qquad \underline{g}^i = \underline{\underline{a}}^{\mathrm{T}} \cdot \underline{h}^i, \qquad \underline{g}_i = \underline{\underline{a}}^{-1} \cdot \underline{h}_i, \qquad \underline{h}^i = \underline{\underline{a}}^{-\mathrm{T}} \cdot \underline{g}^i,$$

and solving for the tensors using the orthogonality relations (3.25) gives

$$\underline{\underline{a}} = \underline{h}_i \underline{g}^i, \qquad \underline{\underline{a}}^{\mathrm{T}} = \underline{g}^i \underline{h}_i, \qquad \underline{\underline{a}}^{-1} = \underline{g}_i \underline{h}^i, \qquad \underline{\underline{a}}^{-\mathrm{T}} = \underline{h}^i \underline{g}_i.$$

4. We summarize the results of this section.

- Any rank-3 tensor $\underline{\underline{a}}$ can be represented as follows:

$$\begin{aligned} \underline{\underline{a}} &= \underline{h}_i \underline{g}^i, & \underline{\underline{a}}^{\mathrm{T}} &= \underline{g}^i \underline{h}_i, \\ \underline{\underline{a}}^{-1} &= \underline{g}_i \underline{h}^i, & \underline{\underline{a}}^{-\mathrm{T}} &= \underline{h}^i \underline{g}_i. \end{aligned} \tag{3.30}$$

Here the $\underline{g}_i$ and $\underline{g}^i$ and also the $\underline{h}_i$ and $\underline{h}^i$ are pairs of reciprocal bases.

- The converse is also true: if a tensor $\underline{\underline{a}}$ can be represented in the form $(3.30)_1$ and both the $\underline{h}_i$ and $\underline{g}^i$ form a basis, then the tensor has rank 3.
- If we solve the equations (3.30) for the four bases, we obtain

$$\begin{aligned} \underline{h}_i &= \underline{\underline{a}} \cdot \underline{g}_i, & \underline{g}^i &= \underline{\underline{a}}^{\mathrm{T}} \cdot \underline{h}^i, \\ \underline{g}_i &= \underline{\underline{a}}^{-1} \cdot \underline{h}_i, & \underline{h}^i &= \underline{\underline{a}}^{-\mathrm{T}} \cdot \underline{g}^i. \end{aligned} \tag{3.31}$$

- For any tensor $\underline{\underline{a}}$, we can freely choose one of the four bases $\underline{g}_i$, $\underline{g}^i$, $\underline{h}_i$, or $\underline{h}^i$, and then the other three are uniquely determined.

Problem 3.10.
Represent the regular tensor with the coordinates

$$a_{ij} = \begin{pmatrix} 2 & 3 & -1 \\ 4 & -2 & 3 \\ 1 & 2 & 1 \end{pmatrix}$$

in the form $\underline{\underline{a}} = \underline{h}_i\, \underline{g}^i$, where the $\underline{g}^i$ are given by their coordinates

$$\underline{g}^1 = (1, 0, -1), \quad \underline{g}^2 = (3, 1, -3), \quad \underline{g}^3 = (1, 2, -2).$$

Hint: The $\underline{h}_i$ can be computed using the Gauss–Jordan algorithm in one step.

3.10.2 Rank 2

1. If $\underline{\underline{a}}$ has rank 2, we choose the basis $\underline{g}_i$ in (3.28) so that $\underline{g}_1$ and $\underline{g}_2$ lie in the row plane of $\underline{\underline{a}}$ and $\underline{g}_3$ is a null vector of $\underline{\underline{a}}$. Then we have $\underline{h}_3 = \underline{0}$, and $\underline{g}^1$ and $\underline{g}^2$ are perpendicular to $\underline{g}_3$, i. e. they also lie in the row plane of $\underline{\underline{a}}$. Then (3.29) reduces to

$$\underline{\underline{a}} = \underline{h}_1\, \underline{g}^1 + \underline{h}_2\, \underline{g}^2. \tag{3.32}$$

We can represent any single singular tensor by four vectors in this way; here the $\underline{g}^1$ and $\underline{g}^2$ are the basis reciprocal to $\underline{g}_1$ and $\underline{g}_2$ in the row plane of the tensor, and $\underline{h}_1$ and $\underline{h}_2$ are the images of $\underline{g}_1$ and $\underline{g}_2$ under the mapping by the tensor and they form a basis in the column plane of the tensor.

2. The converse of (3.32) is also true: if a tensor $\underline{\underline{a}}$ can be written in the form $\underline{\underline{a}} = \underline{h}_1\, \underline{g}^1 + \underline{h}_2\, \underline{g}^2$, and both $\underline{h}_1$ and $\underline{h}_2$ and also $\underline{g}^1$ and $\underline{g}^2$ are not collinear, then $\underline{\underline{a}}$ has rank 2. Then it follows from $\underline{U} = \underline{\underline{a}} \cdot \underline{X}$ that

$$\underline{U} = \underline{h}_1\, \underline{g}^1 \cdot \underline{X} + \underline{h}_2\, \underline{g}^2 \cdot \underline{X}.$$

If we set $\underline{g}^1 \cdot \underline{X} = \alpha^1$ and $\underline{g}^2 \cdot \underline{X} = \alpha^2$, then these two equations form for given $\underline{g}^1, \underline{g}^2, \alpha^1$, and α^2 a system of two linear equations for the three X_i. For each value pair (α^1, α^2) there are infinitely many solutions and, conversely, the value pair (α^1, α^2) can take any values for suitable X_i; for arbitrary $\underline{X}$ the image vector $\underline{U}$ sweeps over the entire plane spanned by $\underline{h}_1$ and $\underline{h}_2$, i. e. $\underline{\underline{a}}$ has rank 2.

3. Transposition of (3.32) gives

$$\underline{\underline{a}}^{\mathrm{T}} = \underline{g}^1 \, \underline{h}_1 + \underline{g}^2 \, \underline{h}_2 .$$

Instead, we can also write $\underline{\underline{a}}^{\mathrm{T}} = \underline{g}^i \, \underline{h}_i$, where i runs only from 1 to 2. Scalar multiplication of this equation from the right by $\underline{h}^j$, the basis reciprocal to $\underline{h}_j$ in the column plane of $\underline{\underline{a}}$, gives

$$\underline{\underline{a}}^{\mathrm{T}} \cdot \underline{h}^j = \underline{g}^i \, \underline{h}_i \cdot \underline{h}^j = \underline{g}^i \, \delta_{ij} = \underline{g}^j .$$

4. We summarize the results of this section.

- Any rank-2 tensor $\underline{\underline{a}}$ can be represented as follows:

$$\underline{\underline{a}} = \underline{h}_1 \, \underline{g}^1 + \underline{h}_2 \, \underline{g}^2, \qquad \underline{\underline{a}}^{\mathrm{T}} = \underline{g}^1 \, \underline{h}_1 + \underline{g}^2 \, \underline{h}_2 . \tag{3.33}$$

  Here $\underline{h}_i$ and $\underline{g}^i$ are linearly independent.
- The converse is also true: if a tensor $\underline{\underline{a}}$ can be represented in the form $(3.33)_1$ and both the $\underline{g}^i$ and the $\underline{h}_i$ are linearly independent, then the tensor has rank 2.
- Solving the equations (3.33) for $\underline{h}_i$ and $\underline{g}^i$ yields

$$\underline{h}_i = \underline{\underline{a}} \cdot \underline{g}_i, \qquad \underline{g}^i = \underline{\underline{a}}^{\mathrm{T}} \cdot \underline{h}^i . \tag{3.34}$$

  Here the $\underline{g}_i$ are the basis reciprocal to the $\underline{g}^i$ in the plane and the $\underline{h}^i$ are the basis reciprocal to the $\underline{h}_i$ in the plane.
- The $\underline{g}_i$ and $\underline{g}^i$ lie in the row plane of $\underline{\underline{a}}$, and the $\underline{h}_i$ and $\underline{h}^i$ lie in the column plane of $\underline{\underline{a}}$.
- For any rank-2 tensor $\underline{\underline{a}}$, we can freely choose one of the four vector pairs $\underline{g}_i$, $\underline{g}^i$, $\underline{h}_i$, or $\underline{h}^i$, and then the other three vector pairs are uniquely determined.

Problem 3.11.

A. Verify that the single singular tensor from Problem 3.7 can be represented in the form $\underline{\underline{a}} = \underline{h}_1 \, \underline{g}^1 + \underline{h}_2 \, \underline{g}^2$, with $\underline{h}_1 = (-1, 2, 1)$ and $\underline{h}_2 = (0, 1, 1)$.

B. Compute $\underline{g}^1$ and $\underline{g}^2$. (This can be done using the Gauss–Jordan algorithm in one step.)

C. Write $\underline{\underline{a}}$ as the sum of $\underline{h}_1 \, \underline{g}^1$ and $\underline{h}_2 \, \underline{g}^2$.

3.10.3 Rank 1

1. Finally, if $\underline{\underline{a}}$ has rank 1, then we choose the basis in (3.28) so that $\underline{g}_1$ lies on the row line of $\underline{\underline{a}}$ and so that $\underline{g}_2$ and $\underline{g}_3$ are null vectors of $\underline{\underline{a}}$. Then $\underline{h}_2 = \underline{h}_3 = \underline{0}$, and $\underline{g}^1$ also lies on the row line of $\underline{\underline{a}}$. Equation (3.29) reduces to

$$\underline{\underline{a}} = \underline{h}_1\, \underline{g}^1, \qquad \underline{\underline{a}}^{\mathrm{T}} = \underline{g}^1\, \underline{h}_1. \tag{3.35}$$

This means that any double singular tensor can be represented as a tensor product of two nonzero vectors, where $\underline{h}_1$ lies on the column line of the tensor, $\underline{g}^1$ lies on the row line of the tensor, and $\underline{h}_1$ is the image of the vector $\underline{g}_1$ reciprocal to $\underline{g}^1$, under the mapping with the tensor. (Two vectors are reciprocal if they are collinear and their scalar product is equal to one.)

2. The converse is also true: if a tensor $\underline{\underline{a}}$ can be written in the form $(3.35)_1$ and $\underline{h}_1$ and $\underline{g}^1$ are nonzero, then the tensor has rank 1. Then from $\underline{U} = \underline{\underline{a}} \cdot \underline{X}$ we have $\underline{U} = \underline{h}_1\, \underline{g}^1 \cdot \underline{X}$, i. e. the image vector $\underline{U}$ is pointing in the direction of $\underline{h}_1$ for all $\underline{X}$.

3. Let $\underline{h}^1$ be the vector reciprocal to $\underline{h}_1$. Then from (3.35) we get by scalar multiplication from the right by $\underline{g}_1$ and $\underline{h}^1$, respectively,

$$\underline{h}_1 = \underline{\underline{a}} \cdot \underline{g}_1, \qquad \underline{g}^1 = \underline{\underline{a}}^{\mathrm{T}} \cdot \underline{h}^1. \tag{3.36}$$

4. We summarize the results.

- Any rank-1 tensor $\underline{\underline{a}}$ can be represented as follows:

$$\underline{\underline{a}} = \underline{h}_1\, \underline{g}^1, \qquad \underline{\underline{a}}^{\mathrm{T}} = \underline{g}^1\, \underline{h}_1. \tag{3.35}$$

- The converse is also true: if a tensor $\underline{\underline{a}}$ can be represented in the form $(3.35)_1$ and $\underline{g}^1$ and $\underline{h}_1$ are nonzero, then the tensor has rank 1.
- If we solve for $\underline{h}_1$ and $\underline{g}^1$, then the equations (3.35) become

$$\underline{h}_1 = \underline{\underline{a}} \cdot \underline{g}_1, \qquad \underline{g}^1 = \underline{\underline{a}}^{\mathrm{T}} \cdot \underline{h}^1. \tag{3.36}$$

  Here $\underline{g}_1$ is the reciprocal vector to $\underline{g}^1$ and $\underline{h}^1$ is the reciprocal vector to $\underline{h}_1$.
- We know $\underline{g}_1$ and $\underline{g}^1$ lie on the row line of $\underline{\underline{a}}$ and $\underline{h}_1$ and $\underline{h}^1$ lie on the column line of $\underline{\underline{a}}$.
- We can freely choose one of the four vectors $\underline{g}_1$, $\underline{g}^1$, $\underline{h}_1$, or $\underline{h}^1$ of a rank-1 tensor $\underline{\underline{a}}$, and then the other three are uniquely determined.

3.11 Eigenvalues and Eigenvectors. The Characteristic Equation

3.11.1 Eigenvalues and Eigenvectors

1. We want to investigate under which conditions the vectors associated with the transformation $U_i = a_{ij} X_j$ are collinear, i. e.

$$U_i = \lambda X_i, \tag{3.37}$$

where the a_{ij} are real, λ is a scalar, and $X_i \neq 0$.

This is the case if

$$a_{ij} X_j = \lambda X_i, \tag{3.38}$$

$$(a_{ij} - \lambda \delta_{ij}) X_j = 0. \tag{3.39}$$

Clearly, this condition is only satisfied if the tensor $(a_{ij} - \lambda \delta_{ij})$ is singular, i. e. if

$$\det (a_{ij} - \lambda \delta_{ij}) = 0, \tag{3.40}$$

and if X_j is a null vector of this singular tensor. The values λ, for which the tensor $(a_{ij} - \lambda \delta_{ij})$ is singular, are called the eigenvalues of the tensor a_{ij}. As scalars, they are clearly invariants of the tensor. The null vectors of $(a_{ij} - \lambda \delta_{ij})$ are called the eigenvectors of a_{ij}, the null directions of $(a_{ij} - \lambda \delta_{ij})$ are called the eigendirections of a_{ij}, and equation (3.39) is called the eigenvalue equation of a_{ij}.

2. In addition to the eigenvalue problem above, we also introduce the eigenvalue problem $a^{T}_{ij} Z_j = \mu Z_i$ for the transposed tensor. We have

$$Z_j a_{ji} = \mu Z_i, \tag{3.41}$$

$$Z_i (a_{ij} - \mu \delta_{ij}) = 0. \tag{3.42}$$

According to the position of the eigenvectors, (3.42) is also called a left-eigenvalue problem and (3.39) a right-eigenvalue problem.

Clearly, (3.40) is the condition determining the eigenvalues of both eigenvalue problems, so the eigenvalues are the same for both problems, i. e.

$$\mu = \lambda. \tag{3.43}$$

The eigenvectors of both problems are also related. We mention here only the simplest relation: right and left eigenvectors, which belong to different eigenvalues, are orthogonal. Let $\underline{X}_m$ be a right eigenvector corresponding to an eigenvalue λ_m and Z_n be a left eigenvector corresponding to an eigenvalue λ_n, i. e. we have

$$\underline{\underline{a}} \cdot \underline{X}_m = \lambda_{\underline{m}} X_m, \quad \underline{Z}_n \cdot \underline{\underline{a}} = \lambda_{\underline{n}} Z_n.$$

Scalar multiplication of the first equation from the left with $\underline{Z}_n$ and of the second equation from the right with $\underline{X}_m$ gives

$$\underline{Z}_n \cdot \underline{\underline{a}} \cdot \underline{X}_m = \lambda_{\underline{m}} \underline{Z}_n \cdot X_m, \quad \underline{Z}_n \cdot \underline{\underline{a}} \cdot \underline{X}_m = \lambda_{\underline{n}} Z_n \cdot \underline{X}_m.$$

The left sides of both equations are the same, so we set the right sides equal and get

$$(\lambda_{\underline{m}} - \lambda_{\underline{n}})\underline{Z}_n \cdot \underline{X}_m = 0.$$

For $\lambda_m \neq \lambda_n$ it follows that

$$\underline{X}_m \cdot \underline{Z}_n = 0 \quad \text{for } m \neq n. \tag{3.44}$$

From now on we will consider only the right-eigenvalue problem.

3.11.2 The Characteristic Equation and Main Invariants

1. The condition for determining the eigenvalues follows from (3.7) and is

$$\begin{aligned} &\det(a_{ij} - \lambda\,\delta_{ij}) \\ &\quad = \frac{1}{6}\,\varepsilon_{ijk}\,\varepsilon_{pqr}(a_{ip} - \lambda\,\delta_{ip})(a_{jq} - \lambda\,\delta_{jq})(a_{kr} - \lambda\,\delta_{kr}) = 0. \end{aligned} \tag{3.45}$$

This is a cubic equation for λ, the so-called characteristic equation of the tensor a_{ij}.[5] Generally, there exist three (not necessarily real) eigenvalues with (at least) one eigendirection corresponding to each eigenvalue.

Expanding the parentheses gives

$$\begin{aligned} &\frac{1}{6}\,\varepsilon_{ijk}\,\varepsilon_{pqr}\,a_{ip}\,a_{jq}\,a_{kr} - \frac{\lambda}{6}\,\varepsilon_{ijk}\,\varepsilon_{pqr}(\delta_{ip}\,a_{jq}\,a_{kr} + \delta_{jq}\,a_{kr}\,a_{ip} + \delta_{kr}\,a_{ip}\,a_{jq}) \\ &\quad + \frac{\lambda^2}{6}\,\varepsilon_{ijk}\,\varepsilon_{pqr}(\delta_{ip}\,\delta_{jq}\,a_{kr} + \delta_{jq}\,\delta_{kr}\,a_{ip} + \delta_{kr}\,\delta_{ip}\,a_{jq}) \\ &\quad - \frac{\lambda^3}{6}\,\varepsilon_{ijk}\,\varepsilon_{pqr}\,\delta_{ip}\,\delta_{jq}\,\delta_{kr} = 0. \end{aligned}$$

For the coefficient of λ^3, we have $\varepsilon_{ijk}\,\varepsilon_{pqr}\,\delta_{ip}\,\delta_{jq}\,\delta_{kr} = \varepsilon_{ijk}\,\varepsilon_{ijk}$, which equals the sum of the squares of all nonzero coordinates of the ε-tensor, hence 6. The coefficients of the remaining powers of λ are clearly scalars and so they are, like the eigenvalues, also invariants of the tensor a_{ij}. We write for the last equation

$$A - A'\lambda + A''\lambda^2 - \lambda^3 = 0, \tag{3.46}$$

5 The left side of this equation is called the characteristic polynomial.

where, using (1.36), (3.11), and (3.7), we have

$$A'' = \frac{1}{2}\,\varepsilon_{ijk}\,\varepsilon_{pqr}\,\delta_{ip}\,\delta_{jq}\,a_{kr} = a_{ii}, \tag{3.47}$$

$$A' = \frac{1}{2}\,\varepsilon_{ijk}\,\varepsilon_{pqr}\,\delta_{ip}\,a_{jq}\,a_{kr} = b_{ii}, \tag{3.48}$$

$$A = \frac{1}{6}\,\varepsilon_{ijk}\,\varepsilon_{pqr}\,a_{ip}\,a_{jq}\,a_{kr} = \det a_{ij}. \tag{3.49}$$

We call these three invariants the first, second, and third main invariant of the tensor a_{ij} (after the power of the coordinates of a_{ij}). According to (1.35), we can also write A' as

$$A' = \frac{1}{2}\,(\delta_{jq}\,\delta_{kr} - \delta_{jr}\,\delta_{kq})\,a_{jq}\,a_{kr}$$

or

$$A' = \frac{1}{2}\,(a_{jj}\,a_{kk} - a_{jk}\,a_{kj}) = \frac{1}{2}\,[\mathrm{tr}^2\,\underline{\underline{a}} - \mathrm{tr}\,(\underline{\underline{a}}\cdot\underline{\underline{a}})]. \tag{3.50}$$

2. For a polar tensor, clearly, the eigenvalues and the main invariants are polar scalars. For an axial tensor, the eigenvalues are, according to (3.39), axial scalars, and then according to (3.46), both A and A'' are axial scalars, while A' is a polar scalar.

3. Using (3.50), we can now derive a simple expression for the cotensor. We have

$$b_{ip} \overset{(3.11)}{=} \frac{1}{2}\,\varepsilon_{ijk}\,\varepsilon_{pqr}\,a_{jq}\,a_{kr}$$

$$\overset{(1.26)}{=} \frac{1}{2}\begin{vmatrix} \delta_{ip} & \delta_{iq} & \delta_{ir} \\ \delta_{jp} & \delta_{jq} & \delta_{jr} \\ \delta_{kp} & \delta_{kq} & \delta_{kr} \end{vmatrix} a_{jq}\,a_{kr} = \frac{1}{2}\begin{vmatrix} \delta_{ip} & \delta_{iq} & \delta_{ir} \\ a_{pq} & a_{qq} & a_{rq} \\ a_{pr} & a_{qr} & a_{rr} \end{vmatrix}$$

$$= \frac{1}{2}\,[\delta_{ip}(a_{qq}\,a_{rr} - a_{qr}\,a_{rq}) + \delta_{iq}(a_{rq}\,a_{pr} - a_{pq}\,a_{rr}) + \delta_{ir}(a_{pq}\,a_{qr} - a_{qq}\,a_{pr})]$$

$$= \frac{1}{2}\,[\delta_{ip}(a_{qq}\,a_{rr} - a_{qr}\,a_{rq}) + a_{ri}\,a_{pr} - a_{pi}\,a_{rr} + a_{pq}\,a_{qi} - a_{qq}\,a_{pi}],$$

$$b_{ip} = A'\delta_{ip} - A''a_{pi} + a_{pq}\,a_{qi}, \qquad \underline{\underline{b}}^{\mathrm{T}} = A'\underline{\underline{\delta}} - A''\underline{\underline{a}} + \underline{\underline{a}}^2. \tag{3.51}$$

3.11.3 Classification of Tensors by the Type of Their Eigenvalues, Eigenvalue Theorems

1. The characteristic equation (3.45) has at least one real root (we assume real coordinates a_{ij}). Thus each tensor with real coordinates has at least one real eigenvalue. The following cases are possible:

I. There are three distinct real eigenvalues.
II. There are three distinct eigenvalues, one is real and two are complex conjugate.
III. There are two distinct real eigenvalues, one has multiplicity 1 and the other has multiplicity 2.
IV. There is only one real eigenvalue, with multiplicity 3.

2. It is easy to show that a symmetric tensor cannot have a complex eigenvalue (cases I, III, and IV). We prove this by contradiction, by assuming the existence of a complex eigenvalue, as well as a complex eigendirection. We have

$$a_{ij}\,(X_j + \mathrm{i}Y_j) = (\lambda + \mathrm{i}\mu)(X_i + \mathrm{i}Y_i).$$

Splitting this equation into real and imaginary parts gives

$$a_{ij}\,X_j = \lambda\,X_i - \mu\,Y_i, \quad a_{ij}\,Y_j = \mu\,X_i + \lambda\,Y_i.$$

We multiply the first equation by Y_i and the second equation by X_i, so we have

$$a_{ij}\,Y_i\,X_j = \lambda\,Y_i\,X_i - \mu\,Y_i^2, \quad a_{ij}\,X_i\,Y_j = \mu\,X_i^2 + \lambda\,X_i\,Y_i.$$

Due to the symmetry of a_{ij} the left sides are equal. Setting the right sides equal gives $\mu(X_i^2 + Y_i^2) = 0$.

Since the eigenvector is nonzero by assumption, it follows that $\mu = 0$, i. e. the eigenvalue must be real.

3. An antimetric tensor has three eigenvalues,

$$\lambda_1 = 0, \qquad \lambda_2 = \mathrm{i}\,\sqrt{A_i^2}, \qquad \lambda_3 = -\mathrm{i}\,\sqrt{A_i^2}, \tag{3.52}$$

where A_i is the vector associated with the tensor according to (3.8), the i before the root is the imaginary unit (case II), and A_i is also an eigenvector corresponding to λ_1. To see this, we write out the characteristic equation (3.46). We showed at the end of Section 3.2 that the determinant of an antimetric tensor is zero. According to (3.48) and (3.12), we have $A' = A_i^2$, and the trace of an antimetric tensor clearly vanishes.

Thus the characteristic equation reads $\lambda(A_i^2 + \lambda^2) = 0$, from which (3.52) follows immediately. The determining equations (3.39) for the eigenvector corresponding to $\lambda_1 = 0$ are $a_{ij}\,X_j = 0$. According to $(3.8)_2$, this gives $\varepsilon_{ijk}\,A_k\,X_j = 0$, and then we immediately recognize that $X_i = A_i$ is a solution, because ε_{ijk} is antimetric and $A_k\,A_j$ is symmetric.

4. A tensor which is neither symmetric nor antimetric can clearly belong to any of the four cases mentioned above.

5. The converse of the characteristic equation (3.46) gives the so-called Vieta formulas

$$\begin{aligned} A &= \lambda_1 \lambda_2 \lambda_3, \\ A' &= \lambda_1 \lambda_2 + \lambda_2 \lambda_3 + \lambda_3 \lambda_1, \\ A'' &= \lambda_1 + \lambda_2 + \lambda_3. \end{aligned} \tag{3.53}$$

A tensor is regular if and only if no eigenvalue is zero, i. e. it is singular if and only if at least one eigenvalue is zero. If only one eigenvalue is zero, the tensor has rank 2, since then clearly $A' \neq 0$ and according to (3.48) also $b_{ij} \neq 0$.[6] We will make further statements for symmetric tensors in Section 3.12.2.

6. If the coordinate matrix of a tensor $\underline{\underline{a}}$ in an appropriate coordinate system is triangular, i. e.

$$\begin{pmatrix} a_{11} & a_{12} & a_{13} \\ 0 & a_{22} & a_{23} \\ 0 & 0 & a_{33} \end{pmatrix},$$

then the diagonal elements of this matrix are the eigenvalues of the tensor.

In this case the coordinate matrix of the tensor $(\underline{\underline{a}} - \lambda \underline{\underline{\delta}})$,

$$\begin{pmatrix} a_{11} - \lambda & a_{12} & a_{13} \\ 0 & a_{22} - \lambda & a_{23} \\ 0 & 0 & a_{33} - \lambda \end{pmatrix},$$

is also triangular, so that the characteristic equation $\det(\underline{\underline{a}} - \lambda \underline{\underline{\delta}}) = 0$ is

$$(a_{11} - \lambda)(a_{22} - \lambda)(a_{33} - \lambda) = 0.$$

7. If in one row or column of the coordinate matrix of a tensor only the diagonal element $a_{\underline{ii}}$ is nonzero, then this diagonal element is an eigenvalue of the tensor, because expanding the determinant in the characteristic equation $\det(a_{ij} - \lambda\, \delta_{ij}) = 0$ with respect to that row or column leads to the factor $(a_{\underline{ii}} - \lambda)$.

If, in particular, in one *column* only the diagonal element is nonzero, then the associated coordinate direction is the eigendirection corresponding to this eigenvalue. To prove this statement, we assume, without loss of generality, that this is the case in the first column and compute the eigenvectors corresponding to the eigenvalue on the diagonal position by solving the system of equations

$$\begin{pmatrix} 0 & a_{12} & a_{13} \\ 0 & a_{22} - a_{11} & a_{23} \\ 0 & a_{32} & a_{33} - a_{11} \end{pmatrix} \begin{pmatrix} X_1 \\ X_2 \\ X_3 \end{pmatrix} = \begin{pmatrix} 0 \\ 0 \\ 0 \end{pmatrix}.$$

6 If two eigenvalues are zero, then $A' = 0$ and hence also $b_{\underline{ii}} = 0$, from which we of course cannot conclude that also $b_{ij} = 0$.

Its nontrivial solutions are the coordinates of the basis vector $\underline{e}_1$, i. e. $X_1 = 1$ and $X_2 = X_3 = 0$.

Clearly, the converse is also true: if a coordinate direction is an eigendirection, then the diagonal element in the associated column of the coordinate matrix is the corresponding eigenvalue, and the remaining elements of this column are zero. If we assume, again without loss of generality, that the x_1-direction is an eigendirection, then from the system of equations

$$\begin{pmatrix} a_{11}-\lambda & a_{12} & a_{13} \\ a_{21} & a_{22}-\lambda & a_{23} \\ a_{31} & a_{32} & a_{33}-\lambda \end{pmatrix} \begin{pmatrix} 1 \\ 0 \\ 0 \end{pmatrix} = \begin{pmatrix} 0 \\ 0 \\ 0 \end{pmatrix}$$

it follows immediately that $a_{11} = \lambda$, $a_{21} = a_{31} = 0$.

3.11.4 Theorems about Eigenvectors

In this section, we derive theorems about eigenvectors of tensors with real coordinates, and we obtain statements about the four classes of tensors, which we distinguished at the beginning of Section 3.11.3.

1. The following statement is fundamental.

Theorem 1. *To each eigenvalue there exists at least one eigendirection, in particular, a real eigenvalue has at least one real eigendirection. A complex eigenvalue cannot have a real eigendirection.*

From (3.39) it follows immediately that each eigenvalue has at least one eigendirection.

To ensure full generality, we assume that both the eigenvalue and the eigendirection in (3.39) are complex, i. e.

$$\underline{\underline{a}} \cdot (\underline{X} + \mathrm{i}\,\underline{Y}) = (\lambda + \mathrm{i}\,\mu)(\underline{X} + \mathrm{i}\,\underline{Y}).$$

Splitting this equation into real and imaginary parts gives

$$\underline{\underline{a}} \cdot \underline{X} = \lambda \underline{X} - \mu \underline{Y}, \quad \underline{\underline{a}} \cdot \underline{Y} = \mu \underline{X} + \lambda \underline{Y}.$$

If we assume that the eigenvalue is real, i. e. $\mu = 0$, it follows that

$$\underline{\underline{a}} \cdot \underline{X} = \lambda \underline{X}, \quad \underline{\underline{a}} \cdot \underline{Y} = \lambda \underline{Y}.$$

These two equations can be satisfied by $\underline{X} \neq \underline{0}$, $\underline{Y} = \underline{0}$, so a real eigenvalue has at least one real eigendirection.

If we assume conversely that the eigenvalue is complex, but the eigendirection is real, i. e. $\mu \neq 0$ and $\underline{Y} = \underline{0}$, then it follows from the imaginary part that $\mu \underline{X} = \underline{0}$ or

$\underline{X} = \underline{0}$. Since $\underline{X}$ and $\underline{Y}$ cannot both be zero, our assumption has led to a contradiction, i. e. a complex eigenvalue cannot have a real eigendirection.

2. The following theorems follow immediately from the fact that the null directions of the tensor $(\underline{\underline{a}} - \lambda\,\underline{\underline{\delta}})$ are the eigendirections of the tensor $\underline{\underline{a}}$ corresponding to the eigenvalue λ.

Theorem 2. *If, for an eigenvalue λ, the tensor $(\underline{\underline{a}} - \lambda\,\underline{\underline{\delta}})$ has rank 2, then there exists only one eigendirection corresponding to this eigenvalue.*

Theorem 3. *If, for an eigenvalue λ, the tensor $(\underline{\underline{a}} - \lambda\,\underline{\underline{\delta}})$ has rank 1, then there exist two distinct (and thus linearly independent) eigendirections corresponding to this eigenvalue.*

All directions which can be represented as linear combinations of distinct eigendirections, i. e. all vectors lying in the plane spanned by the two eigendirections, are eigendirections. There exist infinitely many different eigendirections; we also say they form an eigenplane.

Theorem 4. *If, for an eigenvalue λ, the tensor $(\underline{\underline{a}} - \lambda\,\underline{\underline{\delta}})$ has rank 0, then there exist three linearly independent eigendirections corresponding to this eigenvalue.*

Thus all directions are eigendirections; every direction can be represented as a linear combination of three linearly independent eigendirections. Rank 0 means that the tensor $(\underline{\underline{a}} - \lambda\,\underline{\underline{\delta}})$ must be the zero tensor, i. e. we have $\underline{\underline{a}} = \lambda\,\underline{\underline{\delta}}$; in other words, the tensor $\underline{\underline{a}}$ must be isotropic. Clearly, the converse is also true: if a tensor is isotropic, it has an eigenvalue of multiplicity 3, with three linearly independent eigendirections.

3. We will show in two steps that the eigendirections corresponding to distinct eigenvalues are linearly independent.

Theorem 5. *Two distinct eigenvalues cannot have a shared eigendirection.*

We prove this by contradiction: let us assume that $\underline{q}$ is the common eigendirection corresponding to two eigenvalues λ_1 and λ_2 of the tensor $\underline{\underline{a}}$. Then we have

$$\underline{\underline{a}} \cdot \underline{q} = \lambda_1\,\underline{q} = \lambda_2\,\underline{q}.$$

But since $\underline{q}$ as an eigenvector cannot be the zero vector, it follows from $\lambda_1\,\underline{q} = \lambda_2\,\underline{q}$ that $\lambda_1 = \lambda_2$.

Theorem 6. *The eigendirections corresponding to three distinct eigenvalues are linearly independent.*

We again prove by contradiction. Let λ_1, λ_2, and λ_3 be the three different eigenvalues. Then, according to Theorem 1, there exists at least one eigendirection for each eigenvalue, and, according to Theorem 5, these eigendirections are distinct. Let $\underline{q}_1$, $\underline{q}_2$, and $\underline{q}_3$ be unit vectors in these eigendirections. Then we have

$$\underline{\underline{a}} \cdot \underline{q}_1 = \lambda_1\,\underline{q}_1, \quad \underline{\underline{a}} \cdot \underline{q}_2 = \lambda_2\,\underline{q}_2, \quad \underline{\underline{a}} \cdot \underline{q}_3 = \lambda_3\,\underline{q}_3.$$

We now assume that $\underline{q}_1$, $\underline{q}_2$, and $\underline{q}_3$ are linearly dependent. Then

$$\alpha_i \underline{q}_i = \underline{0}, \quad \alpha_i \neq 0$$

must hold. Let $\alpha_3 \neq 0$. Then we have with $\beta_1 = -\alpha_1/\alpha_3$, $\beta_2 = -\alpha_2/\alpha_3$

$$\underline{q}_3 = \beta_1 \underline{q}_1 + \beta_2 \underline{q}_2.$$

Since $\underline{q}_1$, $\underline{q}_2$, and $\underline{q}_3$ are unit vectors and all are distinct, clearly neither β_1 nor β_2 can be zero. We now substitute the last equation into $\underline{\underline{a}} \cdot \underline{q}_3 = \lambda_3 \underline{q}_3$, and we obtain

$$\underline{\underline{a}} \cdot (\beta_1 \underline{q}_1 + \beta_2 \underline{q}_2) = \lambda_3(\beta_1 \underline{q}_1 + \beta_2 \underline{q}_2),$$

$$\beta_1 \underbrace{\underline{\underline{a}} \cdot \underline{q}_1}_{\lambda_1 \underline{q}_1} + \beta_2 \underbrace{\underline{\underline{a}} \cdot \underline{q}_2}_{\lambda_2 \underline{q}_2} = \beta_1 \lambda_3 \underline{q}_1 + \beta_2 \lambda_3 \underline{q}_2,$$

$$\beta_1(\lambda_1 - \lambda_3) \underline{q}_1 + \beta_2(\lambda_2 - \lambda_3) \underline{q}_2 = \underline{0}.$$

Since $\lambda_1 - \lambda_3 \neq 0$, $\lambda_2 - \lambda_3 \neq 0$, and since $\underline{q}_1$ and $\underline{q}_2$ are distinct eigenvectors, clearly β_1 and β_2 (and thus α_1 and α_2) must be zero, which contradicts the conclusion above that both are nonzero. So the assumption that the $\underline{q}_i$ are linearly dependent is wrong.

4. Next we want to prove an important relation between the multiplicity of an eigenvalue and the number of the corresponding linearly independent eigendirections.

Theorem 7. *The number of the linearly independent eigendirections corresponding to an eigenvalue is at most equal to the multiplicity of the eigenvalue.*

Conversely we conclude the following statement.

Theorem 8. *If an eigenvalue has three linearly independent eigendirections, it has multiplicity 3.*

Theorem 9. *If an eigenvalue has two distinct eigendirections, it can have multiplicity 2 or 3.*

Theorem 10. *If an eigenvalue has only one eigendirection, it can have multiplicity 1, 2, or 3.*

We show first that a tensor $\underline{\underline{a}}$ with an eigenvector of multiplicity 3 can have up to three linearly independent eigendirections.

It is sufficient to give an example for each of these three cases. Let us consider a tensor with the coordinate matrix

$$\begin{pmatrix} \lambda_1 & a_{12} & a_{13} \\ 0 & \lambda_1 & a_{23} \\ 0 & 0 & \lambda_1 \end{pmatrix}.$$

According to Section 3.11.3, No. 6, such a matrix has an eigenvalue λ_1 of multiplicity 3. The coordinate matrix of the tensor $(\underline{\underline{a}} - \lambda_1 \underline{\underline{\delta}})$ is

$$\begin{pmatrix} 0 & a_{12} & a_{13} \\ 0 & 0 & a_{23} \\ 0 & 0 & 0 \end{pmatrix}.$$

For $a_{12} \neq 0$, $a_{23} \neq 0$ this matrix has rank 2 and hence the tensor has, according to Theorem 2, only one eigendirection. For $a_{12} = 0$, $a_{23} \neq 0$ the matrix has rank 1 and hence the tensor has, according to Theorem 3, two distinct eigendirections. For $a_{12} = a_{13} = a_{23} = 0$ the matrix has rank 0 and hence the tensor has, according to Theorem 4, three linearly independent eigendirections.

We now show that an eigenvalue λ_1 of multiplicity 2 can have at most two distinct eigendirections.

The rank of the tensor $(\underline{\underline{a}} - \lambda_1 \underline{\underline{\delta}})$ corresponding to an eigenvalue λ_1 of multiplicity 2 cannot be zero, because otherwise $(\underline{\underline{a}} - \lambda_1 \underline{\underline{\delta}})$ would be the zero tensor, i. e. we would have $\underline{\underline{a}} = \lambda_1 \underline{\underline{\delta}}$, and this tensor has the eigenvalue λ_1 of multiplicity 3. So the rank of $(\underline{\underline{a}} - \lambda_1 \underline{\underline{\delta}})$ must be at least one, and thus the tensor $\underline{\underline{a}}$ has, according to Theorem 2 and Theorem 3, at most two distinct eigendirections.

According to Theorem 1, a real eigenvalue has at least one (real) eigendirection. So it only remains to show that an eigenvalue of multiplicity 2 can also have two distinct eigendirections. An example is a tensor with the coordinate matrix

$$\begin{pmatrix} \lambda_1 & a_{12} & a_{13} \\ 0 & \lambda_1 & a_{23} \\ 0 & 0 & \lambda_2 \end{pmatrix},$$

where $\lambda_2 \neq \lambda_1$. According to Section 3.11.3, No. 6, such a tensor has an eigenvalue λ_1 of multiplicity 2 and an eigenvalue λ_2 of multiplicity 1, and the coordinate matrix of the tensor $(\underline{\underline{a}} - \lambda_1 \underline{\underline{\delta}})$ is given by

$$\begin{pmatrix} 0 & a_{12} & a_{13} \\ 0 & 0 & a_{23} \\ 0 & 0 & \lambda_2 - \lambda_1 \end{pmatrix}.$$

If $a_{12} = 0$, the matrix has rank 1 and thus the tensor has, according to Theorem 3, two distinct eigendirections corresponding to the eigenvalue of multiplicity 2.

Now, let λ_1 be an eigenvalue of $\underline{\underline{a}}$ with multiplicity 1 and let $\underline{q}_1$ be an eigendirection corresponding to λ_1. We choose $\underline{q}_1$ as a unit vector in the x_1-direction of a Cartesian coordinate system, and we further choose two arbitrary unit vectors, which are orthogonal to $\underline{q}_1$ and orthogonal to each other, as the x_2- and x_3-directions. In this coordinate system, the coordinate matrix of $\underline{\underline{a}}$ is, according to Section 3.11.3, No. 7,

$$\begin{pmatrix} \lambda_1 & a_{12}^* & a_{13}^* \\ 0 & a_{22}^* & a_{23}^* \\ 0 & a_{32}^* & a_{33}^* \end{pmatrix},$$

and the characteristic equation is

$$\begin{vmatrix} \lambda_1 - \lambda & a^*_{12} & a^*_{13} \\ 0 & a^*_{22} - \lambda & a^*_{23} \\ 0 & a^*_{32} & a^*_{33} - \lambda \end{vmatrix} = (\lambda_1 - \lambda) \begin{vmatrix} a^*_{22} - \lambda & a^*_{23} \\ a^*_{32} & a^*_{33} - \lambda \end{vmatrix} = 0.$$

For $\lambda = \lambda_1$, the first factor is zero and the equation is satisfied by assumption. If, for $\lambda = \lambda_1$, the second factor were also zero, then λ_1 would not be an eigenvalue of multiplicity 1. Since this is excluded by assumption, the determinant has rank 2, and thus it has only one eigendirection corresponding to λ_1.

5. A tensor with less than three linearly independent eigendirections is called defective, and a tensor with three linearly independent eigendirections is called nondefective.

According to Theorem 7, a defective tensor must have an eigenvalue with a multiplicity of at least 2.

For a nondefective tensor we can choose the $\underline{g}_i$ in (3.28) to be three linearly independent eigenvectors and then we have $\underline{h}_i = \lambda_i \underline{g}_i$. Substituting this into (3.29) gives

$$\underline{\underline{a}} = \lambda_i \underline{g}_i \underline{g}^i, \tag{3.54}$$

where $\underline{g}^i$ is the basis reciprocal to the linearly independent eigenvectors $\underline{g}_i$. Transposing (3.54) gives $\underline{\underline{a}}^{\mathrm{T}} = \lambda_i \underline{g}^i \underline{g}_i$, and applying scalar multiplication from the right by $\underline{g}^j$ gives further

$$\underline{\underline{a}}^{\mathrm{T}} \cdot \underline{g}^j = \lambda_j \underline{g}^j,$$

i. e. the $\underline{g}^j$ are, with (3.42), at the same time left eigenvectors of $\underline{\underline{a}}$.

6. The multiplicity of an eigenvalue as a root of the characteristic equation (3.45) is called the multiplicity of the eigenvalue, or more precisely, its algebraic multiplicity. In addition, we can also define the geometric multiplicity of an eigenvalue: it is the number of linearly independent eigendirections corresponding to this eigenvalue. Using these terms, we can distinguish between defective and nondefective tensors in the following way. If a tensor has an eigenvalue whose geometric multiplicity is less than its geometric multiplicity, then the tensor is defective. Conversely, a tensor is nondefective, if the algebraic multiplicity of each eigenvalue equals its geometric multiplicity.

7. We end this section by summarizing the theorems for the four classes of tensors, which we distinguished at the beginning of the previous section.

I: The tensor has three distinct real eigenvalues.

Then there exist three linearly independent real eigendirections, one for each eigenvalue; the tensor is nondefective.

II: The tensor has three distinct eigenvalues, one is real and two are complex conjugate.

Then there exist three linearly independent eigendirections, the real eigenvalue has a real eigendirection, and the two complex eigenvalues have complex eigendirections; the tensor is nondefective.

III: The tensor has two distinct real eigenvalues, one of multiplicity 1 and one of multiplicity 2.

Then two cases are possible:

a) If the tensor $(\underline{\underline{a}}-\lambda\,\underline{\underline{\delta}})$, with λ being an eigenvalue of multiplicity 2, has rank 2, then there exists one real eigendirection corresponding to this eigenvalue and another real eigendirection (different from the first) corresponding to the second eigenvalue of multiplicity 1; the tensor is defective.

b) If the tensor $(\underline{\underline{a}} - \lambda\,\underline{\underline{\delta}})$, with λ being an eigenvalue of multiplicity 2, has rank 1, then there exist two linearly independent real eigendirections corresponding to this eigenvalue and spanning an eigenplane, and another real eigendirection not lying in the eigenplane and corresponding to the eigenvalue of multiplicity 1; the tensor is nondefective.

IV: The tensor has one real eigenvalue of multiplicity 3.

Then three cases are possible:

a) If the tensor $(\underline{\underline{a}} - \lambda\,\underline{\underline{\delta}})$ has rank 2, then there exists only one real eigendirection; the tensor is defective.

b) If the tensor $(\underline{\underline{a}} - \lambda\,\underline{\underline{\delta}})$ has rank 1, there exist two linearly independent real eigendirections, which span an eigenplane; the tensor is defective.

c) If the tensor $(\underline{\underline{a}} - \lambda\,\underline{\underline{\delta}})$ has rank 0, then the tensor is isotropic and each direction is an eigendirection; the tensor is nondefective.

Problem 3.12.

Consider tensors with the following coordinate matrices:[7]

$$\text{A.}\ \begin{pmatrix} 1 & 1 & 0 \\ 0 & 1 & 1 \\ 0 & 0 & 1 \end{pmatrix}, \quad \text{B.}\ \begin{pmatrix} 1 & 0 & 0 \\ 0 & 1 & 1 \\ 0 & 0 & 1 \end{pmatrix}, \quad \text{C.}\ \begin{pmatrix} 1 & 0 & 0 \\ 0 & 1 & 0 \\ 0 & 0 & 1 \end{pmatrix},$$

$$\text{D.}\ \begin{pmatrix} 1 & 1 & 0 \\ 0 & 1 & 1 \\ 0 & 0 & 2 \end{pmatrix}, \quad \text{E.}\ \begin{pmatrix} 1 & 0 & 0 \\ 0 & 1 & 1 \\ 0 & 0 & 2 \end{pmatrix}, \quad \text{F.}\ \begin{pmatrix} 1 & 0 & 0 \\ 0 & 1 & 0 \\ 0 & 0 & 2 \end{pmatrix}.$$

For each tensor, find the eigenvalues and, for each eigenvalue, find a set of linearly independent eigendirections. In which cases is the tensor nondefective, and in which cases do there exist three mutually orthogonal eigendirections?

---

7 Source: Adalbert Duschek, August Hochrainer: Grundzüge der Tensorrechnung in analytischer Darstellung, Vol. 1: Tensoralgebra, 5th Edition, pp. 113–115. Wien: Springer, 1968.

3.11.5 Eigenvalues and Eigenvectors of Square Matrices

1. The terms eigenvalue and eigenvector carry over to any square matrix of order N. To such a matrix $\underset{\sim}{a}$ we can assign an eigenvalue equation

$$(\underset{\sim}{a} - \lambda \underset{\sim}{I})\underset{\sim}{X} = \underset{\sim}{0}. \tag{3.55}$$

We call the numbers λ, which satisfy this equation, the eigenvalues of the matrix and we call the column matrices $\underset{\sim}{X} \neq \underset{\sim}{0}$, which solve this equation for a particular value of λ, the (right) eigenvectors of the matrix corresponding to the eigenvalue λ.

We can interpret (3.55) as a homogeneous system of linear equations for the N elements of the column matrix $\underset{\sim}{X}$. For this system to have a nontrivial solution, the coefficient determinant must be zero, i. e. the characteristic equation

$$\det(\underset{\sim}{a} - \lambda \underset{\sim}{I}) = 0 \tag{3.56}$$

must hold. Using (1.29), we have for the left side

$$\begin{aligned} &\det(\underset{\sim}{a} - \lambda \underset{\sim}{I}) \\ &= \frac{1}{N!}\, \varepsilon_{ij\ldots k}\, \varepsilon_{pq\ldots r}(a_{ip} - \lambda\, \delta_{ip})(a_{jq} - \lambda\, \delta_{jq}) \cdots (a_{kr} - \lambda\, \delta_{kr}), \\ &ij \ldots kpq \ldots r : N \text{ indices,} \end{aligned} \tag{3.57}$$

which is clearly a polynomial of degree N in λ. We can also write the characteristic equation in the form

$$\alpha - \alpha'\, \lambda + \alpha''\, \lambda^2 + \cdots + (-1)^{N-1}\, \alpha^{(N-1)}\, \lambda^{N-1} + (-1)^N\, \lambda^N = 0. \tag{3.58}$$

Such an equation has N not necessarily distinct and generally complex solutions (roots) $\lambda_1, \lambda_2, \ldots, \lambda_N$, and therefore it can be factored into the form

$$(\lambda - \lambda_1)(\lambda - \lambda_2) \cdots (\lambda - \lambda_N) = 0.$$

By expanding, we obtain for the coefficients $\alpha^{(k)}$ (read: α k-prime)

$$\begin{aligned} \alpha &= \lambda_1\, \lambda_2 \cdots \lambda_N, \\ \alpha' &= \lambda_2\, \lambda_3 \cdots \lambda_N + \lambda_1\, \lambda_3\, \lambda_4 \cdots \lambda_N + \cdots + \lambda_1\, \lambda_2 \cdots \lambda_{N-1}, \\ &\vdots \\ \alpha^{(N-1)} &= \lambda_1 + \lambda_2 + \cdots + \lambda_N. \end{aligned} \tag{3.59}$$

Here the general coefficient $\alpha^{(k)}$ is clearly the sum of all the products with $(N - k)$ different factors, which can be composed of the N roots λ_i, and there are $\binom{N}{k}$ such terms.

By expanding (3.57) and sorting according to the powers of λ, we obtain, alternatively to (3.59), a representation of the coefficients $\alpha^{(k)}$ as functions of the matrix elements a_{ij}.

For the constant term, it follows from (1.29) that

$$\alpha = \frac{1}{N!}\,\varepsilon_{ij\ldots k}\,\varepsilon_{pq\ldots r}\,a_{ip}\,a_{jq}\cdots a_{kr} = \det \underset{\sim}{a}. \tag{3.60}$$

For the linear term, we obtain

$$\alpha' = \frac{1}{N!}\,\varepsilon_{ij\ldots k}\,\varepsilon_{pq\ldots r}(\delta_{ip}\,a_{jq}\cdots a_{kr} + a_{ip}\,\delta_{jq}\cdots a_{kr} + \cdots + a_{ip}\,a_{jq}\cdots \delta_{kr}).$$

Expanding this and exchanging the order of the factors and the corresponding indices of the epsilons, gives, e. g. for the second term,

$$\frac{1}{N!}\,\varepsilon_{ij\ldots k}\,\varepsilon_{pq\ldots r}\,a_{ip}\,\delta_{jq}\cdots a_{kr} = \frac{1}{N!}\,\varepsilon_{ji\ldots k}\,\varepsilon_{qp\ldots r}\,\delta_{jq}\,a_{ip}\cdots a_{kr},$$

but this is equal to the first term, i. e. all N terms are equal, and we have

$$\alpha' = \frac{1}{(N-1)!}\,\varepsilon_{ij\ldots k}\,\varepsilon_{iq\ldots r}\,a_{jq}\cdots a_{kr}. \tag{3.61}$$

According to (1.30), this is, for $i = 1$, the minor of the element a_{11} and analogously, for $i = 2$, the minor of the element a_{22}, etc. This means α' is the sum of the principal minors of the matrix $\underset{\sim}{a}$ (actually, of the cofactors, but on the main diagonal minors and cofactors coincide).

For α'', we have in (1.29) instead of the product $a_{ip}\,a_{jq}\cdots a_{kr}$ the sum of the $\binom{N}{2}$ products, in each of which two factors a are replaced by δ. Analogously to α', we can show that all terms are equal, i. e.

$$\alpha'' = \frac{1}{(N-2)!\,2!}\,\varepsilon_{ijk\ldots l}\,\varepsilon_{ijp\ldots q}\,a_{kp}\cdots a_{lq}.$$

This in turn is the sum of the principal subdeterminants of order $(N-2)$ of the matrix.

For the coefficient of λ^k we get

$$\alpha^{(k)} = \frac{1}{(N-k)!\,k!}\,\varepsilon_{i\ldots jm\ldots n}\,\varepsilon_{i\ldots jp\ldots q}\,a_{mp}\cdots a_{nq}, \quad i\ldots j\text{: } k \text{ indices, } m\ldots n,\, p\ldots q\text{: } N-k \text{ indices,} \tag{3.62}$$

which is the sum of the $\binom{N}{k}$ principal subdeterminants of order $(N-k)$ of the matrix.

For the coefficient $\alpha^{(N-1)}$ of λ^{N-1}, we finally obtain the sum of the elements of the main diagonal; we call this, analogously to the tensors, the trace of the square matrix. We have

$$\alpha^{(N-1)} = a_{ii} =: \operatorname{tr} \underset{\sim}{a}. \tag{3.63}$$

For example, for the easiest case of a square matrix of order 2, the characteristic equation is

$$\alpha - \alpha'\lambda + \lambda^2 = 0, \tag{3.64}$$

and for the coefficients of this equation we have

$$\begin{aligned} \alpha &= \lambda_1 \lambda_2 = \det \underset{\sim}{a} = a_{11}\, a_{22} - a_{12}\, a_{21}, \\ \alpha' &= \lambda_1 + \lambda_2 = \operatorname{tr} \underset{\sim}{a} = a_{11} + a_{22}. \end{aligned} \tag{3.65}$$

2. The theorems from Sections 3.11.3 and 3.11.4 about eigenvalues and eigenvectors carry over, to a large extent, from tensors to square matrices.

A square matrix of order N with real elements has, according to (3.58), exactly N (real and/or complex) eigenvalues, if we count the eigenvalues according to their multiplicity. If N is odd, we always have at least one real eigenvalue; if N is even, all eigenvalues can be complex.

The number of the linearly independent eigendirections corresponding to an eigenvalue λ is determined by the rank defect of the matrix $\underset{\sim}{a} - \lambda \underset{\sim}{I}$. If $\underset{\sim}{a} - \lambda \underset{\sim}{I}$ has the rank defect M, then there exist M linearly independent eigendirections corresponding to this eigenvalue. Because of $\det(\underset{\sim}{a} - \lambda \underset{\sim}{I}) = 0$, the rank defect is at least $M = 1$, i. e. for each eigenvalue, there exists at least one corresponding eigendirection, for a real eigenvalue a real eigendirection, for a complex eigenvalue a complex eigendirection.

If all eigenvalues are real and distinct, then there are N linearly independent eigenvectors, which then form a basis for the N-dimensional vector space of the column matrices of order N.

We can make additional statements, if the matrix $\underset{\sim}{a}$ has a particular shape. In a triangular matrix, the diagonal elements are also eigenvalues; if in the i-th column of $\underset{\sim}{a}$ only the diagonal element is nonzero, then this diagonal element is an eigenvalue of $\underset{\sim}{a}$ and the corresponding eigenvector $\underset{\sim}{X}$ contains only a one in the i-th row, while all the remaining elements are zero.

3.12 Symmetric Tensors

3.12.1 Principal Axes Transformation

1. Every tensor whose coordinate matrix has a diagonal form in a suitable Cartesian coordinate system, i. e. it contains only zeros outside the main diagonal, is clearly symmetric. We will show in this section that, conversely, for every symmetric tensor, there exists (at least) one Cartesian coordinate system in which its coordinate matrix has a diagonal form. We call the coordinate directions of such a coordinate system the principal axes of the tensor, and the transformation of an arbitrary Cartesian coordinate system to the principal axes we call a principal axes transformation. We assume for the rest of this section that all Cartesian coordinate systems and by this also the system of principal axes are right-handed systems; for this we have to choose the order of the principal axes appropriately. With this restriction, we do not need to distinguish between polar and axial tensors for the rest of this section.

2. We will prove that a symmetric tensor always has a system of principal axes by demonstrating how to find such a system of principal axes. According to Section 3.11.3, No. 2, a symmetric tensor always has three (not necessarily distinct) real eigenvalues and according to Section 3.11.4, No. 1, Theorem 1, it has at least one corresponding real eigendirection. Using these facts, we transform the tensor $\underline{\underline{a}}$ in a first step to a Cartesian coordinate system, in which the $\tilde{x}_1$-axis coincides with this eigendirection. According to Section 3.11.3, No. 7 and because of the symmetry, which is, according to Section 2.6, No. 7, invariant under a coordinate transformation, the tensor $\underline{\underline{a}}$ has, in this coordinate system, the coordinate matrix

$$\tilde{a}_{ij} = \begin{pmatrix} \lambda_1 & 0 & 0 \\ 0 & \tilde{a}_{22} & \tilde{a}_{23} \\ 0 & \tilde{a}_{32} & \tilde{a}_{33} \end{pmatrix} . \qquad \text{(a)}$$

The corresponding characteristic equation is

$$\begin{vmatrix} \lambda_1 - \lambda & 0 & 0 \\ 0 & \tilde{a}_{22} - \lambda & \tilde{a}_{23} \\ 0 & \tilde{a}_{32} & \tilde{a}_{33} - \lambda \end{vmatrix} = 0,$$

so the two eigenvalues λ_2 and λ_3 are determined by the equation

$$\begin{vmatrix} \tilde{a}_{22} - \lambda & \tilde{a}_{23} \\ \tilde{a}_{32} & \tilde{a}_{33} - \lambda \end{vmatrix} = 0.$$

The eigenvalue equation corresponding to (a) for λ_2 reads

$$\begin{pmatrix} \lambda_1 - \lambda_2 & 0 & 0 \\ 0 & \tilde{a}_{22} - \lambda_2 & \tilde{a}_{23} \\ 0 & \tilde{a}_{32} & \tilde{a}_{33} - \lambda_2 \end{pmatrix} \begin{pmatrix} \widetilde{X}_1 \\ \widetilde{X}_2 \\ \widetilde{X}_3 \end{pmatrix} = \begin{pmatrix} 0 \\ 0 \\ 0 \end{pmatrix}$$

or, after the multiplication,

$$\begin{aligned} (\lambda_1 - \lambda_2)\widetilde{X}_1 & & &= 0, \\ & (\tilde{a}_{22} - \lambda_2)\widetilde{X}_2 &+ \quad \tilde{a}_{23}\widetilde{X}_3 &= 0, \\ & \tilde{a}_{32}\widetilde{X}_2 &+ (\tilde{a}_{33} - \lambda_2)\widetilde{X}_3 &= 0. \end{aligned} \qquad \text{(b)}$$

If λ_2 and λ_1 are distinct, (b) can only be satisfied if $\widetilde{X}_1 = 0$, i. e. if the eigendirection corresponding to λ_2 is perpendicular to the $\tilde{x}_1$-axis, which is at the same time an eigendirection corresponding to the eigenvalue λ_1. Along the way, we proved an important theorem.

The eigendirections corresponding to two distinct eigenvalues of a symmetric tensor are perpendicular to each other.

It follows immediately from Section 3.11.3, No. 7 that, for three different eigenvalues, the corresponding eigendirections form the basis of a system of principal axes and, according to Section 3.11.4, No. 4, Theorem 7, this principal axes system is unique.

If λ_2 and λ_1 are equal, we have both $\lambda_1 - \lambda_2 = 0$ and

$$(\tilde{a}_{22} - \lambda_1)(\tilde{a}_{33} - \lambda_1) - \tilde{a}_{23}^{(2)} = 0.$$

Then the tensor $\underline{\underline{a}} - \lambda_1 \underline{\underline{\delta}}$ is at least double singular and in addition to the eigendirection $\tilde{X}_1 = 1,\ \tilde{X}_2 = \tilde{X}_3 = 0,$ there is at least one more eigendirection corresponding to the eigenvalue λ_1. For the second eigendirection we can choose $\tilde{X}_1 = 0,$ which then lies in the $\tilde{x}_2, \tilde{x}_3$-plane, i. e. it is perpendicular to the first eigendirection, and both eigendirections span an eigenplane.

Independently of λ_1 and λ_2 being equal or not we can choose the two eigendirections as the $\hat{x}_1$- and $\hat{x}_2$-axes of a new Cartesian coordinate system and transform the tensor $\underline{\underline{a}}$ again. Using the same arguments as before, we have for the coordinate matrix in the new coordinate system

$$\hat{a}_{ij} = \begin{pmatrix} \lambda_1 & 0 & 0 \\ 0 & \lambda_2 & 0 \\ 0 & 0 & \hat{a}_{33} \end{pmatrix}.$$

But then, according to Section 3.11.3, No. 7, the $\hat{x}_3$-axis is also an eigendirection corresponding to the eigenvalue $\lambda_3 = \hat{a}_{33}$; so we showed that every symmetric tensor has at least one system of principal axes and that in any system of principal axes the elements on the main diagonal of the coordinate matrix are eigenvalues and the principal axes are eigendirections.

Thus a symmetric tensor with three different eigenvalues has only one system of principal axes. For an eigenvalue of multiplicity 2, there exist two linearly independent eigendirections, which span an eigenplane, and every Cartesian basis with one basis vector perpendicular to this plane forms a system of principal axes. A tensor with an eigenvalue of multiplicity 3 is isotropic; in this case every Cartesian basis is a system of principal axes.

3. A symmetric tensor is always nondefective and can therefore always be represented in the form (3.54). Since, according to Section 3.9.3, a basis $\underline{q}_i$ of a system of principal axes is identical to its reciprocal basis, we can replace both the $\underline{g}_i$ and $\underline{g}^i$ in (3.54) by $\underline{q}_i$, which gives

$$\underline{\underline{a}} = \lambda_{\underline{i}}\, \underline{q}_i\, \underline{q}_i.$$

4. In our proof of the principal axes transformation, we transformed the coordinate matrix of a symmetric tensor, step-by-step, into diagonal form. For practical computations, it is not necessary to follow this stepwise procedure; we can compute the diagonal form with a single transformation. To this end, we solve the eigenvalue problem $a_{ij} X_j = \lambda X_i$, form a Cartesian basis from the eigenvectors $\underline{q}_i$, and use the coordinates $\overset{j}{q}_i$, as in (2.4), as transformation coefficients $\alpha_{ij} = \overset{j}{q}_i$. Applying the transformation $\hat{a}_{ij} = \alpha_{mi} \alpha_{nj} a_{mn}$ then results in a coordinate representation with respect to the basis $\underline{q}_i$, in which only the diagonal elements are nonzero, so they are the eigenvalues: $\hat{a}_{ij} = \lambda_{\underline{i}} \delta_{ij}$; the order of the eigenvalues on the main diagonal depends on the order in which the basis is constructed from the corresponding eigenvectors.

5. We summarize the results of this section; they are valid under the assumption that all coordinate systems are right-handed systems and so they are valid for both polar and for axial tensors.

- A system of principal axes is a (right-handed) Cartesian coordinate system in which the coordinate matrix of the tensor has a diagonal form.
- Any tensor which has a system of principal axes is symmetric and every symmetric tensor has (at least) one system of principal axes.
- Any principal axis is an eigendirection of the tensor and each triple of mutually orthogonal eigendirections forms a system of principal axes.
- In every system of principal axes the elements on the main diagonal of the coordinate matrix are the eigenvalues of the tensor, i. e.

$$\hat{a}_{ij} = \lambda_{\underline{i}} \delta_{ij} = \begin{pmatrix} \lambda_1 & 0 & 0 \\ 0 & \lambda_2 & 0 \\ 0 & 0 & \lambda_3 \end{pmatrix}. \tag{3.66}$$

- The eigenvalues λ_i are determined by the characteristic equation

$$\det (a_{ij} - \lambda \delta_{ij}) = 0. \tag{3.67}$$

- The transformation coefficients α_{ij} of the principal axes transformation $\hat{a}_{ij} = \alpha_{mi} \alpha_{nj} a_{mn}$ are also the coordinates $\overset{j}{q}_i$ of the three orthonormal eigenvectors $\underline{q}_j$ of the tensor in the original coordinate system, i. e.

$$\alpha_{ij} = \overset{j}{q}_i. \tag{3.68}$$

- The coordinates $\overset{j}{q}_i$ of three orthonormal eigenvectors $\underline{q}_j$ of the tensor in the original coordinate system are determined by the eigenvalue equation

$$(a_{ij} - \lambda_{\underline{k}}\, \delta_{ij})\, \overset{k}{q}_j = 0. \tag{3.69}$$

- Between the coordinates $\hat{a}_{ij}$ of the tensor in a system of principal axes, its eigenvalues λ_i, its coordinates a_{ij} in the original coordinate system, and the coordinates $\overset{j}{q}_i$ of the three orthonormal eigenvectors in the original coordinate system, we have the following relations:

$$\begin{aligned} \hat{a}_{ij} &= \overset{i}{q}_m\, \overset{j}{q}_n\, a_{mn} = \lambda_{\underline{i}}\, \delta_{ij}, \\ a_{ij} &= \overset{m}{q}_i\, \overset{n}{q}_j\, \hat{a}_{mn} = \lambda_{\underline{k}}\, \overset{k}{q}_i\, \overset{k}{q}_j, \\ \underline{\underline{a}} &= \lambda_{\underline{k}}\, \underline{q}_{\underline{k}}\, \underline{q}_{\underline{k}}. \end{aligned} \tag{3.70}$$

- For the number of systems of principal axes we have:
  - If only eigenvalues with multiplicity 1 exist, there is only one system of principal axes.
  - If an eigenvalue of multiplicity 2 exists, then any coordinate system with one axis in the eigendirection corresponding to the eigenvalue with multiplicity 1 is a system of principal axes.
  - If an eigenvalue of multiplicity 3 exists, then every coordinate system is a system of principal axes; the tensor is isotropic.

Problem 3.13.
Consider a tensor which has in its original coordinate system the coordinate matrix

$$t_{ij} = \begin{pmatrix} 5 & 0 & 4 \\ 0 & 9 & 0 \\ 4 & 0 & 5 \end{pmatrix}.$$

Transform the tensor to principal axes, i. e. find a right-handed Cartesian basis (in terms of the coordinates of the original system) in which the coordinate matrix of the tensor has a diagonal form, and write out this coordinate matrix.

3.12.2 Eigenvalues and Rank of a Tensor

From the existence of a system of principal axes (3.66) it further follows that if all eigenvalues are nonzero, the tensor $\underline{\underline{a}}$ has rank 3; if one eigenvalue is zero, it has rank 2; if two eigenvalues are zero, it has rank 1; and if all three eigenvalues are zero, it has rank zero. Thus we have, for a symmetric tensor, a simple relation between the rank of the tensor and the number of its vanishing eigenvalues.

Rank	Eigenvalues
3	$\lambda_{1,2,3} \neq 0$
2	$\lambda_1 = 0, \lambda_{2,3} \neq 0$
1	$\lambda_1 = \lambda_2 = 0, \lambda_3 \neq 0$
0	$\lambda_1 = \lambda_2 = \lambda_3 = 0$

3.12.3 Eigenvalues and Definiteness of a Tensor

1. We consider the quadratic form of an arbitrary tensor $\underline{\underline{a}}$ with the Cartesian coordinates a_{ij},

$$Q = a_{ij} X_i X_j, \tag{3.71}$$

where $\underline{X}$ is an arbitrary nonzero vector. If independently of the choice of $\underline{X}$ the quadratic form Q is always positive, $\underline{\underline{a}}$ is called positive definite; if Q is always negative, $\underline{\underline{a}}$ is called negative definite. If Q is always positive or zero or always negative or zero, $\underline{\underline{a}}$ is called positive semidefinite or negative semidefinite, respectively. Finally, if Q can take both signs, $\underline{\underline{a}}$ is called indefinite. That is, any positive definite tensor is also positive semidefinite, any negative definite tensor is also negative semidefinite, and any tensor which is neither positive semidefinite nor negative semidefinite is indefinite.[8]

2. If we compute the quadratic form (3.71) in the system of principal axes, we get

$$Q = \lambda_1 X_1^2 + \lambda_2 X_2^2 + \lambda_3 X_3^2, \tag{3.72}$$

i. e. the sign of Q depends for $\underline{X} \neq \underline{0}$ only on the sign of the λ_j, and because of Vieta's formulas $(3.53)_1$, the same is true for the determinant of the tensor. Thus, for a symmetric tensor, we find the following relations between the sign of Q and the signs of its eigenvalues and of its determinant.

8 Except for the zero tensor, which is both positive and negative semidefinite, the three sets of positive semidefinite, negative semidefinite, and indefinite tensors are disjoint.

Quadratic Form	Tensor	Eigenvalues	Determinant
$Q > 0$	positive definite	all > 0	> 0
$Q \geqq 0$	positive semidefinite	all $\geqq 0$	$\geqq 0$
$Q < 0$	negative definite	all < 0	< 0
$Q \leqq 0$	negative semidefinite	all $\leqq 0$	$\leqq 0$
Q takes positive and negative values	indefinite	with different sign	no statement possible

Thus positive and negative definite tensors are always regular.

Problem 3.14.
We can generalize the eigenvalue problem of a tensor $\underline{\underline{a}}$ to the case that the eigenvalue λ is associated with a second tensor $\underline{\underline{b}}$, i. e.

$$a_{ij}\,X_j = \lambda\,b_{ij}\,X_j.$$

Let $\underline{\underline{a}}$ and $\underline{\underline{b}}$ be two symmetric tensors. Determine the conditions under which the generalized eigenvalue problem has only real eigenvalues.

3. With the quadratic form $Q = a_{ij}\,X_i\,X_j$ we can interpret the eigendirections of a symmetric tensor $\underline{\underline{a}}$ in a different way than in Section 3.11.1, No. 1. To this end, we assume $\underline{X}$ to be a unit vector and ask for those directions, for which the quadratic form takes its maximal or minimal values. Thus we formulated an extremum problem,

$$Q = a_{ij}\,X_i\,X_j \qquad \longrightarrow \qquad \text{extremum},$$

for the coordinates of the vector $\underline{X}$ with the constraint $X_i\,X_i = 1$ or

$$1 - \delta_{ij}\,X_i\,X_j = 0.$$

A common method for solving such problems is to multiply the left side of the constraint with an initially undetermined factor λ (the so-called Lagrange multiplier) and add this term to the function whose extrema are to be determined, i. e.

$$h = a_{ij}\,X_i\,X_j + \lambda\,(1 - \delta_{ij}\,X_i\,X_j)\,.$$

The necessary condition for the extrema of Q is then that the partial derivatives of the auxiliary function h with respect to the coordinates of $\underline{X}$ must be zero, i. e.

$$\begin{aligned}\frac{\partial h}{\partial X_k} &= (a_{ij} - \lambda\,\delta_{ij})\Big(\underbrace{\frac{\partial X_i}{\partial X_k}}_{\delta_{ik}}\,X_j + X_i\,\underbrace{\frac{\partial X_j}{\partial X_k}}_{\delta_{jk}}\Big)\\ &= (a_{kj} - \lambda\,\delta_{kj})\,X_j + (a_{ik} - \lambda\,\delta_{ik})\,X_i = 0.\end{aligned}$$

From the symmetry of a_{ij} and δ_{ij} it follows that

$$(a_{ij} - \lambda\,\delta_{ij})\,X_j = 0,$$

i. e. we obtained the familiar eigenvalue problem for the tensor $\underline{\underline{a}}$. Thus the quadratic form takes its extrema in the eigendirections of $\underline{\underline{a}}$.

3.12.4 Symmetric Matrices

A symmetric matrix, i. e. a matrix for which $\underset{\sim}{A} = \underset{\sim}{A}^T$, must be square. Furthermore, we restrict ourselves in this section to matrices with real elements, without mentioning it every time.

Several results from symmetric tensors (with real coordinates) carry over to symmetric matrices of order N (with real elements).

1. A fundamental statement for symmetric matrices is the following.

Theorem 1. *A symmetric matrix has only real eigenvalues, and to each eigenvalue corresponds at least one real eigenvector (i. e. an eigenvector where all elements are real numbers).*

We can show, analogously to the corresponding proof for symmetric tensors in Section 3.11.3, No. 2, that a symmetric matrix cannot have complex eigenvalues. Furthermore, from (3.55) it follows that for any real eigenvalue there exists at least one corresponding real eigenvector.

2. We can generalize the concept of orthogonality to an N-dimensional space, using the scalar product of two vectors. If, for the matrix product of two column matrices with N elements, we have $\underset{\sim}{X}^{\mathrm{T}}\,\underset{\sim}{Y} = 0$ (i. e. $X_i\,Y_i = 0$), we call the column matrices $\underset{\sim}{X}$ and $\underset{\sim}{Y}$ mutually orthogonal to each other. From N matrices with mutually orthogonal columns, we can construct an orthogonal matrix $\underset{\sim}{R}$, and then we can assign, according to (1.58), a similar matrix $\underset{\sim}{\tilde{a}} = \underset{\sim}{R}^{\mathrm{T}}\underset{\sim}{a}\,\underset{\sim}{R}$ to a matrix $\underset{\sim}{a}$. Clearly, this is a generalization of the transformation law (2.15) for the coordinates of a second-order tensor. Similar matrices have the same characteristic equation (and thus the same eigenvalues), which in our case follows, using (1.13), from

$$\begin{aligned}\det(\underset{\sim}{\tilde{a}} - \lambda\,\underset{\sim}{E}) &= \det(\underset{\sim}{R}^{\mathrm{T}}\underset{\sim}{a}\,\underset{\sim}{R} - \lambda\,\underset{\sim}{R}^{\mathrm{T}}\underset{\sim}{E}\,\underset{\sim}{R}) = \left(\det\underset{\sim}{R}^{\mathrm{T}}\right)\,\det(\underset{\sim}{a} - \lambda\,\underset{\sim}{E})\,\left(\det\underset{\sim}{R}\right)\\ &= \det(\underset{\sim}{a} - \lambda\,\underset{\sim}{E}) = 0.\end{aligned}$$

With these generalizations we come to the following theorem.

Theorem 2. *A symmetric matrix has at least one system of principal axes consisting of mutually orthogonal eigenvectors. In the system of principal axes the matrix has a diagonal form; the elements on the main diagonal are the eigenvalues of the matrix.*

The principal axes transformation follows the same procedure as for tensors. In the first step we choose a matrix $\underset{\sim}{R}_1$, whose first column is an eigenvector $\underset{\sim}{X}_1$ of $\underset{\sim}{a}$. Because of the symmetry, as in Section 3.12.1, the elements in the first row and the first column of the resulting similar matrix $\underset{\sim}{\tilde{a}}$ are all zero except the element on the diagonal, which is the eigenvalue λ_1 corresponding to the eigenvector $\underset{\sim}{X}_1$, so we have

$$\underset{\sim}{\tilde{a}} = \underset{\sim}{R}_1^{\mathrm{T}} \underset{\sim}{a} \underset{\sim}{R}_1 = \begin{pmatrix} \lambda_1 & 0 & 0 & \dots & 0 \\ 0 & \tilde{a}_{22} & \tilde{a}_{23} & \dots & \tilde{a}_{2N} \\ 0 & \tilde{a}_{23} & \tilde{a}_{33} & \dots & \tilde{a}_{3N} \\ \vdots & & & & \\ 0 & \tilde{a}_{2N} & \tilde{a}_{3N} & \dots & \tilde{a}_{NN} \end{pmatrix}.$$

The first equation in the eigenvalue problem $\underset{\sim}{\tilde{a}} \underset{\sim}{\tilde{X}} = \lambda \underset{\sim}{\tilde{X}}$ of the matrix $\underset{\sim}{\tilde{a}}$ is now decoupled from the remaining equations, i. e. the eigenvector $\underset{\sim}{\tilde{X}}_1$ has the elements $\underset{1\,1}{\tilde{X}} = 1$, $\underset{1\,2}{\tilde{X}} = \dots = \underset{1\,N}{\tilde{X}} = 0$, while further eigenvectors $\underset{\sim}{\tilde{X}}_i$ must satisfy $\underset{i\,1}{\tilde{X}} = 0$ (if $\lambda_i \neq \lambda_1$) or we can choose $\underset{i\,1}{\tilde{X}} = 0$ (if $\lambda_i = \lambda_1$). Thus the eigenvector $\underset{\sim}{\tilde{X}}_1$ is orthogonal to all remaining eigenvectors $\underset{\sim}{\tilde{X}}_i$. Therefore, we can, using the principal submatrix

$$\begin{pmatrix} \tilde{a}_{22} & \tilde{a}_{23} & \dots & \tilde{a}_{2N} \\ \tilde{a}_{23} & \tilde{a}_{33} & \dots & \tilde{a}_{3N} \\ \vdots & & & \\ \tilde{a}_{2N} & \tilde{a}_{3N} & \dots & \tilde{a}_{NN} \end{pmatrix},$$

find a second eigenvalue λ_2, construct an orthogonal matrix $\underset{\sim}{R}_2$ whose first two columns are the eigenvectors $\underset{\sim}{\tilde{X}}_1$ and $\underset{\sim}{\tilde{X}}_2$, and transform the matrix $\underset{\sim}{a}$. This gives

$$\underset{\sim}{\hat{a}} = \underset{\sim}{R}_2^{\mathrm{T}} \underset{\sim}{a} \underset{\sim}{R}_2 = \begin{pmatrix} \lambda_1 & 0 & 0 & \dots & 0 \\ 0 & \lambda_2 & 0 & \dots & 0 \\ 0 & 0 & \hat{a}_{33} & \dots & \hat{a}_{3N} \\ \vdots & & & & \\ 0 & 0 & \hat{a}_{3N} & \dots & \hat{a}_{NN} \end{pmatrix}.$$

After N such steps the matrix is fully diagonalized. If multiple eigenvalues exist, then we have, as in Section 3.12.1, more than one system of principal axes.

3. Any symmetric matrix of order N belongs to one of the five definiteness classes.

Let $\underset{\sim}{a}$ be a symmetric matrix of order N and let $\underset{\sim}{X}$ be a nonzero column matrix with N elements. Then its quadratic form is

$$Q = \underset{\sim}{X}^{\mathrm{T}} \underset{\sim}{a} \underset{\sim}{X}, \qquad Q = X_i \, a_{ij} \, X_j. \tag{3.73}$$

Now, if the number Q is positive, independently of the choice of $\underset{\sim}{X}$, we call the matrix $\underset{\sim}{a}$ positive definite; if it is always negative, we call $\underset{\sim}{a}$ negative definite. If Q is always positive or zero, or always negative or zero, we call $\underset{\sim}{a}$ positive semidefinite or negative semidefinite, respectively. If finally Q can take both signs, we call $\underset{\sim}{a}$ indefinite. With these definitions we come to the following theorem.

Theorem 3. *If a symmetric matrix is positive definite, it has only positive eigenvalues; if it is negative definite, it has only negative eigenvalues; if it is positive semidefinite, it has no negative eigenvalues; and if it is negative semidefinite, it has no positive eigenvalues.*

We can explain this statement as follows. If we choose $\underset{\sim}{X}$ to be an arbitrary eigenvector corresponding to the eigenvalue λ, we have

$$Q = \underset{\sim}{X}^{\mathrm{T}} \underset{\sim}{a} \underset{\sim}{X} = \underset{\sim}{X}^{\mathrm{T}} \lambda \underset{\sim}{X} = \lambda \underset{\sim}{X}^{\mathrm{T}} \underset{\sim}{X}.$$

Since $\underset{\sim}{X}$ has only real elements, $\underset{\sim}{X}^{\mathrm{T}} \underset{\sim}{X}$ is positive, i. e. λ and Q have the same sign.

Since, according to (3.59) and (3.60), the determinant of a symmetric matrix is equal to the product of its N eigenvalues, we further have the following theorem.

Theorem 4. *The determinant of a positive definite symmetric matrix is positive.*

4. A submatrix which is obtained from a symmetric matrix by eliminating a number of columns and rows with the same indices is called a principal submatrix.

Theorem 5. *The principal submatrices of a symmetric matrix have the same definiteness as the matrix itself; for example, if the matrix is positive definite, then all its principal submatrices are also positive definite.*

We prove this statement as follows. If we choose $\underset{\sim}{X}$ in (3.73) such that $X_1 = 0$, then from the N^2 terms on the right side of (3.73) all those with index one vanish, i. e. we obtain the quadratic form of the principal submatrix that we obtained from $\underset{\sim}{a}$ by eliminating the first row and the first column. If $\underset{\sim}{a}$ is positive definite, then Q is also positive for the chosen $\underset{\sim}{X}$, and because this is true for all $\underset{\sim}{X}$ with $X_1 = 0$, the subdeterminant under consideration itself is also positive definite. With analogous considerations for $\underset{\sim}{X}$, in which one or several other elements vanish, we can generalize this to all principal submatrices; and if we follow the same argument for a matrix with a different definiteness, for example a negative semidefinite matrix, we see immediately that our argument remains valid.

Because the main diagonal elements of a matrix are also the principal subdeterminants of order one, we have in particular the following theorem.

Theorem 6. *The main diagonal elements of a positive definite symmetric matrix are all positive; for a negative definite symmetric matrix they are all negative; for a positive semidefinite symmetric matrix they are all positive or zero; and for a negative semidefinite symmetric matrix they are all negative or zero.*

5. The converse of all these theorems is not true, e. g. if all eigenvalues of a symmetric matrix are real, we cannot conclude that the matrix is symmetric; and if the determinant of a symmetric matrix is positive, it does not follow that the matrix is positive definite.

3.13 Orthogonal Polar Tensors

3.13.1 Rotation in a Plane

For the following, we need a matrix which maps a position vector $\underline{X}$ in the x, y-plane by rotating about the z-axis through an angle ϑ to a vector $\underline{U}$. Clearly, these vectors are

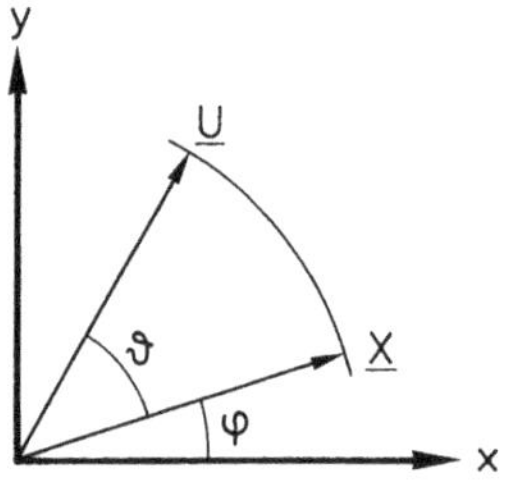

$$
\begin{aligned}
X_1 &= r\cos\varphi, \quad X_2 = r\sin\varphi, \\
U_1 &= r\cos(\varphi+\vartheta) = r(\cos\varphi\cos\vartheta - \sin\varphi\sin\vartheta) \\
&= X_1\cos\vartheta - X_2\sin\vartheta, \\
U_2 &= r\sin(\varphi+\vartheta) = r(\sin\varphi\cos\vartheta + \cos\varphi\sin\vartheta) \\
&= X_2\cos\vartheta + X_1\sin\vartheta,
\end{aligned}
$$

$$
\begin{pmatrix} U_1 \\ U_2 \end{pmatrix} = \begin{pmatrix} \cos\vartheta & -\sin\vartheta \\ \sin\vartheta & \cos\vartheta \end{pmatrix} \begin{pmatrix} X_1 \\ X_2 \end{pmatrix}. \tag{3.74}
$$

3.13.2 Transformation to an Eigendirection

1. Any tensor (with real coordinates), in particular any orthogonal tensor, has (at least) one real eigenvalue and correspondingly (at least) one real eigendirection. Since an orthogonal tensor represents a length preserving mapping, all its real eigenvalues must be, according to (3.37), either +1 or −1.

According to Section 3.11.3, in particular $(3.53)_1$, the following cases can occur:

I. All three eigenvalues are 1, then $\det \underset{\sim}{a} = 1$.
II. All three eigenvalues are −1, then $\det \underset{\sim}{a} = -1$.
III. Two eigenvalues are 1, one is −1, then $\det \underset{\sim}{a} = -1$.

IV. Two eigenvalues are −1, one is 1, then $\det \underset{\sim}{a} = 1$.
V. Two eigenvalues are complex conjugate to each other, one is 1; since the product of two complex conjugate numbers is positive, then $\det \underset{\sim}{a} = 1$.
VI. Two eigenvalues are complex conjugate to each other, one is −1, then $\det \underset{\sim}{a} = -1$.

Thus, in all cases where $\det \underset{\sim}{a} = 1$, one eigenvalue is 1; and in all cases where $\det \underset{\sim}{a} = -1$, one eigenvalue is −1.

2. It is reasonable to assume that in a coordinate system where one basis vector is an eigenvector, an orthogonal tensor has a particularly simple form which is easy to interpret. Therefore, we assume in the following that the basis vector $\underline{e}_3$ of our coordinate system is an eigenvector of the tensor and that the corresponding eigenvalue has the same sign as the determinant of the tensor. The coordinate matrix with respect to such a basis is then, according to Section 3.11.3, No. 7,

$$\begin{pmatrix} a_{11} & a_{12} & 0 \\ a_{21} & a_{22} & 0 \\ a_{31} & a_{32} & \pm 1 \end{pmatrix}.$$

Since the sum of the squares in the last row must be 1, we have $a_{31} = a_{32} = 0$, so

$$\begin{pmatrix} a_{11} & a_{12} & 0 \\ a_{21} & a_{22} & 0 \\ 0 & 0 & \pm 1 \end{pmatrix}.$$

The sum of the squares in the first column is $a_{11}^2 + a_{21}^2 = 1$, which can be satisfied, without loss of generality, by $a_{11} = \cos\vartheta$, $a_{21} = \sin\vartheta$, i. e.

$$\begin{pmatrix} \cos\vartheta & a_{12} & 0 \\ \sin\vartheta & a_{22} & 0 \\ 0 & 0 & \pm 1 \end{pmatrix}.$$

Since the sign of $\det a_{ij}$ must match the sign of a_{33}, it follows that

$$\begin{vmatrix} \cos\vartheta & a_{12} \\ \sin\vartheta & a_{22} \end{vmatrix} = a_{22}\cos\vartheta - a_{12}\sin\vartheta = 1.$$

The sum of the squares in the first row gives $a_{12} = \pm\sin\vartheta$, and from that it follows with the previous equation that $a_{12} = -\sin\vartheta$ and $a_{22} = \cos\vartheta$.

Thus, if $\det \underset{\sim}{a} = 1$, i. e. if the tensor is proper orthogonal, we have for the coordinate matrix of our orthogonal tensor

$$a_{ij} = \begin{pmatrix} \cos\vartheta & -\sin\vartheta & 0 \\ \sin\vartheta & \cos\vartheta & 0 \\ 0 & 0 & +1 \end{pmatrix} \tag{3.75}$$

and if $\det \underset{\sim}{a} = -1$, i. e. if it is improper orthogonal, we have

$$a_{ij} = \begin{pmatrix} \cos\vartheta & -\sin\vartheta & 0 \\ \sin\vartheta & \cos\vartheta & 0 \\ 0 & 0 & -1 \end{pmatrix}. \tag{3.76}$$

3. These formulas are easy to interpret with (3.74). The mapping given by a proper orthogonal tensor does not change the z-coordinate of any position vector and it rotates the projection of the position vector onto the x, y-plane by the angle ϑ; i. e. the mapping of the tensor is a rotation through the angle ϑ about the eigendirection corresponding to its positive eigenvalue. This explains why a proper orthogonal tensor is also called a rotation tensor. The mapping given by an improper orthogonal tensor changes only the sign of the z-coordinate, and it again rotates the projection of the position vector onto the x, y-plane by the angle ϑ; i. e. the mapping of the tensor is a rotation through the angle ϑ about the eigendirection corresponding to its negative eigenvalue, followed by a reflection through the plane orthogonal to this eigendirection.

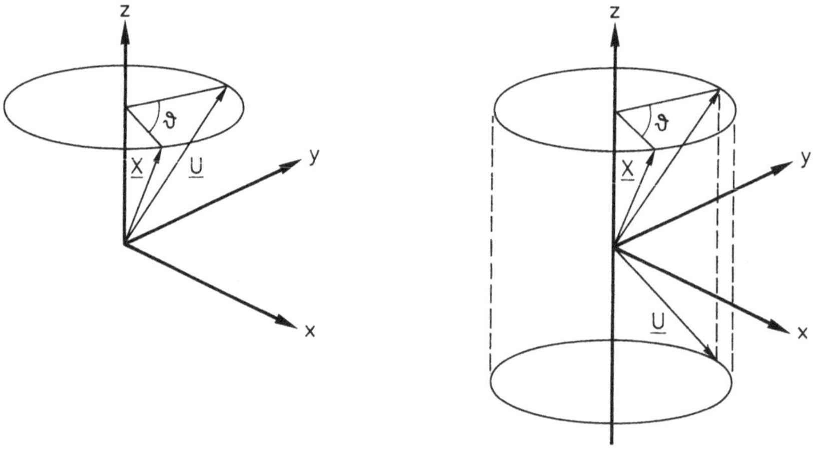

4. Since the mapping by a tensor has to be independent of the coordinate system, this intuitive interpretation implies that an orthogonal tensor must be invariant under a rotation about the eigendirection, i. e. it has the same coordinates in all coordinate systems where the z-axis points in this eigendirection. Hence, such a tensor can be diagonalized only for $\vartheta = 0$ or $\vartheta = \pi$.

Problem 3.15.
Find the eigenvalues of the orthogonal tensor

$$a_{ij} = \begin{pmatrix} \cos\vartheta & -\sin\vartheta & 0 \\ \sin\vartheta & \cos\vartheta & 0 \\ 0 & 0 & \pm 1 \end{pmatrix}.$$

5. We end this section by reviewing the six cases we distinguished earlier, starting with the three cases where the tensor is proper orthogonal:

- $\vartheta = 0$: All eigenvalues are 1. The tensor is the identity tensor, it represents a rotation by 0°, i. e. every vector is mapped onto itself (Case I).
- $0 < \vartheta < \pi$: Two eigenvalues are complex conjugate, one is 1. The tensor represents a rotation by the angle ϑ (Case V).
- $\vartheta = \pi$: Two eigenvalues are –1, one is 1. The tensor represents a rotation by 180° (Case IV).

Next we consider the three cases where the tensor is improper orthogonal:

- $\vartheta = 0$: Two eigenvalues are 1, one is –1. The tensor represents a reflection through the plane orthogonal to the eigendirection of the negative eigenvalue (Case III).
- $0 < \vartheta < \pi$: Two eigenvalues are complex conjugate, one is –1. The tensor represents the same reflection as before, combined with a rotation by the angle ϑ (Case VI).
- $\vartheta = \pi$: All three eigenvalues are –1. The tensor is the negative identity tensor; it is also called the inversion tensor, because it maps any vector to a vector with the same length pointing in the opposite direction (Case II).

Problem 3.16.

A. Find the quantities $\alpha, \beta, \gamma, \delta, \varepsilon, \zeta$ in the matrix

$$\alpha_{ij} = \begin{pmatrix} \alpha & \beta & \delta \\ -\alpha\sqrt{2} & \gamma & \varepsilon \\ \alpha & -\beta & \zeta \end{pmatrix},$$

such that the α_{ij} represent the transformation matrix between two Cartesian coordinate systems (with or without change of orientation).

Hint: Consider first which of the two orthogonality relations (2.6) is more convenient for the computations.

B. How many solutions does the problem have? What additional information do you need, so that the problem has a unique solution? How can the different solutions be interpreted geometrically?

3.13.3 The Orthogonal Tensor as a Function of Rotation Angle and Rotation Axis or Reflection Axis

3.13.3.1 Rotation

1. A proper orthogonal polar tensor (rotation tensor) can be represented by its axis of rotation, which we assume to be given by the axial unit vector $\underline{n}$ and its rotation angle ϑ. To this end, we have to express the rotated vector $\underline{U}$ as a function of $\underline{n}$, ϑ, and the original vector $\underline{X}$. We choose as a basis for this representation:

- the unit vector $\underline{n}$, representing the axis of rotation,
- the projection $\underline{a}$ of the vector $\underline{X}$ onto the plane orthogonal to $\underline{n}$, and
- the vector $\underline{n} \times \underline{X}$.

This basis is orthogonal, but not normalized. To simplify our derivation, we assume that $\underline{X}$ and hence $\underline{U}$ is a polar vector and that the underlying coordinate system is a right-handed system.

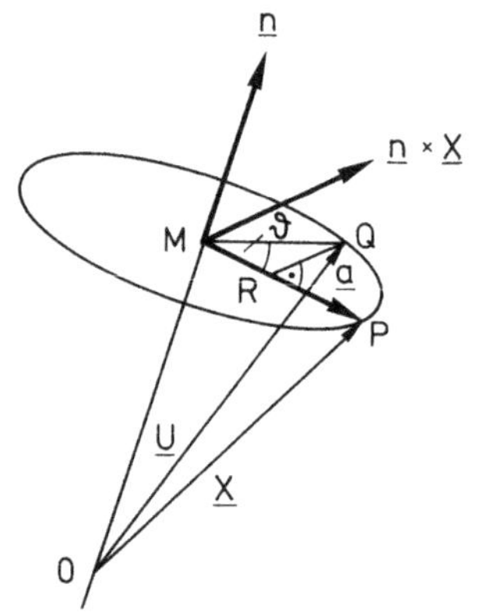

According to the figure, we have

$$\underline{U} = \overrightarrow{OQ} = \overrightarrow{OM} + \overrightarrow{MR} + \overrightarrow{RQ}. \tag{a}$$

We set

$$\overrightarrow{OM} = \alpha\, \underline{n}, \quad \overrightarrow{MR} = \beta\, \underline{a}, \quad \text{and} \quad \overrightarrow{RQ} = \gamma\, \underline{n} \times \underline{X},$$

and now, from the figure, we have to find α, β, γ, and $\underline{a}$ as a function of $\underline{n}$, ϑ, and $\underline{X}$. Since $\underline{n}$ is a unit vector, we have

$$\alpha = OM = OP \cos(\underline{n}, \underline{X}) = X \cos(\underline{n}, \underline{X}) = \underline{n} \cdot \underline{X}.$$

Furthermore,

$$\beta = \frac{MR}{MP} = \frac{MR}{MQ} = \cos\vartheta.$$

For RQ, we have, on the one hand, by taking the length of $\overrightarrow{RQ} = \gamma\, \underline{n} \times \underline{X}$,

$$RQ = \gamma X \sin(\underline{n}, \underline{X}) = \gamma X \frac{MP}{OP} = \gamma\, MP,$$

and on the other hand, from the triangle MRQ,

$$RQ = MQ \sin\vartheta = MP \sin\vartheta.$$

Setting both terms equal gives $\gamma = \sin\vartheta$. Finally,

$$\underline{a} = \overrightarrow{MP} = \overrightarrow{OP} - \overrightarrow{OM} = \underline{X} - \underline{n} \cdot \underline{X}\, \underline{n}.$$

Substituting this into (a) gives

$$\underline{U} = \underline{n} \cdot \underline{X}\, \underline{n} + \cos\vartheta\, (\underline{X} - \underline{n} \cdot \underline{X}\, \underline{n}) + \sin\vartheta\, \underline{n} \times \underline{X}.$$

To obtain this equation in the form $\underline{U} = \underline{\underline{a}} \cdot \underline{X}$, we translate it into coordinate notation:

$$U_i = n_j X_j n_i + \cos\vartheta\, (X_i - n_j X_j n_i) + \sin\vartheta\, \varepsilon_{ijk}\, n_j X_k,$$

from which we get by factoring out X_j

$$U_i = [n_i n_j + (\delta_{ij} - n_i n_j) \cos\vartheta - \varepsilon_{ijk}\, n_k \sin\vartheta]\, X_j,$$

$$a_{ij} = n_i n_j + (\delta_{ij} - n_i n_j) \cos\vartheta - \varepsilon_{ijk}\, n_k \sin\vartheta. \tag{3.77}$$

2. Thus an orthogonal tensor is completely determined by four pieces of information, namely:

- the direction of the rotation axis, represented by the unit vector n_i (two pieces of information, e. g. two coordinates with respect to a Cartesian basis; the third coordinate of n_i is then determined by normalizing the vector to unit length);
- the rotation angle ϑ; and
- the sign of the determinant.

We further see that the coordinates of an orthogonal tensor do not change under a rotation of the coordinate system about its axis of rotation. In all these coordinate systems, the coordinates of n_i and clearly also the scalar ϑ have the same value. Finally, we see that (3.77) is the decomposition of a_{ij} in its symmetric and antimetric parts.

3. We can also write n_i and ϑ as functions of a_{ij}. To this end we contract (3.77), so we have

$$a_{ii} = 1 + 2\cos\vartheta,$$

$$\cos\vartheta = \frac{1}{2}(a_{ii} - 1). \tag{3.78}$$

We further compute $\varepsilon_{ijm}\, a_{ij}$. Here we can replace a_{ij}, according to (2.40), by its antimetric part, so we have

$$\begin{aligned}
\varepsilon_{ijm}\, a_{ij} &= -\,\varepsilon_{ijm}\, \varepsilon_{ijk}\, n_k \sin\vartheta \\
&= -\,(\delta_{jj}\, \delta_{mk} - \delta_{jk}\, \delta_{mj})\, n_k \sin\vartheta \\
&= -\,(3-1)\, \delta_{mk}\, n_k \sin\vartheta, \\
n_m &= -\frac{\varepsilon_{ijm}\, a_{ij}}{2\sin\vartheta}.
\end{aligned} \tag{3.79}$$

Thus we can uniquely assign an angle of rotation between 0° and 180° and an axis of rotation, arbitrarily oriented in space, to any rotation tensor. Rotations by an angle between 0° and –180° correspond to rotations by an angle between 0° and 180° with reversely oriented axis of rotation. Substituting $\vartheta \,||\, -\vartheta$ does not change (3.78) and results with (3.79) in the substitution $n_i \,||\, -n_i$.

3.13.3.2 Reflection

An improper orthogonal tensor, which reflects any vector through the plane orthogonal to $\underline{n}$, can be represented by the unit vector $\underline{n}$. To this end, we have to express the reflected vector $\underline{U}$ as a function of the original vector $\underline{X}$ and the normal vector $\underline{n}$.

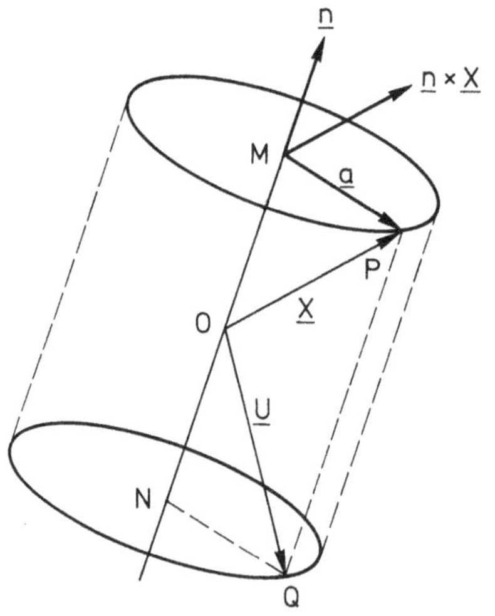

From the figure we have

$$\underline{U} = \overrightarrow{OQ} = \overrightarrow{ON} + \overrightarrow{NQ} = -\overrightarrow{OM} + \overrightarrow{MP}.$$

Following the same considerations as in the last section, we have

$$\overrightarrow{OM} = \underline{n} \cdot \underline{X}\,\underline{n}, \quad \overrightarrow{MP} = \underline{X} - \underline{n} \cdot \underline{X}\,\underline{n},$$

$$\underline{U} = -\underline{n} \cdot \underline{X}\,\underline{n} + \underline{X} - \underline{n} \cdot \underline{X}\,\underline{n} = \underline{X} - 2\,\underline{n} \cdot \underline{X}\,\underline{n}$$

or in coordinate notation

$$U_i = X_i - 2\,n_j\,X_j\,n_i.$$

Factoring out X_j we obtain

$$U_i = (\delta_{ij} - 2\,n_i\,n_j)\,X_j,$$

$$a_{ij} = \delta_{ij} - 2\,n_i\,n_j, \tag{3.80}$$

where a_{ij} is a symmetric tensor which has the eigenvalue −1 with multiplicity 1 and the eigenvalue 1 of multiplicity 2 and $\underline{n}$ is the unit eigenvector corresponding to the eigenvalue −1.

3.13.3.3 Rotary Reflection

A rotary reflection (or improper rotation) is a rotation about the axis $\underline{n}$ by the angle ϑ, combined with a reflection about the plane orthogonal to $\underline{n}$. Then we have

$$\begin{aligned} a_{ij} &= (\delta_{ik} - 2\,n_i\,n_k)[n_k\,n_j + (\delta_{kj} - n_k\,n_j)\cos\vartheta - \varepsilon_{kjm}\,n_m\,\sin\vartheta] \\ &= n_i\,n_j - 2\,n_i\,n_j + (\delta_{ij} - 2\,n_i\,n_j - n_i\,n_j + 2\,n_i\,n_j)\cos\vartheta \\ &\quad - (\varepsilon_{ijm}\,n_m - 2\varepsilon_{kjm}\,n_i\,n_k\,n_m)\sin\vartheta, \end{aligned}$$

$$a_{ij} = -n_i\,n_j + (\delta_{ij} - n_i\,n_j)\cos\vartheta - \varepsilon_{ijm}\,n_m\,\sin\vartheta. \tag{3.81}$$

From (3.81) it follows that $a_{ii} = -1 + 2\cos\vartheta$, $\varepsilon_{ijk}\,a_{ij} = -\varepsilon_{ijk}\,\varepsilon_{ijm}\,n_m\,\sin\vartheta$; hence we obtain the angle and the axis of rotation of a rotary reflection analogously to (3.78) and (3.79) as

$$\cos\vartheta = \frac{1}{2}\,(a_{ii} + 1), \qquad n_m = -\frac{\varepsilon_{ijm}\,a_{ij}}{2\sin\vartheta}. \tag{3.82}$$

Problem 3.17.

One of the solutions to Problem 3.16 is

$$a_{ij} = \begin{pmatrix} \frac{1}{2} & \frac{1}{2}\sqrt{2} & \frac{1}{2} \\ -\frac{1}{2}\sqrt{2} & 0 & \frac{1}{2}\sqrt{2} \\ \frac{1}{2} & -\frac{1}{2}\sqrt{2} & \frac{1}{2} \end{pmatrix}.$$

Find the axis and the angle of rotation and interpret the transformation geometrically.

3.13.4 Rotation and Coordinate Transformation

The equations (2.8)

$$\underline{e}_i = \alpha_{ij}\, \underline{\widetilde{e}}_j, \qquad \underline{\widetilde{e}}_i = \alpha_{ji}\, \underline{e}_j \tag{3.83}$$

represent a mapping between two Cartesian bases, which are rotated to each other.

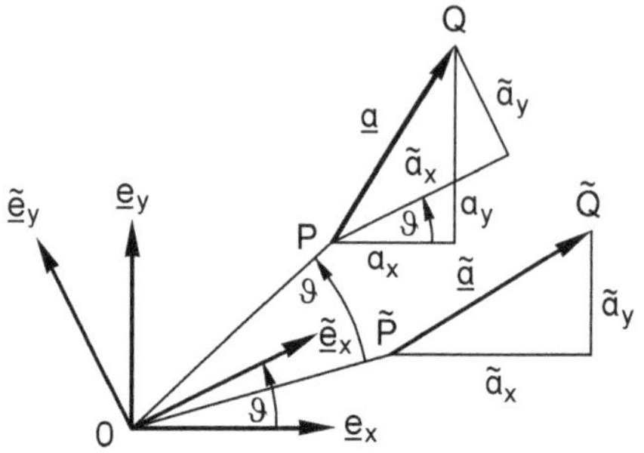

However, the transformation laws (2.17) for tensor coordinates can be interpreted in two ways: either as a transformation law for the coordinates of a tensor between two coordinate systems, which are rotated to each other, or as a transformation law for the coordinates of two tensors in the same coordinate system, which are rotated to each other. For example, in the first interpretation, the equations

$$a_i = \alpha_{ij}\, \widetilde{a}_j, \qquad \widetilde{a}_i = \alpha_{ji}\, a_j \tag{3.84}$$

relate the coordinates of the vector $\underline{a}$, from the figure on the facing page, in the no-tilde coordinate system to the coordinates of this vector in the tilde coordinate system. Then, according to (2.4) $\alpha_{ij} = \overset{j}{\widetilde{e}_i}$ is the transformation matrix whose columns are the basis vectors of the tilde coordinate system, written as a linear combination of the basis vectors of the no-tilde coordinate system. In the second interpretation, the equations (3.84) relate the coordinates of the two vectors $\underline{a}$ and $\underline{\widetilde{a}}$ in the no-tilde coordinate system. Then $\underline{\underline{\alpha}} = \alpha_{ij}\, \underline{e}_i\, \underline{e}_j$ is the rotation tensor, which maps the vector $\underline{\widetilde{a}}$ to the vector $\underline{a}$, i. e. it rotates it through the angle ϑ, which is positive in our case, as can be seen from the figure.

If we interpret $(3.84)_1$ as an equation for the coordinates of the same vector $\underline{a}$ in two different coordinate systems, we cannot translate it into symbolic notation. If we interpret it as an equation for the coordinates of two different vectors, $\underline{a}$ and $\underline{\widetilde{a}}$, in the same coordinate system, then we can translate it as $\underline{a} = \underline{\underline{\alpha}} \cdot \underline{\widetilde{a}}$.

Analogously, the equations

$$x_i = \alpha_{ij}\, \widetilde{x}_j, \qquad \widetilde{x}_i = \alpha_{ji}\, x_j \tag{3.85}$$

either relate the coordinates of the point P, from the figure above, in the no-tilde coordinate system to the coordinates of this point in the tilde coordinate system, or they relate the coordinates of the two points, P and $\widetilde{P}$, in the no-tilde coordinate system. In the first case it is not possible to translate equation $(3.85)_1$ into symbolic notation, in the second case we can translate it into $\underline{x} = \underline{\underline{\alpha}} \cdot \underline{\widetilde{x}}$ with $\underline{\underline{\alpha}}$ as rotation tensor.

3.14 Powers of Tensors. Cayley–Hamilton Theorem

3.14.1 Powers with Integer Exponents

1. Applying the transformation $\underline{U} = \underline{\underline{a}} \cdot \underline{X}$ a second time gives $\underline{V} = \underline{\underline{a}} \cdot \underline{U} = \underline{\underline{a}} \cdot \underline{\underline{a}} \cdot \underline{X}$. This suggests to introduce the square of the tensor $\underline{\underline{a}}$ as a tensor which maps $\underline{X}$ to $\underline{V}$ and to define higher powers of $\underline{\underline{a}}$ analogously. Then we have

$$\begin{aligned}
&\underline{\underline{a}}^2 := \underline{\underline{a}} \cdot \underline{\underline{a}}, &\qquad a_{ij}^{(2)} &:= a_{ik}\, a_{kj}, \\
&\underline{\underline{a}}^3 := \underline{\underline{a}} \cdot \underline{\underline{a}} \cdot \underline{\underline{a}}, &\qquad a_{ij}^{(3)} &:= a_{ik}\, a_{kl}\, a_{lj}, \\
&\text{etc.}, &\qquad &\text{etc.}
\end{aligned} \tag{3.86}$$

Here $a_{ij}^{(2)}$ is the i, j-coordinate of the tensor $\underline{\underline{a}}^2$ and not the square of the i, j-coordinate of the tensor $\underline{\underline{a}}$, i. e. we say “$a$ square $i\,j$”. These powers, where the exponent is a natural number, are defined for arbitrary tensors.

2. For all tensors, except the zero tensor, we define the zeroth power to be the identity tensor; we have

$$\underline{\underline{a}}^0 := \underline{\underline{\delta}}, \qquad a_{ij}^{(0)} := \delta_{ij}. \tag{3.87}$$

3. If the inverse of a tensor exists, i. e. if the original tensor is regular, we define the minus-one power to be the inverse tensor and further powers with negative integer exponents as

$$\begin{aligned}
&\underline{\underline{a}}^{-2} := \underline{\underline{a}}^{-1} \cdot \underline{\underline{a}}^{-1}, &\qquad a_{ij}^{(-2)} &:= a_{ik}^{(-1)}\, a_{kj}^{(-1)}, \\
&\underline{\underline{a}}^{-3} := \underline{\underline{a}}^{-1} \cdot \underline{\underline{a}}^{-1} \cdot \underline{\underline{a}}^{-1}, &\qquad a_{ij}^{(-3)} &:= a_{ik}^{(-1)}\, a_{kl}^{(-1)}\, a_{lj}^{(-1)}, \\
&\text{etc.}, &\qquad &\text{etc.}
\end{aligned} \tag{3.88}$$

4. The n-th power of an eigenvalue of $\underline{\underline{a}}$ is an eigenvalue of $\underline{\underline{a}}^n$, and all powers of $\underline{\underline{a}}$ have the same eigenvectors as $\underline{\underline{a}}$ itself.

We start by proving this theorem for powers with positive integer exponents, using the example

$$b_{ij} = a_{im}\, a_{mn}\, a_{nj}.$$

Let λ be an eigenvalue and let X_j be an eigenvector of a_{ij}, i. e. we have

$$a_{ij}\, X_j = \lambda\, X_i.$$

Then

$$b_{ij}\, X_j = a_{im}\, a_{mn}\, a_{nj}\, X_j = a_{im}\, a_{mn}\, \lambda\, X_n = a_{im}\, \lambda^2\, X_m = \lambda^3\, X_i,$$

i. e. λ^3 is an eigenvalue and X_i is an eigenvector of $b_{ij} = a_{ij}^{(3)}$. The extension of the proof to arbitrary positive integer exponents is clear. We leave the proof for negative integer exponents to the following problem.

Problem 3.18.
Prove that two tensors which are inverses of each other have reciprocal eigenvalues and the same eigenvectors; thus the above theorem is valid for all integer powers of a tensor, as long as negative powers of the tensor exist.

Problem 3.19.
Prove for integer values of n that

$$(\underline{\underline{A}}^n)^{\mathrm{T}} = (\underline{\underline{A}}^{\mathrm{T}})^n. \tag{3.89}$$

Problem 3.20.
The proper orthogonal tensor $\underline{\underline{a}}$ rotates the vector $\underline{X}$ about the axis of rotation $\underline{n}$ through the angle ϑ to the vector $\underline{U}$. Show that the tensor $\underline{\underline{a}}^3$ rotates $\underline{X}$ about the same axis of rotation through the angle $3\,\vartheta$ to a vector $\underline{W}$.

3.14.2 Powers with Real Exponents

1. For positive semidefinite symmetric tensors, we can also define powers with arbitrary real exponents.

A symmetric tensor with eigenvalues λ_k and three orthonormal eigenvectors $\underline{q}_k$ can be represented with (3.70) in the form

$$\underline{\underline{a}} = \lambda_{\underline{k}}\, \underline{q}_k\, \underline{q}_k, \quad a_{ij} = \lambda_{\underline{k}}\, \overset{k}{q}_i\, \overset{k}{q}_j.$$

We now define its n-th power $\underline{\underline{a}}^n$ by

$$\underline{\underline{a}}^n := (\lambda_{\underline{k}})^n\, \underline{q}_k\, \underline{q}_k, \qquad a_{ij}^{(n)} := (\lambda_{\underline{k}})^n\, \overset{k}{q}_i\, \overset{k}{q}_j. \tag{3.90}$$

Clearly, for integer values of n, the power defined in this way is always uniquely determined and identical to the integer power of the tensor defined in Section 3.14.1.

If we restrict ourselves to positive semidefinite tensors, the eigenvalues λ_k are positive or zero. If, in addition, we require that $(\lambda_k)^n$ is also positive or zero, then the n-th power of the tensor is also always uniquely defined by (3.90). (If we allow negative values for $(\lambda_k)^n$, then, for example, $(\lambda_k)^{\frac{1}{2}}$ has two values.)

A tensor defined in this way can be called the n-th power of a_{ij}, if it satisfies the following rules:

$$\underline{\underline{a}}^m \cdot \underline{\underline{a}}^n = \underline{\underline{a}}^{m+n}, \qquad a_{ij}^{(m)}\, a_{jk}^{(n)} = a_{ik}^{(m+n)},$$
$$(\underline{\underline{a}}^m)^n = \underline{\underline{a}}^{mn}, \qquad \left(a_{ij}^{(m)}\right)^n = a_{ij}^{(mn)}. \tag{3.91}$$

Using (3.90) we have

$$a_{ij}^{(m)}\, a_{jk}^{(n)} = (\lambda_{\underline{r}})^m \overset{r}{q}_i \overset{r}{q}_j (\lambda_{\underline{s}})^n \overset{s}{q}_j \overset{s}{q}_k \overset{(3.26)}{=} (\lambda_{\underline{r}})^m (\lambda_{\underline{s}})^n \overset{r}{q}_i \underbrace{\overset{r}{q}_j \overset{s}{q}_j}_{\delta_{rs}} \overset{s}{q}_k$$
$$\overset{(1.18)}{=} (\lambda_{\underline{r}})^m (\lambda_{\underline{r}})^n \overset{r}{q}_i \overset{r}{q}_k = (\lambda_{\underline{r}})^{m+n} \overset{r}{q}_i \overset{r}{q}_k = a_{ik}^{(m+n)},$$

$$\left(a_{ij}^{(m)}\right)^n = \left((\lambda_{\underline{k}})^m \overset{k}{q}_i \overset{k}{q}_j\right)^n \overset{(3.90)}{=} (\lambda_{\underline{k}})^{mn} \overset{k}{q}_i \overset{k}{q}_j = a_{ij}^{(mn)}, \quad \text{Q. E. D.}$$

With the definition (3.90), we assign to any positive semidefinite symmetric tensor $\underline{\underline{a}}$, for any positive real n, exactly one positive semidefinite symmetric tensor $\underline{\underline{a}}^n$, and we call this tensor the n-th power of $\underline{\underline{a}}$, because for all these tensors the rules (3.91) are satisfied.

If we restrict ourselves to positive definite symmetric tensors, our considerations and the formulas (3.90) and (3.91) are also valid for negative real exponents.

2. The definition (3.90) can be generalized to a more general class of tensors, namely the nonnegative tensors. A tensor is called nonnegative, if it is not defective and if its eigenvalues λ_k are positive or zero.

Using (3.54), a nonnegative tensor can be represented in the form

$$\underline{\underline{a}} = \lambda_{\underline{k}}\, \underline{g}_{\underline{k}}\, \underline{g}^k, \qquad a_{ij} = \lambda_{\underline{k}}\, \underset{k}{g}_i\, \overset{k}{g}_j, \tag{3.92}$$

where $\underline{g}^k$ is the basis reciprocal to the eigenvectors $\underline{g}_{\underline{k}}$. The n-th power of $\underline{\underline{a}}$, with $n > 0$, is then defined by

$$\underline{\underline{a}}^n := (\lambda_{\underline{k}})^n\, \underline{g}_{\underline{k}}\, \underline{g}^k, \qquad a_{ij}^{(n)} := (\lambda_{\underline{k}})^n\, \underset{k}{g}_i\, \overset{k}{g}_j; \tag{3.93}$$

if we demand $(\lambda_k)^n$ to be positive or zero, $\underline{\underline{a}}^n$ is defined uniquely by (3.93) and is also nonnegative. It is easy to show that also the powers of a nonnegative tensor as defined by (3.93) satisfy the rules (3.91).

If, in particular, all eigenvalues of a nonnegative tensor are positive, the tensor is called positive. Then the formulas (3.92) and (3.93) also hold for negative exponents.

3. We summarize our definitions of the tensor powers in a table.

Exponent of a tensor power	Admissible tensors
natural number	all tensors
zero	all tensors except the zero tensor
integer number	regular tensors
positive real number	nonnegative tensors
real number	positive tensors

Problem 3.21.

A. Show that the tensor of Problem 3.13 is positive definite.

B. Compute the coordinates of its positive definite square root in the original coordinate system and check your result, i. e. verify that the square of the square root gives the original tensor.

3.14.3 Cayley–Hamilton Theorem

1. Substituting (3.51) into (3.15) gives $A\,\underline{\underline{\delta}} = A'\,\underline{\underline{a}} - A''\,\underline{\underline{a}}^2 + \underline{\underline{a}}^3$, or

$$A\,\underline{\underline{\delta}} - A'\,\underline{\underline{a}} + A''\,\underline{\underline{a}}^2 - \underline{\underline{a}}^3 = \underline{\underline{0}}. \tag{3.94}$$

This equation is called the Cayley–Hamilton theorem. If we replace the powers of the tensor with the powers of the eigenvalues, it has the same form as the characteristic equation (3.46); we also say: a tensor satisfies its characteristic equation. To further generalize (3.94), we scalar multiply by $\underline{\underline{a}}^n$, where n is an integer number, and we obtain

$$A\,\underline{\underline{a}}^n - A'\,\underline{\underline{a}}^{n+1} + A''\,\underline{\underline{a}}^{n+2} - \underline{\underline{a}}^{n+3} = \underline{\underline{0}}. \tag{3.95}$$

This equation is also called the Cayley–Hamilton theorem. Thus, we can write any integer power of a tensor $\underline{\underline{a}}$ as a linear combination of $\underline{\underline{a}}^2$, $\underline{\underline{a}}$, and $\underline{\underline{\delta}}$; and the coefficients in this equation are polynomials of the main invariants of the tensor.

2. For tensors which have at most two distinct eigenvalues λ_1 and λ_2 we can alternatively find all eigenvalues, if we refrain from specifying their multiplicity, from the condition

$$(\lambda_1 - \lambda)(\lambda_2 - \lambda) = 0$$

or

$$\lambda_1 \lambda_2 - (\lambda_1 + \lambda_2)\lambda + \lambda^2 = 0; \tag{3.96}$$

this equation is also called characteristic equation. We multiply this equation by an eigenvector $\underline{X}$,

$$\lambda_1 \lambda_2 \underline{X} - (\lambda_1 + \lambda_2)\lambda \underline{X} + \lambda^2 \underline{X} = \underline{0},$$

and take into account that

$$\underline{\underline{a}} \cdot \underline{X} = \lambda \underline{X}, \quad \underline{\underline{a}}^2 \cdot \underline{X} = \lambda^2 \underline{X},$$

which gives

$$[\lambda_1 \lambda_2 \underline{\underline{\delta}} - (\lambda_1 + \lambda_2)\underline{\underline{a}} + \underline{\underline{a}}^2] \cdot \underline{X} = \underline{0}.$$

If the tensor has three linearly independent eigenvectors, we can substitute for $\underline{X}$ three linearly independent vectors and therefore factor out $\underline{X}$.

Thus, for a nondefective tensor with at most two distinct eigenvalues we found, in addition to (3.94), also the following version of the Cayley–Hamilton theorem:

$$\lambda_1 \lambda_2 \underline{\underline{\delta}} - (\lambda_1 + \lambda_2)\underline{\underline{a}} + \underline{\underline{a}}^2 = \underline{\underline{0}}. \tag{3.97}$$

3.15 Basic Invariants

In addition to the equations (3.47) to (3.49) and (3.53), we can find a third representation for the three main invariants of a tensor. To this end, we take the trace of the Cayley–Hamilton theorem (3.94) and substitute (3.50) for A' and (3.47) for A'', so

$$3A - \frac{1}{2}(\mathrm{tr}^2\,\underline{\underline{a}} - \mathrm{tr}\,\underline{\underline{a}}^2)\,\mathrm{tr}\,\underline{\underline{a}} + \mathrm{tr}\,\underline{\underline{a}}\,\mathrm{tr}\,\underline{\underline{a}}^2 - \mathrm{tr}\,\underline{\underline{a}}^3 = 0,$$

$$A = \frac{1}{6}(\mathrm{tr}^3\,\underline{\underline{a}} - \mathrm{tr}\,\underline{\underline{a}}^2\,\mathrm{tr}\,\underline{\underline{a}} - 2\,\mathrm{tr}\,\underline{\underline{a}}\,\mathrm{tr}\,\underline{\underline{a}}^2 + 2\,\mathrm{tr}\,\underline{\underline{a}}^3).$$

Together with (3.47) and (3.50) we get

$$\begin{aligned} A'' &= \mathrm{tr}\,\underline{\underline{a}}, \\ A' &= \frac{1}{2}(\mathrm{tr}^2\,\underline{\underline{a}} - \mathrm{tr}\,\underline{\underline{a}}^2), \\ A &= \frac{1}{6}(\mathrm{tr}^3\,\underline{\underline{a}} - 3\,\mathrm{tr}\,\underline{\underline{a}}\,\mathrm{tr}\,\underline{\underline{a}}^2 + 2\,\mathrm{tr}\,\underline{\underline{a}}^3). \end{aligned} \tag{3.98}$$

Clearly, $\mathrm{tr}\,\underline{\underline{a}}$, $\mathrm{tr}\,\underline{\underline{a}}^2$, and $\mathrm{tr}\,\underline{\underline{a}}^3$ form, in addition to the main invariants and the eigenvalues, a third set of invariants of a second-order tensor; we call them the basic invariants.

In summary, we have the following expressions.

$$
\begin{aligned}
A'' &= a_{ii} = \lambda_1 + \lambda_2 + \lambda_3 \\
&= \operatorname{tr} \underline{\underline{a}}, \\
A' &= b_{ii} = \lambda_1 \lambda_2 + \lambda_2 \lambda_3 + \lambda_3 \lambda_1 \\
&= \frac{1}{2}(\operatorname{tr}^2 \underline{\underline{a}} - \operatorname{tr} \underline{\underline{a}}^2), \\
A &= \det \underset{\sim}{a} = \lambda_1 \lambda_2 \lambda_3 \\
&= \frac{1}{6}(\operatorname{tr}^3 \underline{\underline{a}} - 3 \operatorname{tr} \underline{\underline{a}} \operatorname{tr} \underline{\underline{a}}^2 + 2 \operatorname{tr} \underline{\underline{a}}^3).
\end{aligned}
\tag{3.99}
$$

If we solve for the basic invariants, we have, after elementary computations, the following formulas.

$$
\begin{aligned}
\operatorname{tr} \underline{\underline{a}} &= \lambda_1 + \lambda_2 + \lambda_3 &&= A'', \\
\operatorname{tr} \underline{\underline{a}}^2 &= \lambda_1^2 + \lambda_2^2 + \lambda_3^2 &&= (A'')^2 - 2A', \\
\operatorname{tr} \underline{\underline{a}}^3 &= \lambda_1^3 + \lambda_2^3 + \lambda_3^3 &&= (A'')^3 - 3A' A'' + 3A.
\end{aligned}
\tag{3.100}
$$

If a tensor has only two distinct eigenvalues, we can, in general, solve the first two equations in (3.99) and the first two equations in (3.100) for the two distinct eigenvalues and then substitute these equations into the third equation of (3.99) or (3.100).

Thus the three main invariants and the three basic invariants are not independent from each other, but only two of them are.

If a tensor has only one eigenvalue (of multiplicity 3), it has only one independent main invariant and only one independent basic invariant.

The three different sets of invariants (eigenvalues, main invariants, and basic invariants) usually do not form a complete set of invariants; a general second-order tensor has seven independent invariants. We will come back to this topic in Chapter 5.

3.16 Polar Decomposition of a Tensor

1. In this section we consider the following theorem. Any regular tensor F_{ij} can be written in the form

$$
F_{ij} = V_{ik} R_{kj} = R_{ik} U_{kj}, \tag{3.101}
$$

where R_{ij} is orthogonal, U_{ij} and V_{ij} are symmetric and positive definite, and R_{ij}, U_{ij}, and V_{ij} are uniquely determined. This decomposition is called polar, because it resembles the decomposition of a complex number into polar coordinates, i. e. into modulus and argument: U_{ij} and V_{ij} are positive definite tensors, similarly as the modulus of a complex number is positive, the determinant of R_{ij} is 1, similarly as the exponential factor of a complex number has modulus 1, and R_{ij} can be interpreted as a rotation or as a rotary reflection, respectively (compare Section 3.13.3).

2. We prove the decomposition theorem (3.101) in six steps.

I. We first show that the scalar product of any tensor with its transposed tensor gives a positive semidefinite symmetric tensor. If the original tensor is regular, then the scalar product is even positive definite and thus with $(3.53)_1$ also regular.

Both are easy to see. We note that, for arbitrary F_{ij}, $F_{ij}F^{\mathrm{T}}_{jk} = F_{ij}F_{kj}$ is symmetric. Furthermore $A := F_{ij}\,F_{kj}\,X_i\,X_k$ is the square of the vector $F_{ij}\,X_i$ and thus it is always greater than or equal to zero, i. e. $F_{ij}\;F_{kj}$ is positive semidefinite.

If, in particular, F_{ij} is regular, it does not have a null direction, i. e. for $X_i \neq 0$ we have $F_{ij}\;X_i \neq 0$, and hence $A > 0$. Then, according to Section 3.12.3, $F_{ij}\;F_{kj}$ is positive definite, and its eigenvalues are all positive.

II. Next we show that any regular tensor F_{ij} can be represented in the form

$$F_{ij} = V_{ik}\;R_{kj}, \tag{3.102}$$

where V_{ij} is symmetric and positive definite and R_{ij} is orthogonal.

This is also easy to see. If F_{ij} is regular, then, according to step I, the product $F_{ik}\;F^{\mathrm{T}}_{kj}$ is symmetric and positive definite, i. e. with Section 3.14.2 there exists a symmetric and positive definite square root

$$V_{ij} := \sqrt{F_{ik}\;F^{\mathrm{T}}_{kj}}, \qquad V^{(2)}_{ij} = V_{ik}\;V_{kj} = F_{ik}\;F^{\mathrm{T}}_{kj}. \tag{3.103}$$

Here $\sqrt{F_{ik}\;F^{\mathrm{T}}_{kj}}$ is the i,j-coordinate of the tensor $(\underline{\underline{F}}\cdot\underline{\underline{F}}^{\mathrm{T}})^{\frac{1}{2}}$ and not the square root of the i,j-coordinate of the tensor $(\underline{\underline{F}}\cdot\underline{\underline{F}}^{\mathrm{T}})$.

Since a positive definite symmetric tensor is regular, $V^{(-1)}_{ij}$ exists and hence the tensor

$$R_{ij} := V^{(-1)}_{ik}\;F_{kj} \tag{3.104}$$

also exists. Multiplying the last equation by V_{mi} gives

$$V_{mi}\;R_{ij} = V_{mi}\;V^{(-1)}_{ik}\;F_{kj} = \delta_{mk}\;F_{kj} = F_{mj},$$

i. e. the tensors defined by (3.103) and (3.104) satisfy equation (3.102).

It remains to show that R_{ij} is orthogonal. For this we exploit that the inverse of a (regular) symmetric tensor is again symmetric. We can show this as follows. Let $\underline{\underline{V}} = \underline{\underline{V}}^{\mathrm{T}}$. Then we have $(\underline{\underline{V}}^{-1})^{\mathrm{T}} \overset{(1.54)}{=} (\underline{\underline{V}}^{\mathrm{T}})^{-1} = \underline{\underline{V}}^{-1}$. Now we compute

$$\begin{aligned}
R_{ik}\;R^{\mathrm{T}}_{kj} &\overset{(3.104)}{=} V^{(-1)}_{im}\;F_{mk}\left(V^{(-1)}_{kn}\;F_{nj}\right)^{\mathrm{T}} \\
&= V^{(-1)}_{im}\;F_{mk}\;F^{\mathrm{T}}_{kn}\left(V^{(-1)}_{nj}\right)^{\mathrm{T}} \\
&\overset{(3.103)}{=} V^{(-1)}_{im}\;V_{mk}\;V_{kn}\;V^{(-1)}_{nj} = \delta_{ik}\;\delta_{kj} = \delta_{ij}.
\end{aligned}$$

So the transpose of the tensor R_{ij} is equal to the inverse of R_{ij}, i. e. R_{ij} is orthogonal.

III. Uniqueness of the decomposition is easy to prove by contradiction.

Let $F_{ij} = v_{ik}\, r_{kj}$ be a second decomposition. Then we have $V_{ik}\, R_{kj} = v_{ik}\, r_{kj}$ and, if we transpose both sides, $R^{\mathrm{T}}_{ik}\, V_{kj} = r^{\mathrm{T}}_{ik}\, v_{kj}$. This gives

$$\begin{aligned} V^{(2)}_{in} &= V_{ik}\, V_{kn} = V_{ik}\, \delta_{km}\, V_{mn} = V_{ik}\, R_{kj}\, R^{\mathrm{T}}_{jm}\, V_{mn} \\ &= v_{ik}\, r_{kj}\, r^{\mathrm{T}}_{jm}\, v_{mn} = v_{ik}\, \delta_{km}\, v_{mn} = v_{ik}\, v_{kn} = v^{(2)}_{in}\,. \end{aligned}$$

Now, since V_{ij} is symmetric and positive definite, $V^{(2)}_{ij}$ is also symmetric and positive definite: $V^{(2)}_{ij} = V_{ik}\, V_{kj} = V_{ik}\, V_{jk}$ is symmetric, and $A := V^{(2)}_{ij}\, X_i\, X_j = V_{ik}\, V_{jk}\, X_i\, X_j$ is the square of the vector $V_{ik}\, X_i$, and since V_{ij} is regular, V_{ij} has no null direction, i. e. the square is always positive.

From the fact that a symmetric and positive definite tensor has only one symmetric and positive definite square root, it follows that $V_{ij} = v_{ij}$ and hence $V_{ik} R_{kj} = V_{ik}\, r_{kj}$. Multiplying by $V^{(-1)}_{mi}$ then gives $R_{mj} = r_{mj}$.

IV. Next we show (also to practice symbolic notation) that any regular tensor $\underline{\underline{F}}$ can also be represented as

$$\underline{\underline{F}} = \underline{\underline{S}} \cdot \underline{\underline{U}}, \tag{3.105}$$

where $\underline{\underline{S}}$ is orthogonal and $\underline{\underline{U}}$ is symmetric and positive definite. To this end we introduce

$$\underline{\underline{U}} := \sqrt{\underline{\underline{F}}^{\mathrm{T}} \cdot \underline{\underline{F}}}, \qquad \underline{\underline{U}}^2 = \underline{\underline{U}} \cdot \underline{\underline{U}} = \underline{\underline{F}}^{\mathrm{T}} \cdot \underline{\underline{F}}, \tag{3.106}$$

and

$$\underline{\underline{S}} := \underline{\underline{F}} \cdot \underline{\underline{U}}^{-1}. \tag{3.107}$$

Analogously to step II, $\underline{\underline{U}}$ is again symmetric and positive definite, and if we scalar multiply (3.107) from the right by $\underline{\underline{U}}$, we get $\underline{\underline{S}} \cdot \underline{\underline{U}} = \underline{\underline{F}}$, i. e. the two tensors $\underline{\underline{U}}$ and $\underline{\underline{S}}$ satisfy equation (3.105). It remains to show that $\underline{\underline{S}}$ is orthogonal. This follows, with $(\underline{\underline{U}}^{-1})^{\mathrm{T}} = \underline{\underline{U}}^{-1}$ and $\underline{\underline{S}}^{\mathrm{T}} \overset{(2.35)}{=} \underline{\underline{U}}^{-1} \cdot \underline{\underline{F}}^{\mathrm{T}}$, from

$$\underline{\underline{S}}^{\mathrm{T}} \cdot \underline{\underline{S}} = \underline{\underline{U}}^{-1} \cdot \underline{\underline{F}}^{\mathrm{T}} \cdot \underline{\underline{F}} \cdot \underline{\underline{U}}^{-1} \overset{(3.106)}{=} \underline{\underline{U}}^{-1} \cdot \underline{\underline{U}} \cdot \underline{\underline{U}} \cdot \underline{\underline{U}}^{-1} = \underline{\underline{\delta}}.$$

V. We show (again by contradiction) that also this second decomposition is unique.

Let $\underline{\underline{F}} = \underline{\underline{s}} \cdot \underline{\underline{u}}$ be a second decomposition. Then we have $\underline{\underline{S}} \cdot \underline{\underline{U}} = \underline{\underline{s}} \cdot \underline{\underline{u}}$ and $\underline{\underline{U}} \cdot \underline{\underline{S}}^{\mathrm{T}} = \underline{\underline{u}} \cdot \underline{\underline{s}}^{\mathrm{T}}$. Furthermore, we have

$$\underline{\underline{U}}^2 = \underline{\underline{U}} \cdot \underline{\underline{\delta}} \cdot \underline{\underline{U}} = \underline{\underline{U}} \cdot \underline{\underline{S}}^{\mathrm{T}} \cdot \underline{\underline{S}} \cdot \underline{\underline{U}} = \underline{\underline{u}} \cdot \underline{\underline{s}}^{\mathrm{T}} \cdot \underline{\underline{s}} \cdot \underline{\underline{u}} = \underline{\underline{u}} \cdot \underline{\underline{\delta}} \cdot \underline{\underline{u}} = \underline{\underline{u}}^2,$$

and since $\underline{\underline{U}}^2$ and $\underline{\underline{u}}^2$ are both symmetric and positive definite, both have only one symmetric and positive definite square root, i. e. we have $\underline{\underline{U}} = \underline{\underline{u}}$ and hence $\underline{\underline{S}} \cdot \underline{\underline{U}} = \underline{\underline{s}} \cdot \underline{\underline{U}}$. Scalar multiplication by $\underline{\underline{U}}^{-1}$ from the right gives $\underline{\underline{S}} = \underline{\underline{s}}$.

VI. It remains to show that $\underline{\underline{R}} = \underline{\underline{S}}$.

From $\underline{\underline{F}} = \underline{\underline{V}} \cdot \underline{\underline{R}}$ it follows that $\underline{\underline{F}} = \underline{\underline{R}} \cdot \underline{\underline{R}}^{\mathrm{T}} \cdot \underline{\underline{V}} \cdot \underline{\underline{R}}$. Since $\underline{\underline{V}}$ is symmetric and positive definite, $\underline{\underline{T}} := \underline{\underline{R}}^{\mathrm{T}} \cdot \underline{\underline{V}} \cdot \underline{\underline{R}}$ is also symmetric and positive definite, i. e. in coordinate notation $T_{ij} := R^{\mathrm{T}}_{im} V_{mn} R_{nj} = R_{mi} R_{nj} V_{mn}$. So we can interpret V_{ij} and T_{ij} with (2.17) as Cartesian coordinates of the same tensor in two different coordinate systems, which are related by the transformation R_{ij}. Because the decomposition $\underline{\underline{F}} = \underline{\underline{S}} \cdot \underline{\underline{U}}$ is unique, it follows that $\underline{\underline{S}} = \underline{\underline{R}}$ and $\underline{\underline{U}} = \underline{\underline{R}}^{\mathrm{T}} \cdot \underline{\underline{V}} \cdot \underline{\underline{R}}$.

3. Scalar multiplication of the last equation by $\underline{\underline{R}}$ from the left and by $\underline{\underline{R}}^{\mathrm{T}}$ from the right allows us to solve for $\underline{\underline{V}}$: $\underline{\underline{R}} \cdot \underline{\underline{U}} \cdot \underline{\underline{R}}^{\mathrm{T}} = \underline{\underline{V}}$. Then we obtain a relation between $\underline{\underline{V}}$ and $\underline{\underline{U}}$ which reads in coordinate notation

$$U_{ij} = R^{\mathrm{T}}_{im} V_{mn} R_{nj}, \quad V_{ij} = R_{im} U_{mn} R^{\mathrm{T}}_{nj}$$

or

$$V_{ij} = R_{im} R_{jn} U_{mn}, \qquad U_{ij} = R_{mi} R_{nj} V_{mn}. \tag{3.108}$$

So we can again interpret V_{ij} and U_{ij} as the Cartesian coordinates of the same tensor in two Cartesian coordinate systems, which are related by the transformation matrix R_{ij}.

Thus both tensors have the same eigenvalues, and the unit eigenvectors $\overset{k}{\underline{q}}_V$ and $\overset{k}{\underline{q}}_U$ corresponding to the eigenvalue λ_k are related by

$$(\overset{k}{q}_V)_i = R_{ij} (\overset{k}{q}_U)_j, \qquad (\overset{k}{q}_U)_i = R_{ji} (\overset{k}{q}_V)_j. \tag{3.109}$$

Since the R_{ij} are the coordinates of the tensor $\underline{\underline{R}}$, we can write equations (3.108) and (3.109) also in symbolic notation. For example, for $(3.108)_1$ and $(3.109)_1$ we have

$$\underline{\underline{V}} = \underline{\underline{R}} \cdot \underline{\underline{U}} \cdot \underline{\underline{R}}^{\mathrm{T}}, \qquad \overset{k}{\underline{q}}_V = \underline{\underline{R}} \cdot \overset{k}{\underline{q}}_U. \tag{3.110}$$

The tensor $\underline{\underline{R}}$ rotates the tensor $\underline{\underline{U}}$ to the tensor $\underline{\underline{V}}$ and it rotates the eigendirections $\overset{k}{\underline{q}}_U$ of $\underline{\underline{U}}$ to the eigendirections $\overset{k}{\underline{q}}_V$ of $\underline{\underline{V}}$.

Problem 3.22.
Find, for the tensor

$$a_{ij} = \begin{pmatrix} 2 & 2 & 1 \\ 1 & -2 & 2 \\ 2 & -1 & -2 \end{pmatrix},$$

the two factors of the polar decomposition $a_{ij} = V_{ik} R_{kj}$ and check your result, i. e. verify that $V_{ik} R_{kj} = a_{ij}$.

4 Tensor Analysis in Curvilinear Coordinates

4.1 Curvilinear Coordinates

4.1.1 Curvilinear Coordinate Systems

1. Until now, we described a point P in space using a Cartesian coordinate system. Such a coordinate system consists of an origin O and an orthonormal basis $\underline{e}_i$ at this origin, and we described the point P by the three coordinates x_i of the position vector $\underline{x} = \overrightarrow{OP} = x_i\,\underline{e}_i$.

Let us consider a set of transformation equations (one-to-one, except at singular points),

$$u^i = u^i(x_j), \qquad x_i = x_i(u^j). \tag{4.1}$$

Then every point in space is uniquely described by the three variables u^i. Thus we can also call the u^i point coordinates, i. e. each set of transformation equations (4.1), together with a Cartesian coordinate system, defines again a coordinate system. All these coordinate systems are called curvilinear coordinate systems.[1]

As an example we consider cylindrical coordinates. They are described by the transformation equations

$$x_1 = u^1 \cos u^2, \quad x_2 = u^1 \sin u^2, \quad x_3 = u^3.$$

We obtain conversely

$$u^1 = \sqrt{x_1^2 + x_2^2}, \quad u^2 = \arctan\frac{x_2}{x_1}, \quad u^3 = x_3$$

or in the usual notation $x_1 = x,\ x_2 = y,\ u^1 = R,\ u^2 = \varphi,\ x_3 = u^3 = z$

$$\begin{aligned} x &= R\cos\varphi, & y &= R\sin\varphi, & z &= z, \\ R &= \sqrt{x^2+y^2}, & \varphi &= \arctan\frac{y}{x}, & z &= z. \end{aligned} \tag{4.2}$$

2. From $\mathrm{d}x_i = (\partial x_i/\partial u^j)\,\mathrm{d}u^j$ it follows that $\partial x_i/\partial x_k = \delta_{ik} = (\partial x_i/\partial u^j)\,(\partial u^j/\partial x_k)$, and analogously from $\mathrm{d}u^i = (\partial u^i/\partial x_j)\mathrm{d}x_j$ it follows that $\partial u^i/\partial u^k = \delta_{ik} = (\partial u^i/\partial x_j)(\partial x_j/\partial u^k)$, i. e. the partial derivatives of the transformation equations satisfy the orthogonality relations

$$\frac{\partial x_i}{\partial u^j}\frac{\partial u^j}{\partial x_k} = \delta_{ik}, \qquad \frac{\partial u^i}{\partial x_j}\frac{\partial x_j}{\partial u^k} = \delta_{ik}. \tag{4.3}$$

1 Since the transformation equations (2.9) between two Cartesian coordinate systems are special cases of (4.1), Cartesian coordinate systems are special cases of curvilinear coordinate systems.

https://doi.org/10.1515/9783110404265-004

3. Computing the determinant of both sides of (4.3) and using the multiplication theorem for determinants (1.13), for the two Jacobians

$$\frac{\partial(x_1, x_2, x_3)}{\partial(u^1, u^2, u^3)} := \begin{vmatrix} \frac{\partial x_1}{\partial u^1} & \frac{\partial x_1}{\partial u^2} & \frac{\partial x_1}{\partial u^3} \\ \frac{\partial x_2}{\partial u^1} & \frac{\partial x_2}{\partial u^2} & \frac{\partial x_2}{\partial u^3} \\ \frac{\partial x_3}{\partial u^1} & \frac{\partial x_3}{\partial u^2} & \frac{\partial x_3}{\partial u^3} \end{vmatrix}$$

and

$$\frac{\partial(u^1, u^2, u^3)}{\partial(x_1, x_2, x_3)} := \begin{vmatrix} \frac{\partial u^1}{\partial x_1} & \frac{\partial u^1}{\partial x_2} & \frac{\partial u^1}{\partial x_3} \\ \frac{\partial u^2}{\partial x_1} & \frac{\partial u^2}{\partial x_2} & \frac{\partial u^2}{\partial x_3} \\ \frac{\partial u^3}{\partial x_1} & \frac{\partial u^3}{\partial x_2} & \frac{\partial u^3}{\partial x_3} \end{vmatrix},$$

we obtain the relation

$$\frac{\partial(x_1, x_2, x_3)}{\partial(u^1, u^2, u^3)} \frac{\partial(u^1, u^2, u^3)}{\partial(x_1, x_2, x_3)} = 1. \tag{4.4}$$

Since we can interchange the rows and columns of a determinant, it does not matter if we take the row or the column index as the numerator index (unlike with the coordinate matrices $\partial a_i/\partial x_j$ of the gradient of a vector, where it does matter).

4. We can show that the coordinate transformation (4.1) is one-to-one at all the points where the Jacobian is neither zero nor infinite. The transformation is called regular at these points and singular at all other points.

In the example of the cylindrical coordinates we have

$$\frac{\partial(x, y, z)}{\partial(R, \varphi, z)} = \begin{vmatrix} \cos\varphi & -R\sin\varphi & 0 \\ \sin\varphi & R\cos\varphi & 0 \\ 0 & 0 & 1 \end{vmatrix} = R,$$

which gives with (4.4)

$$\frac{\partial(R, \varphi, z)}{\partial(x, y, z)} = \frac{1}{R}.$$

Thus cylindrical coordinates are singular at all points where $R = \sqrt{x^2 + y^2} = 0$, i. e. on the z-axis. At these points the transformation (4.2) is indeed not one-to-one, since the value of φ is not unique at these points.

4.1.2 Coordinate Surfaces and Coordinate Curves

The surfaces with u^i = const are called coordinate surfaces. Except at singular points, they cover the space simply, i. e. every point lies on exactly one coordinate surface of each of the three families of coordinate surfaces. Each pair of coordinate surfaces intersects along a coordinate curve, so that, again except for singular points, three coordinate curves intersect at each point.

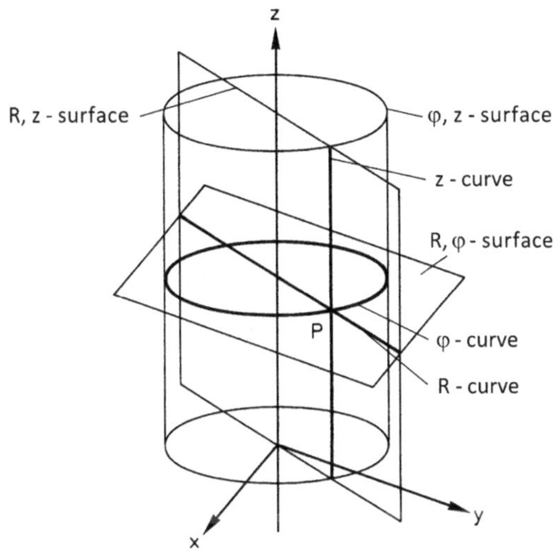

In the example of the cylindrical coordinates, the equations for the coordinate surfaces are $R = \sqrt{x^2 + y^2}$ = const, or simpler, $x^2 + y^2$ = const, $\varphi = \arctan(y/x)$ = const, or simpler, y/x = const, z = const.

The surfaces with $x^2 + y^2$ = const are cylinders which are coaxial with the z-axis; the surfaces y/x = const are planes through the z-axis; and the surfaces z = const are planes perpendicular to the z-axis.

On the coordinate surfaces $R = \sqrt{x^2 + y^2}$ = const, only φ and z vary; on the coordinate surfaces $\varphi = \arctan(y/x)$ = const, only R and z vary; and on the coordinate surfaces z = const, only R and φ vary. On the curve of intersection of the surfaces R = const and φ = const, only z varies; on the curve of intersection of the surfaces φ = const and z = const, only R varies; and on the curve of intersection of the surfaces z = const and R = const, only φ varies.

4.1.3 Holonomic Bases

1. At every point, the differentials $\mathrm{d}\,x_i$ and $\mathrm{d}\,u^i$ of the two coordinate systems are related by

$$\mathrm{d}\,x_1 = \frac{\partial x_1}{\partial u^1}\,\mathrm{d}\,u^1 + \frac{\partial x_1}{\partial u^2}\,\mathrm{d}\,u^2 + \frac{\partial x_1}{\partial u^3}\,\mathrm{d}\,u^3,$$

$$\mathrm{d}\,x_2 = \frac{\partial x_2}{\partial u^1}\,\mathrm{d}\,u^1 + \frac{\partial x_2}{\partial u^2}\,\mathrm{d}\,u^2 + \frac{\partial x_2}{\partial u^3}\,\mathrm{d}\,u^3,$$

$$\mathrm{d}\,x_3 = \frac{\partial x_3}{\partial u^1}\,\mathrm{d}\,u^1 + \frac{\partial x_3}{\partial u^2}\,\mathrm{d}\,u^2 + \frac{\partial x_3}{\partial u^3}\,\mathrm{d}\,u^3.$$

On the curve of intersection of the surfaces $u^2 = \text{const}$ and $u^3 = \text{const}$ we have at any point $\mathrm{d}\,u^2 = \mathrm{d}\,u^3 = 0$. Thus the polar vector with the Cartesian coordinates $(\partial x_1/\partial u^1,\ \partial x_2/\partial u^1,\ \partial x_3/\partial u^1)$ is a (nonnormalized) tangent vector to this coordinate curve. We consider it as the vector $\underline{g}_1$ of a basis. Then the basis defined by

$$\underline{g}_i = \frac{\partial x_j}{\partial u^i}\,\underline{e}_j \tag{4.5}$$

with the Cartesian coordinates

$$\underset{i\,j}{g} = \frac{\partial x_j}{\partial u^i} \tag{4.6}$$

consists of vectors tangent to the coordinate curves of the u^i-coordinates at every point. This basis is called the natural or covariant basis of the u^i-coordinates at the point under consideration.

2. Clearly, the (nonnormalized) polar vector with the Cartesian coordinates $(\partial u^1/\partial x_1,\ \partial u^1/\partial x_2,\ \partial u^1/\partial x_3)$ is perpendicular to the surface $u^1(x_1, x_2, x_3) = \text{const}$ at the point (x_1, x_2, x_3). We consider this vector as the vector $\underline{g}^1$ of another basis. The basis defined by

$$\underline{g}^i = \frac{\partial u^i}{\partial x_j}\,\underline{e}_j \tag{4.7}$$

with the Cartesian coordinates

$$\overset{i}{g}_j = \frac{\partial u^i}{\partial x_j} \tag{4.8}$$

has thus the property that its vectors are perpendicular to the coordinate surfaces of the u^i-coordinates at each point. We call this basis the contravariant basis of the u^i-coordinates at this point.

3. The covariant and the contravariant bases are not the only bases which we can associate to a point in a curvilinear coordinate system. However, computations with these bases are particularly simple, as we can easily convince ourselves, because they are reciprocal bases. Both of these bases are called holonomic bases.

4. If we contract (4.5) with $\partial u^i/\partial x_k$ and (4.7) with $\partial x_k/\partial u^i$ and then solve for the Cartesian bases, we get

$$\frac{\partial u^i}{\partial x_k}\underline{g}_i = \frac{\partial x_j}{\partial u^i}\frac{\partial u^i}{\partial x_k}\underline{e}_j \overset{(4.3)}{=} \underline{e}_k, \qquad \frac{\partial x_k}{\partial u^i}\underline{g}^i = \frac{\partial x_k}{\partial u^i}\frac{\partial u^i}{\partial x_j}\underline{e}_j \overset{(4.3)}{=} \underline{e}_k.$$

Thus we have

$$\underline{e}_i = \frac{\partial u^j}{\partial x_i}\underline{g}_j = \frac{\partial x_i}{\partial u^j}\underline{g}^j. \tag{4.9}$$

5. A system of transformation equations (4.1) associates a curvilinear coordinate system to any given Cartesian coordinate system, i. e. we can associate a triple of curvilinear coordinates and a pair of holonomic bases to each point in space.

6. We have

$$\frac{\partial(x_1, x_2, x_3)}{\partial(u^1, u^2, u^3)} = \det\left(\frac{\partial x_i}{\partial u^j}\right) \overset{(4.6)}{=} \det\left(g_j{}_i\right) \overset{(2.45)}{=} [\underline{g}_1, \underline{g}_2, \underline{g}_3],$$

$$\frac{\partial(u^1, u^2, u^3)}{\partial(x_1, x_2, x_3)} = \det\left(\frac{\partial u^i}{\partial x_j}\right) \overset{(4.8)}{=} \det\left(g^i_j\right) \overset{(2.45)}{=} [\underline{g}^1, \underline{g}^2, \underline{g}^3],$$

$$\frac{\partial(x_1, x_2, x_3)}{\partial(u^1, u^2, u^3)} = [\underline{g}_1, \underline{g}_2, \underline{g}_3], \qquad \frac{\partial(u^1, u^2, u^3)}{\partial(x_1, x_2, x_3)} = [\underline{g}^1, \underline{g}^2, \underline{g}^3]. \tag{4.10}$$

Both triple products have, according to (4.4), the same sign. If the $\underline{e}_i$ form a right-handed system and if the triple products are positive in this right-handed system, the $\underline{g}_i$ and the $\underline{g}^i$ also form right-handed systems. If the triple products are negative in this right-handed system, the $\underline{g}_i$ and the $\underline{g}^i$ form left-handed systems (see Section 2.10.3).

Since we can change the orientation of a basis by interchanging two basis vectors, we can, without significant loss of generality, avoid many sign rules and case distinctions if we restrict ourselves to right-handed systems. So for the rest of Chapter 4 we make the following assumption.

The order of the basis vectors $\underline{e}_i$ of the Cartesian coordinate system x_i, in which the curvilinear coordinate system $u^i(x_j)$ is embedded, and the order of the basis vectors $\underline{g}_i$ and $\underline{g}^i$ of its holonomic bases are such that all three bases form right-handed systems.

Then the Jacobians (4.10) are positive, axial tensors are also independent of the coordinate system, and their coordinates transform in the same way as the coordinates of polar tensors. Thus we do not need to distinguish between polar and axial tensors.

4.1.4 Rectilinear and Cartesian Coordinate Systems

The transformation matrices $\partial x_i/\partial u^j$ and $\partial u^i/\partial x_j$ are, according to (4.6) and (4.8), the Cartesian coordinates of the two reciprocal bases for the u^i-coordinates. These matrices are in general functions of position and thus the two bases are also different at different positions.

The transformation matrices, and hence also the reciprocal bases, are clearly independent of position, if and only if the transformation equations (4.1) have the form

$$u^i = A^{ij}\, x_j + B^i, \qquad x_i = A_{ij}\, u^j + B_i \tag{4.11}$$

and if the transformation coefficients A^{ij}, A_{ij}, B^i, and B_i are independent of position.

Since the reciprocal bases are independent of position, all coordinate curves are straight lines and all coordinate surfaces are planes. Equally named coordinate curves are parallel at all points (i. e. all u^1 lines are parallel, all u^2 lines are parallel, and all u^3 lines are parallel). Equally named coordinate surfaces are, at all points, parallel planes. Such coordinate systems are called rectilinear and thus rectilinear coordinate systems are a special case of curvilinear coordinate systems. (For cylindrical coordinates, all R-curves and all z-curves are lines, but the φ-curves are not. Thus cylindrical coordinates do not form a rectilinear coordinate system.)

The transformation coefficients of a rectilinear coordinate system satisfy the relations

$$\begin{aligned} A^{ij} &= \frac{\partial u^i}{\partial x_j} = \overset{i}{g}_j, & A_{ij} &= \frac{\partial x_i}{\partial u^j} = \underset{j}{g}{}_i, \\ A^{ij} A_{jk} &= \delta_{ik}, & A_{ij} A^{jk} &= \delta_{ik}, \\ B^i &= -A^{ij} B_j, & B_i &= -A_{ij} B^j. \end{aligned} \tag{4.12}$$

The transformation matrices A^{ij} and A_{ij} are inverses of each other and hence regular.

If furthermore the transformation matrices are special orthogonal, the u^i form a Cartesian coordinate system, the two bases $\underline{g}_i$ and $\underline{g}^i$ are orthonormal and coincide, and the equations (4.11) simplify to the transformation equations (2.9).

Problem 4.1.

Consider an oblique, rectilinear coordinate system in the plane, sharing the origin with a Cartesian coordinate system and with axes at angles α and β with respect to the axes of this Cartesian coordinate system.

A. Write out the transformation equations $x_i = x_i(u^j)$ and $u^i = u^i(x_j)$ for the coordinates of a point P.

B. Find the Cartesian coordinates of the holonomic bases. Which basis vectors are unit vectors?

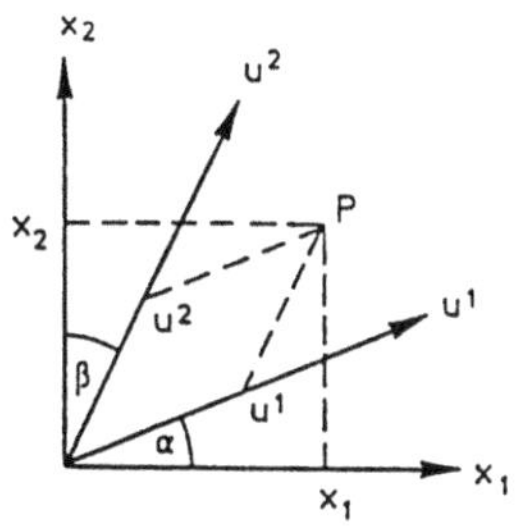

4.1.5 Orthogonal Coordinate Systems

A coordinate system in which the coordinate curves at a point are mutually orthogonal and hence the covariant basis vectors at each point are also mutually orthogonal is called an orthogonal coordinate system. Cylindrical coordinates are an example of an orthogonal coordinate system. According to Section 3.9.3, in an orthogonal coordinate system, the coordinate surfaces at a point and hence the contravariant basis vectors at every point are also orthogonal. The homologous vectors of both bases have the same direction and their lengths are reciprocal. The direction and length of the basis vectors can vary from point to point, as for example in cylindrical coordinates.

Problem 4.2.
Find the Cartesian coordinates of the covariant basis and of the contravariant basis in cylindrical coordinates.

4.2 Holonomic Tensor Coordinates

4.2.1 Introduction

1. Just as we can represent a tensor, according to (2.28), by coordinates with respect to a Cartesian basis, i. e.

$$\underline{a} = a_i\, \underline{e}_i, \quad \underline{\underline{a}} = a_{ij}\, \underline{e}_i\, \underline{e}_j, \quad \text{etc.},$$

we can represent the tensor also with respect to the two holonomic bases of a curvilinear coordinate system. Clearly, already a vector has different coordinates with respect to the covariant basis and to the contravariant basis. The coordinates with respect to the covariant basis are called contravariant coordinates and are denoted by an upper index, the coordinates with respect to the contravariant basis are called covariant coordinates and are denoted by a lower index. We have the representations

$$\underline{a} = a^i\, \underline{g}_i = a_i\, \underline{g}^i.$$

For a second-order tensor, we have, in addition to the covariant basis $\underline{g}_i\,\underline{g}_j$ and the contravariant basis $\underline{g}^i\,\underline{g}^j$, also mixed bases, i. e. we have the four representations

$$\underline{\underline{a}} = a^{ij}\,\underline{g}_i\,\underline{g}_j = a^i{}_j\,\underline{g}_i\,\underline{g}^j = a_i{}^j\,\underline{g}^i\,\underline{g}_j = a_{ij}\,\underline{g}^i\,\underline{g}^j,$$

where we call e. g. the $a^i{}_j$ contravariant–covariant coordinates and the corresponding basis $\underline{g}_i\,\underline{g}^j$ covariant–contravariant basis.

Lower indices of bases and tensor coordinates are called covariant and upper indices are called contravariant.

We obtain the following representations of a tensor with respect to the two holonomic bases of a curvilinear coordinate system.

$$\begin{aligned} \underline{a} &= a^i\,\underline{g}_i = a_i\,\underline{g}^i, \\ \underline{\underline{a}} &= a^{ij}\,\underline{g}_i\,\underline{g}_j = a^i{}_j\,\underline{g}_i\,\underline{g}^j = a_i{}^j\,\underline{g}^i\,\underline{g}_j = a_{ij}\,\underline{g}^i\,\underline{g}^j, \\ &\text{etc.} \end{aligned} \tag{4.13}$$

Clearly, these equations generalize the representation equations (2.28) of a tensor with respect to a Cartesian basis.

All bases of a tensor defined in (4.13) are called holonomic bases, and all corresponding coordinates are called holonomic coordinates.

In the case of nonrectilinear coordinates, the bases change from position to position and a tensor field depends on position both through the coordinates and the bases.

2. In addition to coordinates, we can again introduce components. We write a vector $\underline{a}$ with respect to, for example, the covariant basis $\underline{g}_i$, i. e.

$$\underline{a} = a^1\,\underline{g}_1 + a^2\,\underline{g}_2 + a^3\,\underline{g}_3,$$

and we call $a^1\,\underline{g}_1$ the vector component corresponding to a^1. Again, vector components are vectors, but the vector coordinates are not; they are signed (real) numbers. In general, the holonomic basis vectors of a curvilinear coordinate system are not unit vectors, so a part of the length of a vector component is contained in the coordinate and another part in the basis. Thus, contrary to Cartesian coordinates, the magnitude of a vector coordinate and the magnitude of the corresponding vector component are generally not equal.

3. Using the orthogonality relations (3.24) between reciprocal bases we obtain from (4.13), by an appropriate scalar multiplication, the converse equations

$$
\begin{aligned}
a^i &= \underline{a} \cdot \underline{g}^i, \\
a_i &= \underline{a} \cdot \underline{g}_i, \\
a^{ij} &= \underline{\underline{a}} \cdot\cdot\, \underline{g}^i \underline{g}^j = \underline{g}^i \cdot \underline{\underline{a}} \cdot \underline{g}^j, \\
a^i{}_j &= \underline{\underline{a}} \cdot\cdot\, \underline{g}^i \underline{g}_j = \underline{g}^i \cdot \underline{\underline{a}} \cdot \underline{g}_j, \\
a_i{}^j &= \underline{\underline{a}} \cdot\cdot\, \underline{g}_i \underline{g}^j = \underline{g}_i \cdot \underline{\underline{a}} \cdot \underline{g}^j, \\
a_{ij} &= \underline{\underline{a}} \cdot\cdot\, \underline{g}_i \underline{g}_j = \underline{g}_i \cdot \underline{\underline{a}} \cdot \underline{g}_j, \\
a^{ijk} &= \underline{\underline{\underline{a}}} \cdots \underline{g}^i \underline{g}^j \underline{g}^k = \underline{g}^i \cdot \underline{\underline{\underline{a}}} \cdot\cdot\, \underline{g}^j \underline{g}^k = \underline{g}^i \underline{g}^j \cdot\cdot\, \underline{\underline{\underline{a}}} \cdot \underline{g}^k = \underline{g}^i \underline{g}^j \underline{g}^k \cdots \underline{\underline{\underline{a}}},
\end{aligned}
\tag{4.14}
$$

etc.

4. Clearly we could also write a position vector with respect to the holonomic bases of a curvilinear coordinate system, i. e.

$$\underline{x} = x^i \underline{g}_i = x_i \underline{g}^i .$$

However, in general neither the x^i nor the x_i depend in a simple manner on the curvilinear position coordinates u^i, so they are not used.

5. In equations where both Cartesian and holonomic curvilinear coordinates appear, we find it useful to distinguish the Cartesian coordinates from the covariant coordinates by denoting e. g. the Cartesian coordinates of a vector as $\hat{a}_i$, or written out as (a_x, a_y, a_z).

Problem 4.3.
A tensor $\underline{\underline{T}}$ and a basis $\underline{g}_i$ are given by their Cartesian coordinates, i. e.

$$\hat{T}_{ij} = \begin{pmatrix} 1 & -1 & 2 \\ 2 & 2 & -1 \\ -1 & -2 & 1 \end{pmatrix}, \qquad \begin{aligned} \underline{g}_1 &= (1, 0, 0), \\ \underline{g}_2 &= (1, 1, 0), \\ \underline{g}_3 &= (1, 1, 1). \end{aligned}$$

Find the covariant coordinates T^{ij} of the tensor $\underline{\underline{T}}$ in the basis $\underline{g}_i \underline{g}_j$.

4.2.2 Transformations Between Two Curvilinear Coordinate Systems

1. Let two curvilinear coordinate systems be given by their transformation equations with respect to the same Cartesian coordinate system, i. e.

$$\begin{aligned} u^i &= u^i(x_j), & x_i &= x_i(u^j), \\ \tilde{u}^i &= \tilde{u}^i(x_j), & x_i &= x_i(\tilde{u}^j). \end{aligned} \tag{4.15}$$

By eliminating this Cartesian coordinate system from the above equations, we obtain analogous transformation equations between the two curvilinear Coordinate systems themselves, i. e.

$$u^i = u^i(\tilde{u}^j), \qquad \tilde{u}^i = \tilde{u}^i(u^j). \tag{4.16}$$

2. Similarly as in Section 4.1.1, we obtain analogous orthogonality relations to (4.3) by writing out the differential of the first equation and taking the appropriate derivative. We have

$$\frac{\partial u^i}{\partial \tilde{u}^j}\frac{\partial \tilde{u}^j}{\partial u^k} = \delta_{ik}. \tag{4.17}$$

Because the tilde and the no-tilde coordinate systems are on an equal footing, it is trivial to write down the equation corresponding to the second equation in (4.3).

3. From (4.9) it follows that

$$\underline{e}_i = \frac{\partial u^j}{\partial x_i}\underline{g}_j = \frac{\partial \tilde{u}^j}{\partial x_i}\underline{\tilde{g}}_j = \frac{\partial x_i}{\partial u^j}\underline{g}^j = \frac{\partial x_i}{\partial \tilde{u}^j}\underline{\tilde{g}}^j,$$

$$\frac{\partial u^j}{\partial x_i}\underline{g}_j = \frac{\partial \tilde{u}^j}{\partial x_i}\underline{\tilde{g}}_j, \qquad \frac{\partial x_i}{\partial u^j}\underline{g}^j = \frac{\partial x_i}{\partial \tilde{u}^j}\underline{\tilde{g}}^j.$$

Contraction with $\partial x_i/\partial \tilde{u}^m$ and $\partial \tilde{u}^m/\partial x_i$, respectively, gives

$$\underbrace{\frac{\partial u^j}{\partial x_i}\frac{\partial x_i}{\partial \tilde{u}^m}}_{\frac{\partial u^j}{\partial \tilde{u}^m}}\underline{g}_j = \underbrace{\frac{\partial \tilde{u}^j}{\partial x_i}\frac{\partial x_i}{\partial \tilde{u}^m}}_{\delta_{jm}}\underline{\tilde{g}}_j, \qquad \underbrace{\frac{\partial \tilde{u}^m}{\partial x_i}\frac{\partial x_i}{\partial u^j}}_{\frac{\partial \tilde{u}^m}{\partial u^j}}\underline{g}^j = \underbrace{\frac{\partial \tilde{u}^m}{\partial x_i}\frac{\partial x_i}{\partial \tilde{u}^j}}_{\delta_{mj}}\underline{\tilde{g}}^j,$$

$$\underline{\tilde{g}}_m = \frac{\partial u^j}{\partial \tilde{u}^m}\underline{g}_j, \qquad \underline{\tilde{g}}^m = \frac{\partial \tilde{u}^m}{\partial u^j}\underline{g}^j. \tag{4.18}$$

Since the two coordinate systems are on an equal footing, we can exchange the tilde and the no-tilde quantities in these equations.

The two equations (4.18) are clearly transformation equations between the covariant and the contravariant bases of two curvilinear coordinate systems. They generalize the transformation equations (2.8) between the bases of two Cartesian coordinate systems. The matrices $\partial u^j/\partial\tilde{u}^m$ and $\partial\tilde{u}^m/\partial u^j$ in these equations are, according to (4.17), inverses of each other and therefore regular; they are called the transformation matrices between the two curvilinear coordinate systems. Clearly they generalize the (orthogonal) transformation matrices α_{ij} and $\alpha_{ji} = \alpha_{ij}^{\mathrm{T}}$ between two Cartesian coordinate systems, which are also inverses of each other.

4. From the equations (4.18) we immediately obtain the transformation equations for tensor coordinates of the same type. For example, we have

$$\underline{a} = a^i\,\underline{g}_i = \tilde{a}^i\,\underline{\tilde{g}}_i = \tilde{a}^i\,\frac{\partial u^j}{\partial\tilde{u}^i}\,\underline{g}_j \quad \text{or} \quad a^i = \frac{\partial u^i}{\partial\tilde{u}^k}\,\tilde{a}^k.$$

Hence the transformation equations between the holonomic coordinates of the same type for a tensor in two curvilinear coordinate systems are

$$\begin{aligned}
\tilde{a}^i &= \frac{\partial\tilde{u}^i}{\partial u^m}\,a^m, & \tilde{a}_i &= \frac{\partial u^m}{\partial\tilde{u}^i}\,a_m,\\
\tilde{a}^{ij} &= \frac{\partial\tilde{u}^i}{\partial u^m}\,\frac{\partial\tilde{u}^j}{\partial u^n}\,a^{mn}, & \tilde{a}^i{}_j &= \frac{\partial\tilde{u}^i}{\partial u^m}\,\frac{\partial u^n}{\partial\tilde{u}^j}\,a^m{}_n,\\
\tilde{a}_i{}^j &= \frac{\partial u^m}{\partial\tilde{u}^i}\,\frac{\partial\tilde{u}^j}{\partial u^n}\,a_m{}^n, & \tilde{a}_{ij} &= \frac{\partial u^m}{\partial\tilde{u}^i}\,\frac{\partial u^n}{\partial\tilde{u}^j}\,a_{mn},
\end{aligned} \tag{4.19}$$

etc.

These equations generalize the transformation equations (2.17) for the coordinates of a tensor in two Cartesian coordinate systems (when both are right-handed systems).

5. Thus a covariant index of a tensor coordinate transforms like a covariant index of a basis and "oppositely" to a contravariant index of a tensor coordinate or a basis. This is the case, because contraction of a covariant index with a contravariant index (and vice versa) results in an index-free expression, which is as such independent of a coordinate system, and this does not depend on whether or not these indices are indices of a tensor or of a basis; $a_i\,b^i = c$, $a_i\,\underline{g}^i = \underline{a}$, $\underline{g}_i\,\underline{g}^i = \underline{\underline{\delta}}$. (This property of covariant and contravariant indices explains the names covariant and contravariant: covariant means changing similarly, contravariant means changing oppositely.)

6. The transformation law for tensor coordinates depends only on the number of the upper and lower indices, but not on their order; according to (4.19) we have

$$\tilde{a}^i{}_j = \frac{\partial\tilde{u}^i}{\partial u^m}\,\frac{\partial u^n}{\partial\tilde{u}^j}\,a^m{}_n, \quad \tilde{a}_j{}^i = \frac{\partial\tilde{u}^i}{\partial u^m}\,\frac{\partial u^n}{\partial\tilde{u}^j}\,a_n{}^m.$$

Problem 4.4.
Compute the transformation equations between the Cartesian coordinates and the holonomic cylindrical coordinates of a vector.

Problem 4.5.

A. For the oblique coordinate system from Problem 4.1, compute the transformation equations $A^i = A^i(\widehat{A}_j)$, $\widehat{A}_i = \widehat{A}_i(A^j)$, and $A_i = A_i(\widehat{A}_j)$ between the holonomic and the Cartesian vector coordinates.

B. Add the representations $\underline{A} = A_i \underline{g}^i$ and $\underline{A} = A^i \underline{g}_i$ to the graphs (in each case as addition of two vectors). Show that the contravariant vector coordinates can be constructed by parallel projection and that the covariant vector coordinates can be constructed by perpendicular projection onto the coordinate lines.

Hint: First verify geometrically the computed representations $\widehat{A}_i = \widehat{A}_i(A^j)$ and $A_i = A_i(\widehat{A}_j)$ using auxiliary lines.

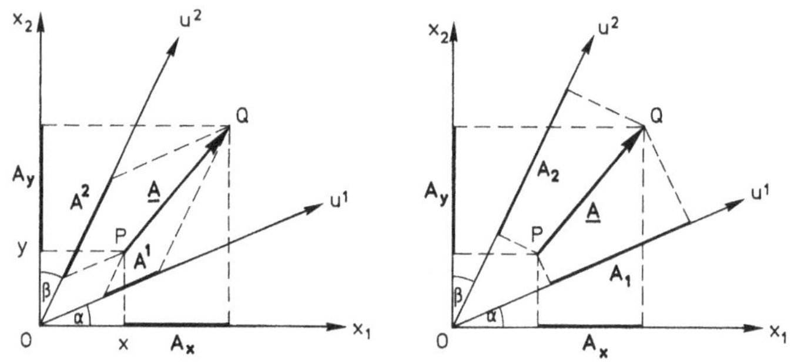

4.2.3 The Summation Convention

The distinction between upper and lower indices in holonomic bases and tensor coordinates requires that we adjust the summation convention and its consequent rules for the use of running indices. This follows from the representation equations (4.13) and (4.14) and from the transformation equations (4.18) and (4.19).

We modify the summation convention for equations in which, besides tensors, only holonomic tensor coordinates, holonomic bases, and transformation matrices appear, i. e. for tensor equations, transformation equations, and representation equations (compare Section 2.7.4) we have the following rule.

If a running index appears twice in a term, once as a lower index and once as an upper index, summation over the values of the index from one to three is implied, without explicitly writing the summation sign. In a transformation matrix, the index in the “numerator” counts as an upper index and the index in the “denominator” counts as a lower index.

This has the following implications for running indices (see Section 1.1, No. 4):

- A running index can appear only once or twice in a term. If it appears twice, one must be in upper position and one in lower position.
- Each running index must take the values one, two, three. (In Chapters 2 to 4 we restrict ourselves to the three-dimensional space of our intuition.)
- All terms of an equation must match in their free indices, also with respect to their positions, but not with respect to their order.

In the rare cases where we do not want to apply the summation convention, we again underline the corresponding running index.

4.2.4 The δ-Tensor

4.2.4.1 Holonomic Coordinates

1. For the holonomic coordinates of the δ-tensor a special notation is used, and we write

$$\underline{\underline{\delta}} = g^{ij}\,\underline{g}_i\,\underline{g}_j = \delta^i_j\,\underline{g}_i\,\underline{g}^j = \delta^j_i\,\underline{g}^i\,\underline{g}_j = g_{ij}\,\underline{g}^i\,\underline{g}^j. \tag{4.20}$$

Conversely, using (4.14), we obtain for the coordinates

$$\begin{aligned}
g^{ij} &= \underline{\underline{\delta}}\cdot\cdot\,\underline{g}^i\,\underline{g}^j = \underline{g}^i\cdot\underline{g}^j,\\
\delta^i_j &= \underline{\underline{\delta}}\cdot\cdot\,\underline{g}^i\,\underline{g}_j = \underline{g}^i\cdot\underline{g}_j \overset{(3.24)}{=} \delta_{ij},\\
\delta^j_i &= \underline{\underline{\delta}}\cdot\cdot\,\underline{g}_i\,\underline{g}^j = \underline{g}_i\cdot\underline{g}^j \overset{(3.24)}{=} \delta_{ij},\\
g_{ij} &= \underline{\underline{\delta}}\cdot\cdot\,\underline{g}_i\,\underline{g}_j = \underline{g}_i\cdot\underline{g}_j,
\end{aligned}$$

or, using the fact that the scalar product of two vectors does not depend on the order of its factors, we write the following.

$$\begin{aligned}
\delta^j_i &= \delta^i_j = \underline{g}^i\cdot\underline{g}_j = \underline{g}_i\cdot\underline{g}^j = \delta_{ij},\\
g^{ij} &= g^{ji} = \underline{g}^i\cdot\underline{g}^j,\\
g_{ij} &= g_{ji} = \underline{g}_i\cdot\underline{g}_j.
\end{aligned} \tag{4.21}$$

Here (in accordance with the definition (1.19) of the generalized Kronecker symbols), δ^i_j and δ^j_i have the same meaning as δ_{ij}; so we will set one index of the Kronecker symbol as upper index, if this is required to satisfy the rules for computations with running indices for curvilinear coordinates. Because the order of the indices does not matter in the Kronecker symbol, often both indices are set on top of each other (unlike for mixed tensor coordinates, where the order does matter).

2. The g^{ij} and g_{ij} are also called metric coefficients ("measure coefficients"). This makes sense, because we need them to compute the length of a vector (to "measure" its length) from the three holonomic coordinates; if, for example, the covariant coordinates a_i of a vector $\underline{a}$ are given, then we compute its length as

$$a = \sqrt{\underline{a} \cdot \underline{a}} = \sqrt{a_i \, \underline{g}^i \cdot a_j \, \underline{g}^j} = \sqrt{a_i \, a_j \, \underline{g}^i \cdot \underline{g}^j} = \sqrt{a_i \, a_j \, g^{ij}},$$

i. e. we need both the covariant coordinates a_i and the contravariant metric coefficients. This explains why the δ-tensor is called metric tensor (a name we had already mentioned earlier).

3. Using (4.21), we can also interpret the coordinates of the δ-tensor geometrically; they are the scalar products of two basis vectors. We compute for the mixed coordinates in every curvilinear coordinate system, according to the definition (3.21) of reciprocal bases or the orthogonality relations (3.24), the same values, namely 1 if both indices are equal and 0 if they are not equal. The covariant coordinates and the contravariant coordinates of the δ-tensor, however, depend on the coordinate system; coordinates with the same indices are the square of the length of the corresponding basis vector, e. g. we have

$$g_{11} = |\underline{g}_1|^2.$$

Coordinates with different indices are the product of the lengths of the corresponding two basis vectors and the cosine of the enclosed angle, e. g.

$$g^{13} = |\underline{g}^1||\underline{g}^3| \cos(\underline{g}^1, \underline{g}^3).$$

Problem 4.6.
Compute the covariant and contravariant cylindrical coordinates of the δ-tensor.

4.2.4.2 Properties of the Metric Coefficients

1. The metric coefficients of a curvilinear coordinate system at a point (or shorter, the metric of a curvilinear coordinate system) are defined by the lengths of the three basis vectors and the angles enclosed by them, i. e. by six parameters. All coordinate systems which have the same values for these six parameters (at every point) have the same metric coefficients. Geometrically, these are coordinate systems whose bases can be obtained from each other by a rotation or by a rotary reflection.

2. Now we investigate some properties of the metric coefficients. What conditions must a square matrix of order 3 satisfy, so that it can be interpreted as a matrix of metric coefficients?

The most obvious property or condition is that the metric coefficients are symmetric, i. e.

$$g_{ij} = g_{ji}, \qquad g^{ij} = g^{ji}. \tag{4.22}$$

Next, from the condition

$$a^i a^j g_{ij} = a_i a_j g^{ij} = \underline{a} \cdot \underline{a} > 0,$$

which is true for any vector $\underline{a}$ (except the zero vector), it follows that the matrices of the metric coefficients g_{ij} and g^{ij} are positive definite. From this, according to Theorem 5 and Theorem 4 from Section 3.12.4, it follows that the determinant and all principal subdeterminants (e. g. for a matrix of order 3 the principal minors and the elements on the main diagonal) are positive.

First, the diagonal elements must be positive, i. e.

$$g_{\underline{ii}} > 0, \qquad g^{\underline{ii}} > 0. \tag{4.23}$$

This we see also from their geometric interpretation as the square of the length of the basis vectors. Secondly, the principal minors must be positive, i. e. the subdeterminants of order 2, which are obtained by dropping a row and its associated column, so we must have

$$\begin{vmatrix} g_{\underline{ii}} & g_{ij} \\ g_{ji} & g_{\underline{jj}} \end{vmatrix} > 0, \qquad \begin{vmatrix} g^{\underline{ii}} & g^{ij} \\ g^{ji} & g^{\underline{jj}} \end{vmatrix} > 0, \quad i \neq j. \tag{4.24}$$

Lastly, the determinant of the metric coefficients itself must also be positive, and this must be true independently of the orientation of the basis vectors. We have e. g.

$$\det g_{ij} = \begin{vmatrix} g_{11} & g_{12} & g_{13} \\ g_{21} & g_{22} & g_{23} \\ g_{31} & g_{32} & g_{33} \end{vmatrix} = \begin{vmatrix} \underline{g}_1 \cdot \underline{g}_1 & \underline{g}_1 \cdot \underline{g}_2 & \underline{g}_1 \cdot \underline{g}_3 \\ \underline{g}_2 \cdot \underline{g}_1 & \underline{g}_2 \cdot \underline{g}_2 & \underline{g}_2 \cdot \underline{g}_3 \\ \underline{g}_3 \cdot \underline{g}_1 & \underline{g}_3 \cdot \underline{g}_2 & \underline{g}_3 \cdot \underline{g}_3 \end{vmatrix}$$

$$\stackrel{(2.46)}{=} [\underline{g}_1, \underline{g}_2, \underline{g}_3][\underline{g}_1, \underline{g}_2, \underline{g}_3],$$

which is clearly always positive. Writing it as a square, we have

$$\det g_{ij} = [\underline{g}_1, \underline{g}_2, \underline{g}_3]^2 > 0, \quad \det g^{ij} = [\underline{g}^1, \underline{g}^2, \underline{g}^3]^2 > 0. \tag{4.25}$$

3. Because the two holonomic bases are reciprocal, one is uniquely determined by the other, and thus the contravariant metric coefficients are uniquely determined by the covariant metric coeficients and vice versa. Using this, we obtain relations between the two types of metric coefficients. We have

$$g_{ij}\, g^{jk} = \underline{g}_i \cdot \underline{g}_j\, \underline{g}^j \cdot \underline{g}^k \overset{(3.25)}{=} \underline{g}_i \cdot \underline{\underline{\delta}} \cdot \underline{g}^k = \underline{g}_i \cdot \underline{g}^k = \delta_i^k,$$

$$g_{ij}\, g^{jk} = \delta_i^k, \tag{4.26}$$

i. e. the matrices of the two types of metric coefficients are inverses of each other, and one can be computed from the other using the Gauss–Jordan algorithm.

4. Clearly we have

$$\det g_{ij} \det g^{ij} \overset{(1.13)}{=} \det (g_{ij}\, g^{jk}) \overset{(4.26)}{=} 1.$$

If we introduce the symbol g for the determinant of g_{ij}, we have

$$g := \det g_{ij}, \quad \det g^{ij} = \frac{1}{g}, \quad \det g_{ij} \det g^{ij} = 1. \tag{4.27}$$

5. Using (2.46) and (4.25), we further have

$$[\underline{g}_1, \underline{g}_2, \underline{g}_3][\underline{g}_1, \underline{g}_2, \underline{g}_3] = \begin{vmatrix} \underline{g}_1 \cdot \underline{g}_1 & \underline{g}_1 \cdot \underline{g}_2 & \underline{g}_1 \cdot \underline{g}_3 \\ \underline{g}_2 \cdot \underline{g}_1 & \underline{g}_2 \cdot \underline{g}_2 & \underline{g}_2 \cdot \underline{g}_3 \\ \underline{g}_3 \cdot \underline{g}_1 & \underline{g}_3 \cdot \underline{g}_2 & \underline{g}_3 \cdot \underline{g}_3 \end{vmatrix} = g$$

and accordingly

$$[\underline{g}^1, \underline{g}^2, \underline{g}^3]^2 = \frac{1}{g}.$$

So we obtain, together with (4.10),

$$\begin{aligned} [\underline{g}_1, \underline{g}_2, \underline{g}_3] &= \frac{\partial(x_1, x_2, x_3)}{\partial(u^1, u^2, u^3)} = \sqrt{\det g_{ij}} = \sqrt{g}, \\ [\underline{g}^1, \underline{g}^2, \underline{g}^3] &= \frac{\partial(u^1, u^2, u^3)}{\partial(x_1, x_2, x_3)} = \sqrt{\det g^{ij}} = \sqrt{\frac{1}{g}}. \end{aligned} \tag{4.28}$$

6. From the geometric interpretation of the metric coefficients as scalar products of basis vectors, it follows immediately that for orthogonal coordinate systems all metric coefficients with different indices are zero, and that homologous covariant and contravariant metric coefficients are reciprocal, i. e.

$$g_{ij} = g_{\underline{ii}}\, \delta_{ij}, \qquad g^{ij} = g^{\underline{ii}}\, \delta_{ij}, \qquad g^{\underline{ii}}\, g_{\underline{ii}} = 1. \tag{4.29}$$

For Cartesian coordinates we have

$$g_{ij} = g^{ij} = \delta_{ij}. \tag{4.30}$$

Problem 4.7.
Find the transformation equation for $g := \det g_{ij}$, when changing from one curvilinear coordinate system to another.

Hint: Start with the transformation equations (4.19) for the covariant tensor coordinates g_{ij}.

4.2.5 Raising and Lowering Indices

Clearly we have

$$g^{ij}\,\underline{g}_j = \underline{g}^i \cdot \underline{g}^j\,\underline{g}_j = \underline{g}^i \cdot \underline{\underline{\delta}} = \underline{g}^i,$$
$$g_{ij}\,\underline{g}^j = \underline{g}_i \cdot \underline{g}_j\,\underline{g}^j = \underline{g}_i \cdot \underline{\underline{\delta}} = \underline{g}_i.$$

It follows further from

$$\underline{a} = a^i\,\underline{g}_i = a_i\,\underline{g}^i,$$

after scalar multiplication by $\underline{g}^j$ and $\underline{g}_j$, respectively, that

$$a^i\,\underline{g}_i \cdot \underline{g}^j = a_i\,\underline{g}^i \cdot \underline{g}^j, \quad a^i\,\delta_i^j = a_i\,g^{ij}, \quad a^j = g^{ji}\,a_i,$$
$$a^i\,\underline{g}_i \cdot \underline{g}_j = a_i\,\underline{g}^i \cdot \underline{g}_j, \quad a^i\,g_{ij} = a_i\,\delta_j^i, \quad a_j = g_{ji}\,a^i.$$

The same applies to higher-order tensors; for example, from

$$\underline{\underline{a}} = a^{ij}\,\underline{g}_i\,\underline{g}_j = a^i{}_j\,\underline{g}_i\,\underline{g}^j = a_i{}^j\,\underline{g}^i\,\underline{g}_j = a_{ij}\,\underline{g}^i\,\underline{g}^j$$

it follows after scalar multiplication by $\underline{g}^k$ from the right that

$$a^{ij}\,\underline{g}_i\,\underline{g}_j \cdots \underline{g}^k = a^i{}_j\,\underline{g}_i\,\underline{g}^j \cdot \underline{g}^k \quad \text{and} \quad a_i{}^j\,\underline{g}^i\,\underline{g}_j \cdot \underline{g}^k = a_{ij}\,\underline{g}^i\,\underline{g}^j \cdot \underline{g}^k,$$
$$a^{ij}\,\underline{g}_i\,\delta_j^k = a^i{}_j\,\underline{g}_i\,g^{jk} \quad \text{and} \quad a_i{}^j\,\underline{g}^i\,\delta_j^k = a_{ij}\,\underline{g}^i\,g^{jk},$$
$$a^{ik} = g^{kj}\,a^i{}_j \quad \text{and} \quad a_i{}^k = g^{kj}\,a_{ij}.$$

Thus we can raise or lower an index in a basis or in tensor coordinates by contracting with the corresponding metric coefficient, i. e.

$$
\begin{aligned}
\underline{g}^i &= g^{im}\,\underline{g}_m,\\
\underline{g}_i &= g_{im}\,\underline{g}^m,\\
a^i &= g^{im}\,a_m,\\
a_i &= g_{im}\,a^m,\\
a^{ij} &= g^{im}\,a_m{}^j = g^{jn}\,a^i{}_n = g^{im}\,g^{jn}\,a_{mn},\\
a^i{}_j &= g^{im}\,a_{mj} = g_{jn}\,a^{in} = g^{im}\,g_{jn}\,a_m{}^n,\\
a_i^{\ j} &= g_{im}\,a^{mj} = g^{jn}\,a_{in} = g_{im}\,g^{jn}\,a^m{}_n,\\
a_{ij} &= g_{im}\,a^m{}_j = g_{jn}\,a_i{}^n = g_{im}\,g_{jn}\,a^{mn},\\
&\text{etc.}
\end{aligned}
\tag{4.31}
$$

4.2.6 The ε-Tensor

4.2.6.1 Holonomic Coordinates

1. We denote the holonomic coordinates of the ε-tensor by e. There are eight different versions of e:

$$
\underline{\underline{\underline{\varepsilon}}} = e^{ijk}\,\underline{g}_i\,\underline{g}_j\,\underline{g}_k = e^{ij}{}_k\,\underline{g}_i\,\underline{g}_j\,\underline{g}^k = \cdots = e_{ijk}\,\underline{g}^i\,\underline{g}^j\,\underline{g}^k. \tag{4.32}
$$

From (4.14) and (2.45) we obtain the converse equations

$$
\begin{aligned}
e^{ijk} &= \underline{\underline{\underline{\varepsilon}}} \cdots \underline{g}^i\,\underline{g}^j\,\underline{g}^k = [\underline{g}^i, \underline{g}^j, \underline{g}^k],\\
e^{ij}{}_k &= \underline{\underline{\underline{\varepsilon}}} \cdots \underline{g}^i\,\underline{g}^j\,\underline{g}_k = [\underline{g}^i, \underline{g}^j, \underline{g}_k],\\
&\vdots\\
e_{ijk} &= \underline{\underline{\underline{\varepsilon}}} \cdots \underline{g}_i\,\underline{g}_j\,\underline{g}_k = [\underline{g}_i, \underline{g}_j, \underline{g}_k].
\end{aligned}
$$

According to (2.46), we can write the square of these triple products as determinant as follows:

$$
[\underline{g}^i, \underline{g}^j, \underline{g}^k][\underline{g}^{\underline{i}}, \underline{g}^{\underline{j}}, \underline{g}^{\underline{k}}] = \begin{vmatrix} g^{i\underline{i}} & g^{i\underline{j}} & g^{i\underline{k}} \\ g^{j\underline{i}} & g^{j\underline{j}} & g^{j\underline{k}} \\ g^{k\underline{i}} & g^{k\underline{j}} & g^{k\underline{k}} \end{vmatrix},
$$

$$
[\underline{g}^i, \underline{g}^j, \underline{g}_k][\underline{g}^{\underline{i}}, \underline{g}^{\underline{j}}, \underline{g}_{\underline{k}}] = \begin{vmatrix} g^{i\underline{i}} & g^{i\underline{j}} & \delta^i_{\underline{k}} \\ g^{j\underline{i}} & g^{j\underline{j}} & \delta^j_{\underline{k}} \\ \delta^{\underline{i}}_k & \delta^{\underline{j}}_k & g_{k\underline{k}} \end{vmatrix},
$$

etc.

Thus we obtain the converse equations to (4.32) as follows.

$$
\begin{aligned}
e^{ijk} &= [\underline{g}^i, \underline{g}^j, \underline{g}^k] = \pm\sqrt{\begin{vmatrix} g^{i\underline{i}} & g^{i\underline{j}} & g^{i\underline{k}} \\ g^{j\underline{i}} & g^{j\underline{j}} & g^{j\underline{k}} \\ g^{k\underline{i}} & g^{k\underline{j}} & g^{k\underline{k}} \end{vmatrix}}, \\
e^{ij}{}_k &= [\underline{g}^i, \underline{g}^j, \underline{g}_k] = \pm\sqrt{\begin{vmatrix} g^{i\underline{i}} & g^{i\underline{j}} & \delta^i_{\underline{k}} \\ g^{j\underline{i}} & g^{j\underline{j}} & \delta^j_{\underline{k}} \\ \delta^{\underline{i}}_k & \delta^{\underline{j}}_k & g_{k\underline{k}} \end{vmatrix}}, \\
&\vdots \\
e_{ijk} &= [\underline{g}_i, \underline{g}_j, \underline{g}_k] = \pm\sqrt{\begin{vmatrix} g_{i\underline{i}} & g_{i\underline{j}} & g_{i\underline{k}} \\ g_{j\underline{i}} & g_{j\underline{j}} & g_{j\underline{k}} \\ g_{k\underline{i}} & g_{k\underline{j}} & g_{k\underline{k}} \end{vmatrix}}.
\end{aligned}
\tag{4.33}
$$

All eight versions can be obtained from each other by raising or lowering single indices and by replacing g with δ as necessary.

2. The sign of a holonomic coordinate of the ε-tensor changes, if we change the order of two indices, e. g. i and j. The sign does not change, if we raise or lower an index. For example,

$$
\begin{aligned}
[\underline{g}_1, \underline{g}_2, \underline{g}^3] &\overset{(2.45)}{=} \varepsilon_{ijk} \underset{1}{g}_i \underset{2}{g}_j \overset{3}{g}_k \overset{(3.22)}{=} \varepsilon_{ijk} \underset{1}{g}_i \underset{2}{g}_j \frac{\varepsilon_{pqk} \underset{1}{g}_p \underset{2}{g}_q}{\varepsilon_{lmn} \underset{1}{g}_l \underset{2}{g}_m \underset{3}{g}_n} \\
&= \frac{(\delta_{ip}\,\delta_{jq} - \delta_{iq}\,\delta_{jp}) \underset{1}{g}_i \underset{2}{g}_j \underset{1}{g}_p \underset{2}{g}_q}{\varepsilon_{lmn} \underset{1}{g}_l \underset{2}{g}_m \underset{3}{g}_n} = \frac{\underset{1}{g}_i \underset{1}{g}_i \underset{2}{g}_j \underset{2}{g}_j - \underset{1}{g}_i \underset{2}{g}_i \underset{1}{g}_j \underset{2}{g}_j}{\varepsilon_{lmn} \underset{1}{g}_l \underset{2}{g}_m \underset{3}{g}_n} = \frac{\begin{vmatrix} g_{11} & g_{12} \\ g_{21} & g_{22} \end{vmatrix}}{[\underline{g}_1, \underline{g}_2, \underline{g}_3]}.
\end{aligned}
$$

Since the numerator is, according to (4.24), always positive, $[\underline{g}_1, \underline{g}_2, \underline{g}^3]$ has the same sign as $[\underline{g}_1, \underline{g}_2, \underline{g}_3]$.

4.2.6.2 Properties of the Holonomic Coordinates

1. Using (4.33) and (4.28), we have

$$
e_{123} = [\underline{g}_1, \underline{g}_2, \underline{g}_3] = \sqrt{g}, \quad e^{123} = [\underline{g}^1, \underline{g}^2, \underline{g}^3] = \frac{1}{\sqrt{g}}, \quad e^{123}\, e_{123} = 1.
$$

Since a triple product is zero if two of its vectors are equal and since it changes the sign when we change the order of two of its vectors, we have for the entirely covariant coordinates and the entirely contravariant coordinates of the ε-tensor[2]

$$e_{ijk} = \sqrt{g}\,\varepsilon_{ijk}, \qquad e^{ijk} = \frac{1}{\sqrt{g}}\,\varepsilon_{ijk}, \qquad e^{ijk} = \frac{1}{g}\,e_{ijk}. \tag{4.34}$$

2. The holonomic coordinates of the ε-tensor only change their sign if we exchange two indices, while keeping their positions (upper or lower). The sign does not change when the indices are changed cyclically, while keeping their positions, for example, we have

$$e^{ij}{}_k = e^{j}{}_k{}^{i} = e_k{}^{ij} = -e_k{}^{ji} = -e^{ji}{}_k = -e^{i}{}_k{}^{j}.$$

3. Holonomic coordinates of the ε-tensor are zero:
- if at least two indices in the same positions (both either lower or upper) are equal (because then two vectors in the corresponding triple product are equal),
- for orthogonal coordinate systems, also, if at least two indices in different positions are equal (because then two homologous vectors of the corresponding reciprocal bases are collinear).

4. We can interpret (1.26) as an equation between the Cartesian coordinates of the ε-tensor and the Cartesian coordinates of the δ-tensor, i. e. as a tensor equation in Cartesian coordinates. Its translation[3] e. g. into contravariant curvilinear coordinates gives

$$e^{ijk}\,e^{pqr} = \begin{vmatrix} g^{ip} & g^{iq} & g^{ir} \\ g^{jp} & g^{jq} & g^{jr} \\ g^{kp} & g^{kq} & g^{kr} \end{vmatrix}, \tag{4.35}$$

where indices can be lowered or raised and indices in mixed positions can be set both in upper or lower position. Thus we obtain the translated Grassman identity (1.35).

$$e^{ijk}\,e^{pq}{}_k = \left(g^{ip}\,g^{jq} - g^{iq}\,g^{jp}\right). \tag{4.36}$$

Here, again we can lower free indices.

5. We have

$$\underline{g}^i \overset{(3.23)}{=} \frac{1}{2}\,\varepsilon_{ijk}\,\frac{\underline{g}_j \times \underline{g}_k}{[\underline{g}_1, \underline{g}_2, \underline{g}_3]} \overset{(4.28)}{=} \frac{1}{2}\,\frac{\varepsilon_{ijk}}{\sqrt{g}}\,\underline{g}_j \times \underline{g}_k \overset{(4.34)}{=} \frac{1}{2}\,e^{ijk}\,\underline{g}_j \times \underline{g}_k,$$

2 The rule from Section 4.2.3, according to which free indices must be in the same position in all terms of an equation, applies only to equations which contain besides tensors only holonomic tensor coordinates, holonomic bases, and transformation coefficients. This is not the case for an equation with an ε_{ijk}.

3 See Section 4.2.8.5.

which we can solve for $\underline{g}_i \times \underline{g}_j$ by contraction with e_{ipq} as follows:

$$\underline{g}^i = \frac{1}{2}\, e^{ijk}\, \underline{g}_j \times \underline{g}_k, \qquad \underline{g}_i \times \underline{g}_j = e_{ijk}\, \underline{g}^k. \tag{4.37}$$

Problem 4.8.
How many holonomic coordinates does the ε-tensor have? How many of them are not zero in cylindrical coordinates? Find the nonzero holonomic cylindrical coordinates of the ε-tensor.

Problem 4.9.
Complete the expression $e^{ijk}\, e^{m}{}_{nk}$ to obtain the Grassmann identity.

4.2.7 Isotropic Tensors

The δ-tensor, the ε-tensor, and combinations of both are, according to Section 2.8.3, characterized by the property that their Cartesian coordinates are equal in all Cartesian coordinate systems. We called them isotropic tensors.

Clearly, holonomic curvilinear coordinates of isotropic tensors do not have this property: the metric coefficients can have different values in different curvilinear coordinate systems. However, since the metric coefficients depend only on the length and the relative position of the basis vectors, which do not change under a rotation of the basis (at the point under consideration), and since the holonomic curvilinear coordinates of isotropic tensors depend only on the metric coefficients, the following property holds. The holonomic curvilinear coordinates of isotropic tensors are invariant under a rotation of the basis. (Note that the bases of all Cartesian coordinate systems at a point can be obtained from each other by a rotation or a rotary reflection.)

4.2.8 Tensor Algebra in Holonomic Coordinates

We can derive rules for computations with holonomic coordinates from the rules for computations in symbolic notation or (e. g. when operations are only defined in coordinate notation) from the rules for computations in Cartesian coordinates. Thus we find as a third notation, in addition to the symbolic notation and the notation in Cartesian coordinates, the notation of an equation in holonomic curvilinear coordinates, with analogous translation rules.

4.2.8.1 Equality, Addition, and Subtraction

For example, from $\underline{a} = \underline{b}$ it follows that $a^i\,\underline{g}_i = b^i\,\underline{g}_i$ and also that $a_i\,\underline{g}^i = b_i\,\underline{g}^i$. Since the decomposition of two vectors $\underline{a}$ and $\underline{b}$ with respect to a basis $\underline{g}_i$ or $\underline{g}^i$ is unique, we can cancel out the basis and obtain the equations for coordinates

$$a^i = b^i \quad \text{and} \quad a_i = b_i.$$

Analogously, from $\underline{\underline{a}} = \underline{\underline{b}}$ it follows that

$$a^{ij} = b^{ij}, \quad a^i{}_j = b^i{}_j, \quad a_i^{\ j} = b_i^{\ j}, \quad a_{ij} = b_{ij}.$$

Thus two tensors are equal if we have in symbolic notation, in coordinate notation in Cartesian coordinates, and in coordinate notation in holonomic curvilinear coordinates

$$\begin{array}{lll}
a = A, & a = A, & a = A, \\
\underline{a} = \underline{B}, & \hat{a}_i = \hat{B}_i, & a^i = B^i, \\
 & & a_i = B_i, \\
\underline{\underline{a}} = \underline{\underline{C}}, & \hat{a}_{ij} = \hat{C}_{ij}, & a^{ij} = C^{ij}, \\
 & & a^i{}_j = C^i{}_j, \\
 & & a_i^{\ j} = C_i^{\ j}, \\
 & & a_{ij} = C_{ij}, \\
\text{etc.} & &
\end{array} \tag{4.38}$$

Analogously, from $\underline{a} \pm \underline{b} = \underline{c}$ it follows that

$$a^i \pm b^i = c^i \quad \text{and} \quad a_i \pm b_i = c_i,$$

and, in general, for addition and subtraction of tensors in the three notations

$$\begin{array}{lll}
a \pm b = A, & a \pm b = A, & a \pm b = A, \\
\underline{a} \pm \underline{b} = \underline{B}, & \hat{a}_i \pm \hat{b}_i = \hat{B}_i, & a^i \pm b^i = B^i, \\
 & & a_i \pm b_i = B_i, \\
\underline{\underline{a}} \pm \underline{\underline{b}} = \underline{\underline{C}}, & \hat{a}_{ij} \pm \hat{b}_{ij} = \hat{C}_{ij}, & a^{ij} \pm b^{ij} = C^{ij}, \\
 & & a^i{}_j \pm b^i{}_j = C^i{}_j, \\
 & & a_i^{\ j} \pm b_i^{\ j} = C_i^{\ j}, \\
 & & a_{ij} \pm b_{ij} = C_{ij}, \\
\text{etc.} & &
\end{array} \tag{4.39}$$

Thus, we can write an equation, whose terms are vectors in holonomic curvilinear coordinates, in two equivalent versions: either in contravariant coordinates or in covariant coordinates; an equation whose terms are second-order tensors can be written in four equivalent versions. In general, we can write an equation whose terms are tensors of order n in 2^n equivalent versions (corresponding to the number of holonomic curvilinear coordinates).

4.2.8.2 Transposed, Symmetric, and Antimetric Tensors

1. If we assume that one of the two coordinate systems in the transformation equations (4.19) is a Cartesian coordinate system, we have

$$a^{ij} = \frac{\partial u^i}{\partial x_m}\frac{\partial u^j}{\partial x_n}\hat{a}_{mn} = \frac{\partial u^j}{\partial x_n}\frac{\partial u^i}{\partial x_m}\hat{a}^{\mathrm{T}}_{nm} = (a^{\mathrm{T}})^{ji},$$

$$a^i{}_j = \frac{\partial u^i}{\partial x_m}\frac{\partial x_n}{\partial u^j}\hat{a}_{mn} = \frac{\partial x_n}{\partial u^j}\frac{\partial u^i}{\partial x_m}\hat{a}^{\mathrm{T}}_{nm} = (a^{\mathrm{T}})_j{}^i,$$

etc.

The transpose of a tensor in Cartesian coordinates is given by (2.23), and with the equations above we have in holonomic curvilinear coordinates

$$\hat{a}^{\mathrm{T}}_{ij} = \hat{a}_{ji}, \qquad \begin{aligned} (a^{\mathrm{T}})^{ij} &= a^{ji}, \\ (a^{\mathrm{T}})^i{}_j &= a_j{}^i, \\ (a^{\mathrm{T}})_i{}^j &= a^j{}_i, \\ (a^{\mathrm{T}})_{ij} &= a_{ji}, \end{aligned} \tag{4.40}$$

i. e. we transpose a tensor by exchanging the two free indices of the holonomic curvilinear tensor coordinates while keeping their positions.

2. According to (2.26) and with (4.39), we have for the symmetric part of a second-order tensor

$$\hat{a}_{(ij)} = \frac{1}{2}(\hat{a}_{ij} + \hat{a}_{ji}), \qquad \begin{aligned} a^{(ij)} &= \frac{1}{2}(a^{ij} + a^{ji}), \\ a^{(i}{}_{j)} &= \frac{1}{2}(a^i{}_j + a_j{}^i), \\ a_{(i}{}^{j)} &= \frac{1}{2}(a_i{}^j + a^j{}_i), \\ a_{(ij)} &= \frac{1}{2}(a_{ij} + a_{ji}) \end{aligned} \tag{4.41}$$

and for the antimetric part of a second-order tensor

$$\hat{a}_{[ij]} = \frac{1}{2}(\hat{a}_{ij} - \hat{a}_{ji}), \qquad \begin{aligned} a^{[ij]} &= \frac{1}{2}(a^{ij} - a^{ji}), \\ a^{[i}{}_{j]} &= \frac{1}{2}(a^{i}{}_{j} - a_{j}{}^{i}), \\ a_{[i}{}^{j]} &= \frac{1}{2}(a_{i}{}^{j} - a^{j}{}_{i}), \\ a_{[ij]} &= \frac{1}{2}(a_{ij} - a_{ji}). \end{aligned} \tag{4.42}$$

4.2.8.3 Tensor Multiplication

For example, from $\underline{\underline{a}}\,\underline{b} = \underline{\underline{\underline{c}}}$ it follows that

$$a^{ij}\,\underline{g}_i\,\underline{g}_j\,b^k\,\underline{g}_k = a^{ij}\,b^k\,\underline{g}_i\,\underline{g}_j\,\underline{g}_k = c^{ijk}\,\underline{g}_i\,\underline{g}_j\,\underline{g}_k \quad \text{or} \quad a^{ij}\,b^k = c^{ijk}$$

and

$$a^{i}{}_{j}\,\underline{g}_i\,\underline{g}^j\,b_k\,\underline{g}^k = a^{i}{}_{j}\,b_k\,\underline{g}_i\,\underline{g}^j\,\underline{g}^k = c^{i}{}_{jk}\,\underline{g}_i\,\underline{g}^j\,\underline{g}^k \quad \text{or} \quad a^{i}{}_{j}\,b_k = c^{i}{}_{jk}.$$

Six more equivalent versions are possible, because three indices i, j, and k can be arranged into $2^3 = 8$ different combinations of upper and lower indices. We note that in computations with these tensor products we can pull the coordinates before the bases (because they are numbers), but we cannot change the order of the bases, because tensor products of vectors are not commutative, i. e. $\underline{g}_i\,\underline{g}_j \neq \underline{g}_j\,\underline{g}_i$. Furthermore, to be able to cancel out bases, they must match not only in their order, but also in the positions of their indices, i. e. $\underline{g}_i\,\underline{g}^j \neq \underline{g}_i\,\underline{g}_j$. As a result of these two conditions, after canceling out the bases, the free indices in a tensor equation match in both their order and their positions, i. e.

$$\begin{array}{lll}
a\,b = A, & a\,b = A, & a\,b = A, \\
a\,\underline{b} = \underline{B}, & a\,\hat{b}_i = \hat{B}_i, & a\,b^i = B^i, \\
 & & a\,b_i = B_i, \\
\underline{a}\,b = \underline{C}, & \hat{a}_i\,b = \hat{C}_i, & a^i\,b = C^i, \\
 & & a_i\,b = C_i, \\
\underline{a}\,\underline{b} = \underline{\underline{D}}, & \hat{a}_i\,\hat{b}_j = \hat{D}_{ij}, & a^i\,b^j = D^{ij}, \\
 & & a^i\,b_j = D^{i}{}_{j}, \\
 & & a_i\,b^j = D_{i}{}^{j}, \\
 & & a_i\,b_j = D_{ij},
\end{array} \tag{4.43}$$

etc.

4.2.8.4 Contraction and Its Special Cases

For example, from $\underline{\underline{a}} \cdot \underline{b} = \underline{c}$ it follows analogously that

$$a^{ij}\,\underline{g}_i\,\underline{g}_j \cdot b_k\,\underline{g}^k = a^{ij}\,b_k\,\underline{g}_i\,\underline{g}_j \cdot \underline{g}^k = a^{ij}\,b_k\,\underline{g}_i\,\delta_j^k = a^{ij}\,b_j\,\underline{g}_i = c^i\,\underline{g}_i$$

or

$$a^{ij}\,b_j = c^i.$$

The other three possible positions of the indices j and k are

$$a^i{}_j\,b^j = c^i, \quad a^{ij}\,\underbrace{b^k\,g_{jk}}_{b_j} = c^i, \quad a^i{}_j\,\underbrace{b_k\,g^{jk}}_{b^j} = c^i.$$

Since, according to (4.31), we have $b^k\,g_{jk} = b_j$ and $b_k\,g^{jk} = b^j$, we essentially have for a contravariant i the two notations

$$a^{ij}\,b_j = c^i \quad \text{and} \quad a^i{}_j\,b^j = c^i.$$

Since the right sides and thus also the left sides are equal (unlike in the equivalent formulations for indices in different positions), we interpret both equations as different notations of the same version, i. e. of the contravariant version of the vector equation $\underline{\underline{a}} \cdot \underline{b} = \underline{c}$. In addition we have the covariant version $a_i{}^j\,b_j = a_{ij}\,b^j = c_i$. We note that summation indices must be in different positions, but it does not matter which of the two indices is in upper position and which is in lower position;

$$\begin{aligned}
&\underline{a} \cdot \underline{b} = A, && \hat{a}_l\,\hat{b}_l = A, && a_i\,b^i = a^i\,b_i = A,\\
&\underline{a} \cdot \underline{\underline{b}} = \underline{B}, && \hat{a}_i\,\hat{b}_{ij} = \hat{B}_j, && a_i\,b^{ij} = a^i\,b_i{}^j = B^j,\\
& && && a_i\,b^i{}_j = a^i\,b_{ij} = B_j,
\end{aligned} \tag{4.44}$$

etc.

If a contraction cannot be written as scalar product, we have analogously

$$\begin{aligned}
&\hat{a}_{ijk}\,\hat{b}_j = \hat{A}_{ik}, \qquad && a^{ijk}\,b_j = a^i{}_j{}^k\,b^j = A^{ik},\\
& && a^{ij}{}_k\,b_j = a^i{}_{jk}\,b^j = A^i{}_k,\\
& && a_i{}^{jk}\,b_j = a_{ij}{}^k\,b^j = A_i{}^k,\\
& && a_i{}^j{}_k\,b_j = a_{ijk}\,b^j = A_{ik}.
\end{aligned} \tag{4.45}$$

This can be proven using the transformation law (4.19).

We abstain from writing out the analogous equations for multiple scalar products, vector products, and triple products.

4.2.8.5 Summary

1. Tensor equations in holonomic curvilinear coordinates must satisfy the three rules from Section 4.2.3.

2. To translate a tensor-algebraic equation from Cartesian coordinates into holonomic curvilinear coordinates, every free index must be replaced by either a covariant index or a contravariant index (i. e. the same index in each term must be replaced by the same type of holonomic index) and we have to replace each pair of summation indices with two holonomic indices in different positions. This means that we have to replace δ_{ij} with either g^{ij}, δ^i_j, δ^j_i, or g_{ij} and $\varepsilon_{...}$ with $e^{...}_{...}$.

To translate from curvilinear coordinates into Cartesian coordinates, we have to write all upper indices as lower indices, and we have to replace g^{ij} and g_{ij} with δ_{ij} and $e^{...}_{...}$ with $\varepsilon_{...}$.

3. Furthermore, when translating from symbolic notation to holonomic curvilinear coordinates and vice versa, the order of the free indices must match in every term.

4. Thus the equations (4.31) on raising and lowering of indices are simply translations of identities such as $\hat{a}_i = \delta_{ij}\,\hat{a}_j$ into curvilinear coordinates, where the index of the vector coordinate is set in different positions on the left and on the right side.

Problem 4.10.
Translate (in one possible way) into holonomic curvilinear coordinates

A. the equations from Problem 2.5;
B. $\underline{a} \times (\underline{b} \times \underline{c}) = \underline{a} \cdot \underline{c}\,\underline{b} - \underline{a} \cdot \underline{b}\,\underline{c}$.

4.3 Physical Bases and Tensor Coordinates

1. Consider a vector $\underline{a}$ and the two bases of a curvilinear coordinate system. Then we can always decompose $\underline{a}$ uniquely into its components with respect to the two bases $\underline{g}_i$ and $\underline{g}^i$, i. e.

$$\underline{a} = a^1\,\underline{g}_1 + a^2\,\underline{g}_2 + a^3\,\underline{g}_3 = a_1\,\underline{g}^1 + a_2\,\underline{g}^2 + a_3\,\underline{g}^3.$$

Since in general the basis vectors are not unit vectors, one part of the length of the components $a^i\,\underline{g}_i$ and $a_i\,\underline{g}^i$ is contained in the coordinates and the other part in the basis vectors. If the vector, and thus its components, represent a physical quantity, such as force, then one part of the physical dimension is in the coordinates and the other part is in the basis vectors. This is often inconvenient, so we define for each holonomic basis another basis, whose basis vectors have the same directions, but are unit vectors. We call this basis the corresponding physical basis and denote this basis and its corresponding coordinates by a star. Thus we have

$$\underline{g}^*{}_i = \frac{\underline{g}_i}{\sqrt{g_{\underline{ii}}}}, \qquad \underline{g}^{*i} = \frac{\underline{g}^i}{\sqrt{g^{\underline{ii}}}}, \tag{4.46}$$

and for a vector $\underline{a}$

$$\underline{a} = a^i \underline{g}_i = a^{*i} \underline{g}^*{}_i = a_i \underline{g}^i = a^*{}_i \underline{g}^{*i}. \tag{4.47}$$

The so-defined physical coordinates a^{*i} and $a^*{}_i$ clearly represent the length of the corresponding components, and they have the same physical dimension as the vector itself. We can generalize (4.47) analogously to higher-order tensors, by substituting (4.46) for the relation between the physical coordinates and the holonomic coordinates, and we obtain

$$\begin{aligned} a^{*i} &= a^i \sqrt{g_{\underline{ii}}}, & a^*{}_i &= a_i \sqrt{g^{\underline{ii}}}, \\ a^{*ij} &= a^{ij} \sqrt{g_{\underline{ii}}} \sqrt{g_{\underline{jj}}}, & a^{*i}{}_j &= a^i{}_j \sqrt{g_{\underline{ii}}} \sqrt{g^{\underline{jj}}}, \\ a^*{}_i{}^j &= a_i{}^j \sqrt{g^{\underline{ii}}} \sqrt{g_{\underline{jj}}}, & a^*{}_{ij} &= a_{ij} \sqrt{g^{\underline{ii}}} \sqrt{g^{\underline{jj}}}, \end{aligned} \tag{4.48}$$

etc.

2. That some of the indices in these equations are underlined, i. e. that the summation convention cannot be applied, already indicates that we pay a price for the dimensional equality of the physical coordinates. The powerful elegance of tensor calculus, which is expressed in the summation convention, exists only for holonomic coordinates. In practice, we therefore perform computations as long as possible in holonomic coordinates, and convert only the final result into physical coordinates, if necessary.

3. For example, in general the two physical bases $\underline{g}^*{}_i$ and $\underline{g}^{*i}$, corresponding to the reciprocal bases $\underline{g}_i$ and $\underline{g}^i$, are not reciprocal, i. e. $\underline{g}^*{}_i$ and $\underline{g}^{*i}$ do not satisfy the orthogonality relations (3.24) and (3.25).

However, for an important group of curvilinear coordinates, namely the orthogonal coordinates, this is still the case. Here, the homologous vectors of the two holonomic bases have the same direction, i. e. the corresponding physical bases coincide. We then write

$$\underline{g}^*{}_i = \underline{g}^{*i} =: \underline{g}_{<i>} \tag{4.49}$$

and, for example, for a vector

$$\underline{a} = a_{<i>} \underline{g}_{<i>}. \tag{4.50}$$

Thus the physical bases corresponding to an orthogonal coordinate system are orthonormal at every point, but in general their direction changes from point to point. This explains why orthogonal physical coordinates are also called locally Cartesian coordinates.

This change of direction differentiates orthonormal bases from Cartesian bases. For tensor *algebra* it does not matter if tensors or bases depend on the position. Consequently, all tensor-*algebraic* equations have the same form in orthogonal physical coordinates and in Cartesian coordinates. To translate into orthogonal physical coordinates, we only have to replace the lower indices of the Cartesian coordinates by indices in acute brackets.

Problem 4.11.

A. Compute the Cartesian coordinates of the physical basis in cylindrical coordinates.

B. Compute the transformation equations between the Cartesian coordinates and the physical cylindrical coordinates of a vector.

   Hint: Using the results of Problems 4.2 and 4.4, the desired equations can be found immediately or with only few intermediate steps.

C. What is the value of the physical cylindrical coordinates of the δ-tensor and of the ε-tensor? (No computations necessary.)

D. Translate into orthogonal physical coordinates:
   $(\underline{\underline{a}} \cdot \underline{\underline{b}})^{\mathrm{T}} = \underline{\underline{c}}, \quad \alpha_{ik}\, b^i{}_j = c_{jk}, \quad \underline{a} \times (\underline{b} \times \underline{c}) = \underline{a} \cdot \underline{c}\, \underline{b} - \underline{a} \cdot \underline{b}\, \underline{c}.$

4.4 Tensor-Analytical Operations

Let us again consider tensor fields, i. e. we assume that bases and tensors and hence also the tensor coordinates are functions of the three curvilinear coordinates u^i of a point in space.

We introduce an abbreviated notation for the partial derivative of a quantity G with respect to the curvilinear coordinates u^i.

$$G_{,i} := \frac{\partial G}{\partial u^i}. \tag{4.51}$$

Here G can be the position vector, a basis, a tensor, or a tensor coordinate. For the total differential $\mathrm{d}\,G$ of a quantity G, we have the following expression.

$$\mathrm{d}\,G = G_{,i}\,\mathrm{d}\,u^i. \tag{4.52}$$

The partial derivative of a quantity is, like the quantity itself, also a function of position. The total differential of a quantity depends on the coordinates u^i and also on their differentials $\mathrm{d}\,u^i$. For small $\mathrm{d}\,u^i$, i. e. for nearby points with the position vectors

$\underline{x} + \mathrm{d}\,\underline{x}$ and $\underline{x}$, the differential of a tensor or of a tensor coordinate is, up to second-order terms in the $\mathrm{d}\,u^i$, equal to the difference of the tensor (or the tensor coordinate) between the two points, i. e.

$$\mathrm{d}\,G = G(\underline{x} + \mathrm{d}\,\underline{x}) - G(\underline{x}). \tag{4.53}$$

4.4.1 Partial Derivative and Differential of the Position Vector

Differentiating $\underline{x} = x_j\,\underline{e}_j$ with respect to u^i gives, since the Cartesian basis is independent of position, $\underline{x}_{,i} = (\partial x_j/\partial u^i)\,\underline{e}_j$, which, according to (4.5), is $\underline{g}_i$;

$$\underline{x}_{,i} = \underline{g}_i, \tag{4.54}$$

i. e. the partial derivative of the position vector with respect to the curvilinear coordinates gives the covariant basis. From (4.52) it further follows that $\mathrm{d}\,\underline{x} = \underline{g}_i\,\mathrm{d}\,u^i$, i. e. the differentials of the curvilinear coordinates u^i are the contravariant coordinates of the vector $\mathrm{d}\,\underline{x}$. (The position vector $\underline{x}$, as we know, is not a vector; the differential $\mathrm{d}\,\underline{x}$ of the position vector, however, is a vector.) This property of the coordinate differentials explains why we write the index as an upper index. In Cartesian and in curvilinear coordinates, we have

$$\mathrm{d}\,\underline{x} = \mathrm{d}\,x_i\,\underline{e}_i = \mathrm{d}\,u^i\,\underline{g}_i. \tag{4.55}$$

4.4.2 Partial Derivative and Total Differential of the Holonomic Bases, Christoffel Symbols

1. Let us compute the partial derivatives of the two bases with respect to the curvilinear coordinates u^i and write the result with respect to the original basis.

For the covariant basis we get

$$\underline{g}_{i,j} \overset{(4.5)}{=} \frac{\partial^2 x_k}{\partial u^i\,\partial u^j}\,\underline{e}_k \overset{(4.9)}{=} \frac{\partial^2 x_k}{\partial u^i\,\partial u^j}\,\frac{\partial u^m}{\partial x_k}\,\underline{g}_m.$$

Here, the expressions

$$\Gamma^m_{ij} := \frac{\partial^2 x_k}{\partial u^i\,\partial u^j}\,\frac{\partial u^m}{\partial x_k} \tag{4.56}$$

are called Christoffel symbols (of the second kind). Note that they are symmetric in the lower indices.

$$\Gamma_{ij}^{m} = \Gamma_{ji}^{m}. \tag{4.57}$$

However, they are not tensor coordinates, since they do not satisfy the transformation law for single contravariant and twice covariant tensor coordinates, as we will see in Problem 4.13. If the u^i are rectilinear, for example, Cartesian coordinates, the Christoffel symbols are zero.

To derive the partial derivative of the contravariant basis, we differentiate $\underline{g}^i \cdot \underline{g}_j = \delta_j^i$ with respect to u^k. We have

$$\underline{g}^i{}_{,k} \cdot \underline{g}_j + \underline{g}^i \cdot \underline{g}_{j,k} = 0,$$

$$\underline{g}^i{}_{,k} \cdot \underline{g}_j = -\Gamma_{jk}^{m}\, \underline{g}_m \cdot \underline{g}^i = -\Gamma_{jk}^{m}\, \delta_m^i = -\Gamma_{jk}^{i}.$$

Tensor multiplication with $\underline{g}^j$ gives

$$\underline{g}^i{}_{,k} \cdot \underline{g}_j\, \underline{g}^j = \underline{g}^i{}_{,k} \cdot \underline{\underline{\delta}} = \underline{g}^i{}_{,k} = -\Gamma_{jk}^{i}\, \underline{g}^j.$$

Thus, the partial derivatives of the bases with respect to the curvilinear coordinates are

$$\underline{g}_{i,j} = \Gamma_{ij}^{m}\, \underline{g}_m, \qquad \underline{g}^i{}_{,j} = -\Gamma_{jm}^{i}\, \underline{g}^m. \tag{4.58}$$

2. The total differential of a holonomic basis is given by (4.52) as

$$\begin{aligned} \mathrm{d}\underline{g}_i &= \underline{g}_{i,j}\, \mathrm{d}u^j = \Gamma_{ij}^{m}\, \mathrm{d}u^j\, \underline{g}_m, \\ \mathrm{d}\underline{g}^i &= \underline{g}^i{}_{,j}\, \mathrm{d}u^j = -\Gamma_{jm}^{i}\, \mathrm{d}u^j\, \underline{g}^m. \end{aligned} \tag{4.59}$$

For small $\mathrm{d}u^i$, they are the difference between the corresponding basis vectors of two neighboring points.

4.4.3 Christoffel Symbols and Metric Coefficients

An alternative to (4.56) is to express the Christoffel symbols as a function of the metric coefficients.

To do this we differentiate $\underline{g}_i = g_{ij}\, \underline{g}^j$ with respect to u^k. Then we obtain

$$\underline{g}_{i,k} = g_{ij,k}\, \underline{g}^j + g_{ij}\, \underline{g}^j{}_{,k},$$

$$\Gamma_{ik}^{m}\, \underline{g}_m = g_{ij,k}\, \underline{g}^j - g_{ij}\, \Gamma^j_{km}\, \underline{g}^m.$$

Scalar multiplication by $\underline{g}^n$ gives

$$\Gamma^m_{ik}\, \underline{g}_m \cdot \underline{g}^n = g_{ij,k}\, \underline{g}^j \cdot \underline{g}^n - g_{ij}\, \Gamma^j_{km}\, \underline{g}^m \cdot \underline{g}^n,$$

and with (4.21) we obtain

$$\Gamma^n_{ik} = g_{ij,k}\, g^{jn} - g_{ij}\, g^{mn}\, \Gamma^j_{km}.$$

Multiplying by g_{np} gives

$$g_{np}\, \Gamma^n_{ik} + g_{ij}\, g^{mn}\, g_{np}\, \Gamma^j_{km} = g_{ij,k}\, g^{jn}\, g_{np}$$

and (4.26) yields

$$g_{pj}\, \Gamma^j_{ik} + g_{ij}\, \Gamma^j_{kp} = g_{ip,k}.$$

Cyclic permutations of the free indices i, p, and k give

$$g_{kj}\, \Gamma^j_{pi} + g_{pj}\, \Gamma^j_{ik} = g_{pk,i},$$
$$g_{ij}\, \Gamma^j_{kp} + g_{kj}\, \Gamma^j_{pi} = g_{ki,p}.$$

If we multiply the first of the last three equations by $-\frac{1}{2}$, the other two by $+\frac{1}{2}$, and add all three equations, we get

$$g_{kj}\, \Gamma^j_{ip} = \frac{1}{2}(g_{pk,i} + g_{ki,p} - g_{ip,k}).$$

Multiplication by g^{km} finally gives (independently of the underlying Cartesian coordinate system)

$$\Gamma^m_{ip} = \frac{1}{2}\, g^{mk}(g_{ki,p} + g_{kp,i} - g_{ip,k}). \tag{4.60}$$

Problem 4.12.
Compute the Christoffel symbols in cylindrical coordinates.

Problem 4.13.
Derive the transformation law for Christoffel symbols when changing between two curvilinear coordinate systems.

4.4.4 Partial Derivatives of Tensors. Partial and Covariant Derivatives of Tensor Coordinates

1. Let us now compute the partial derivatives of tensors with respect to the curvilinear coordinates u^i and write them with respect to the original basis or bases. We obtain

$$\begin{aligned}
\underline{a}_{,k} &= (a^i \underline{g}_i)_{,k} = a^i{}_{,k}\,\underline{g}_i + a^i\,\underline{g}_{i,k} = a^i{}_{,k}\,\underline{g}_i + a^i\,\Gamma^m_{ik}\,\underline{g}_m = (a^i{}_{,k} + \Gamma^i_{mk}\,a^m)\,\underline{g}_i \\
&= (a_i\,\underline{g}^i)_{,k} = a_{i,k}\,\underline{g}^i + a_i\,\underline{g}^i{}_{,k} = a_{i,k}\,\underline{g}^i - a_i\,\Gamma^i_{km}\,\underline{g}^m \\
&= (a_{i,k} - \Gamma^m_{ik}\,a_m)\,\underline{g}^i, \\
\underline{\underline{a}}_{,k} &= (a^{ij}\,\underline{g}_i\,\underline{g}_j)_{,k} = a^{ij}{}_{,k}\,\underline{g}_i\,\underline{g}_j + a^{ij}\,\underline{g}_{i,k}\,\underline{g}_j + a^{ij}\,\underline{g}_i\,\underline{g}_{j,k} \\
&= a^{ij}{}_{,k}\,\underline{g}_i\,\underline{g}_j + a^{ij}\,\Gamma^m_{ik}\,\underline{g}_m\,\underline{g}_j + a^{ij}\,\underline{g}_i\,\Gamma^m_{jk}\,\underline{g}_m \\
&= (a^{ij}{}_{,k} + \Gamma^i_{mk}\,a^{mj} + \Gamma^j_{mk}\,a^{im})\,\underline{g}_i\,\underline{g}_j \\
&= (a^i{}_j\,\underline{g}_i\,\underline{g}^j)_{,k} = \cdots,
\end{aligned}$$

etc.

The coefficients of the original bases in these equations are called covariant derivatives or absolute derivatives of the curvilinear tensor coordinates and they are denoted by a vertical line; i. e. we have

$$\begin{aligned}
a|_k &:= a_{,k}, \\
a^i|_k &:= a^i{}_{,k} + \Gamma^i_{mk}\,a^m, \\
a_i|_k &:= a_{i,k} - \Gamma^m_{ik}\,a_m, \\
a^{ij}|_k &:= a^{ij}{}_{,k} + \Gamma^i_{mk}\,a^{mj} + \Gamma^j_{mk}\,a^{im}, \\
a^i{}_j|_k &:= a^i{}_{j,k} + \Gamma^i_{mk}\,a^m{}_j - \Gamma^m_{jk}\,a^i{}_m, \\
a_i{}^j|_k &:= a_i{}^j{}_{,k} - \Gamma^m_{ik}\,a_m{}^j + \Gamma^j_{mk}\,a_i{}^m, \\
a_{ij}|_k &:= a_{ij,k} - \Gamma^m_{ik}\,a_{mj} - \Gamma^m_{jk}\,a_{im},
\end{aligned} \tag{4.61}$$

etc.

Using these covariant derivatives of tensor coordinates, we can write the partial derivatives of tensors as follows.

$$\begin{aligned}
a_{,k} &= a|_k, \\
\underline{a}_{,k} &= a^i|_k\,\underline{g}_i = a_i|_k\,\underline{g}^i, \\
\underline{\underline{a}}_{,k} &= a^{ij}|_k\,\underline{g}_i\,\underline{g}_j = a^i{}_j|_k\,\underline{g}_i\,\underline{g}^j = a_i{}^j|_k\,\underline{g}^i\,\underline{g}_j = a_{ij}|_k\,\underline{g}^i\,\underline{g}^j,
\end{aligned} \tag{4.62}$$

etc.

2. The partial derivatives of tensors clearly depend on the coordinate system. Using the chain rule, we obtain the transformation law for derivatives with respect to the curvilinear coordinates

$$\frac{\partial \mathscr{A}}{\partial \tilde{u}^i} = \frac{\partial u^m}{\partial \tilde{u}^i} \frac{\partial \mathscr{A}}{\partial u^m}, \tag{4.63}$$

i. e. the index of the partial derivative of a tensor transforms like a covariant index.

Since, on both sides of the equations (4.62), the index k must transform similarly, the covariant derivative of the holonomic coordinates of a tensor of n-th order gives the holonomic coordinates of a tensor of order $n+1$, i. e. the covariant derivative adds another covariant index. On the other hand, the partial derivative of a tensor coordinate with respect to the curvilinear coordinates is not a tensor coordinate.

4.4.5 The Total Differential of Tensors. The Total and the Absolute Differential of Tensor Coordinates

1. The total differential of a tensor is given by (4.52) as

$$\mathrm{d}\mathscr{A} = \mathscr{A}_{,i}\,\mathrm{d}u^i. \tag{4.64}$$

As we would expect, it is independent of the coordinates; the index of the partial derivative of a tensor is covariant, the index of the coordinate differentials is contravariant.

2. For tensor coordinates we must distinguish two different differentials (similar to two different derivatives). The total differential is, according to (4.52),

$$\mathrm{d}\,a^{i...j}_{m...n} = a^{i...j}_{m...n,k}\,\mathrm{d}u^k. \tag{4.65}$$

This represents, up to second-order terms in $\mathrm{d}\,u^k$, the increase of the tensor coordinates $\mathrm{d}\,a^{i...j}_{m...n}$ between two neighboring points. Since the partial derivative of a tensor coordinate is not a tensor coordinate, also the total differential of a tensor coordinate is not a tensor coordinate.

The expression analogous to (4.65), but with the covariant derivative of the tensor coordinates in place of the partial derivative, is called the absolute differential, to distinguish it from the total differential, and it is denoted by $\delta a^{i...j}_{m...n}$.

$$\delta a^{i...j}_{m...n} = a^{i...j}_{m...n}|_k\,\mathrm{d}u^k. \tag{4.66}$$

The absolute differential of a holonomic tensor coordinate is, just like the covariant derivative, also a holonomic tensor coordinate, and it is a tensor coordinate of the same order and type as the original coordinate.

3. The total differential of, for example, a vector, is

$$\mathrm{d}\,\underline{a} \overset{(4.64)}{=} \underline{a}_{,k}\,\mathrm{d}u^k \overset{(4.62)}{=} a^i|_k\,\underline{g}_i\,\mathrm{d}u^k \overset{(4.66)}{=} \delta a^i\,\underline{g}_i.$$

In general, we have

$$\begin{aligned}
&\mathrm{d}\,a = \delta a,\\
&\mathrm{d}\,\underline{a} = \delta a^i\,\underline{g}_i = \delta a_i\,\underline{g}^i,\\
&\mathrm{d}\,\underline{\underline{a}} = \delta a^{ij}\,\underline{g}_i\,\underline{g}_j = \delta a_i^{\ j}\,\underline{g}^i\,\underline{g}_j = \delta a^i_{\ j}\,\underline{g}_i\,\underline{g}^j = \delta a_{ij}\,\underline{g}^i\,\underline{g}^j,\\
&\text{etc.}
\end{aligned} \tag{4.67}$$

Thus the absolute differential of tensor coordinates are the coordinates of the total differential of a tensor with respect to the holonomic bases. They are zero if and only if the total differential of the tensor is zero, i. e. if the tensor is equal at the two neighboring points. In particular, for a vector, this means geometrically that the two vectors can be obtained from each other by a parallel translation.

We translate differentials of tensor coordinates from Cartesian coordinates into holonomic curvilinear coordinates, according to (4.67), as follows.

We replace the total differential of a Cartesian tensor coordinate by the absolute differential of a holonomic curvilinear tensor coordinate, e. g.

$$\mathrm{d}\,\underline{a} = \mathrm{d}\,\hat{a}_i\,\underline{e}_i = \delta a^i\,\underline{g}_i = \delta a_i\,\underline{g}^i.$$

4. The following computation shows the relation between the total differentials $\mathrm{d}\,\underline{a}$, $\mathrm{d}\,a^i$, the differentials $\mathrm{d}\,u^i$, and the absolute differentials δa^i, using a vector as an example:

$$\mathrm{d}\,\underline{a} = \mathrm{d}\,(a^i\,\underline{g}_i) = \underbrace{\mathrm{d}\,a^i}_{a^i_{\ ,k}\,\mathrm{d}u^k}\,\underline{g}_i + a^i\,\underbrace{\mathrm{d}\,\underline{g}_i}_{\Gamma^j_{ik}\,\underline{g}_j\,\mathrm{d}u^k} = \underbrace{(a^i_{\ ,k} + \Gamma^i_{jk}\,a^j)}_{a^i|_k}\,\underline{g}_i\,\mathrm{d}u^k,$$

$$\mathrm{d}\,\underline{a} = \underbrace{\underbrace{\mathrm{d}u^k\,(a^i_{\ ,k}}_{\mathrm{d}\,a^i} + \Gamma^i_{jk}\,a^j)}_{\delta a^i}\,\underline{g}_i. \qquad \overbrace{(a^i_{\ ,k} + \Gamma^i_{jk}\,a^j)}^{a^i|_k}, \quad \overbrace{(a^i_{\ ,k} + \Gamma^i_{jk}\,a^j)\,\underline{g}_i}^{\underline{a}_{,k}} \tag{4.68}$$

4.4.6 Derivatives with Respect to a Parameter

If the curvilinear coordinates u^i depend on a parameter t, then the derivatives of a tensor coordinate with respect to this parameter are given by (4.65) and (4.66) as

$$\begin{aligned}\frac{\mathrm{d}\,a^{i\ldots j}_{m\ldots n}}{\mathrm{d}t} &= a^{i\ldots j}_{m\ldots n,k}\,\frac{\mathrm{d}u^k}{\mathrm{d}t}\,,\\ \frac{\delta a^{i\ldots j}_{m\ldots n}}{\mathrm{d}t} &= a^{i\ldots j}_{m\ldots n}|_k\,\frac{\mathrm{d}u^k}{\mathrm{d}t}\,.\end{aligned} \tag{4.69}$$

Only the second of these two derivatives with respect to a parameter is again a tensor coordinate, and it is a tensor coordinate of the same order and type as the original tensor coordinate.

From (4.67), we analogously obtain

$$\begin{aligned}\frac{\mathrm{d}\underline{a}}{\mathrm{d}t} &= \frac{\delta a^i}{\mathrm{d}t}\,\underline{g}_i = \frac{\delta a_i}{\mathrm{d}t}\,\underline{g}^i\,,\\ \frac{\mathrm{d}\underline{\underline{a}}}{\mathrm{d}t} &= \frac{\delta a^{ij}}{\mathrm{d}t}\,\underline{g}_i\,\underline{g}_j = \frac{\delta a^i{}_j}{\mathrm{d}t}\,\underline{g}_i\,\underline{g}^j = \frac{\delta a_i{}^j}{\mathrm{d}t}\,\underline{g}^i\,\underline{g}_j = \frac{\delta a_{ij}}{\mathrm{d}t}\,\underline{g}^i\,\underline{g}^j\,,\end{aligned} \tag{4.70}$$

etc.

4.4.7 Gradient

1. The partial derivative (4.62) of a tensor is not independent of the coordinates; it has a covariant index. Thus it is not a tensor. However, we can make it a tensor, if we multiply it by the corresponding contravariant basis. Depending on whether we multiply it from the left or from the right, we obtain the left or the right gradient.

Using the chain rule, we obtain for the del operator[4] (2.47)

$$\underline{\nabla} := \frac{\partial}{\partial x_k}\,\underline{e}_k = \frac{\partial}{\partial u^i}\frac{\partial u^i}{\partial x_k}\,\underline{e}_k \overset{(4.7)}{=} \frac{\partial}{\partial u^i}\,\underline{g}^i\,,$$

$$\underline{\nabla} = \frac{\partial}{\partial u^i}\,\underline{g}^i\,. \tag{4.71}$$

(The partial derivative of a tensor with respect to the curvilinear coordinates transforms, according to (4.63), like a covariant vector coordinate.)

Thus, according to (2.48), we have for the (right) gradient

$$\operatorname{grad}\mathscr{A} = \mathscr{A}\,\underline{\nabla} = \mathscr{A}_{,k}\,\underline{g}^k \tag{4.72}$$

or, using (4.62), written out for tensors of different order, we obtain the following formulas.

4 Here e. g. $\underline{a}\,\underline{\nabla} = (\partial\underline{a}/\partial u^i)\,\underline{g}^i = \underline{a}_{,i}\,\underline{g}^i$; i. e. we do not compute the partial derivative of $\underline{g}^i$!

$$
\begin{aligned}
&\operatorname{grad} a = a|_k \, \underline{g}^k, \\
&\operatorname{grad} \underline{a} = a^i|_k \, \underline{g}_i \, \underline{g}^k = a_i|_k \, \underline{g}^i \, \underline{g}^k, \\
&\operatorname{grad} \underline{\underline{a}} = a^{ij}|_k \, \underline{g}_i \, \underline{g}_j \, \underline{g}^k = a^i{}_j|_k \, \underline{g}_i \, \underline{g}^j \, \underline{g}^k \\
&\qquad\quad = a_i{}^j|_k \, \underline{g}^i \, \underline{g}_j \, \underline{g}^k = a_{ij}|_k \, \underline{g}^i \, \underline{g}^j \, \underline{g}^k,
\end{aligned}
\tag{4.73}
$$

etc.

In other words, the covariant derivatives of the coordinates of a tensor are the coordinates of the gradient of this tensor, and we translate partial derivatives into holonomic curvilinear coordinates as follows.

We replace the partial derivative of a Cartesian tensor coordinate with the covariant derivative of a holonomic curvilinear tensor coordinate.

2. Since the δ-tensor and the ε-tensor are independent of position, their gradients must be zero and thus, according to (4.73), the covariant derivatives of their coordinates must be zero, i. e.

$$
\begin{aligned}
g^{ij}|_k &= g_{ij}|_k = 0, \\
e^{ijk}|_m &= e^{ij}{}_k|_m = \cdots = e_{ijk}|_m = 0.
\end{aligned}
\tag{4.74}
$$

(The first of these two equations is also called the Ricci theorem.) This implies that, in general, the partial derivatives of the coordinates of the δ-tensor and the ε-tensor are nonzero. We can compute them from the corresponding covariant derivative and the definition (4.61). For the partial derivatives of the δ-tensor we again obtain the relations (4.60), which we derived earlier.

4.4.8 Divergence and Curl

1. The (right) divergence of a tensor follows from (2.56) as

$$
\operatorname{div} \mathscr{A} = \mathscr{A} \cdot \underline{\nabla} = \mathscr{A}_{,k} \cdot \underline{g}^k. \tag{4.75}
$$

Using (4.62), we have e. g. $\operatorname{div} \underline{a} = a^i|_k \, \underline{g}_i \cdot \underline{g}^k \overset{(4.21)}{=} a^i|_k \, \delta_i^k = a^k|_k$, and for tensors of various orders we write the following expressions.

$$
\begin{aligned}
&\operatorname{div} \underline{a} = a^k|_k, \\
&\operatorname{div} \underline{\underline{a}} = a^{ik}|_k \, \underline{g}_i = a_i{}^k|_k \, \underline{g}^i,
\end{aligned}
\tag{4.76}
$$

etc.

2. Analogously, we have for the (right) curl of a tensor, according to (2.59),

$$\operatorname{curl} \mathscr{A} = \mathscr{A} \otimes \underline{\nabla} = -\mathscr{A} \times \underline{\nabla} = -\mathscr{A}_{,k} \times \underline{g}^k . \tag{4.77}$$

With (4.62), we have e. g. $\operatorname{curl} \underline{a} = -a^i|_k \, \underline{g}_i \times \underline{g}^k \overset{(4.37)}{=} -a^i|_k \, e_i^{\ kj} \, \underline{g}_j$, and for tensors of various orders we obtain the following formulas.

$$\begin{aligned} \operatorname{curl} \underline{a} &= a^i|_k \, e_i^{\ jk} \, \underline{g}_j = a_i|_k \, e^{ijk} \, \underline{g}_j = a^i|_k \, e_{ij}^{\ \ k} \, \underline{g}^j = a_i|_k \, e^{i\ k}_{\ j} \, \underline{g}^j , \\ \operatorname{curl} \underline{\underline{a}} &= a^{mi}|_k \, e_i^{\ jk} \, \underline{g}_m \underline{g}_j = a^m{}_i|_k \, e^{ijk} \, \underline{g}_m \underline{g}_j = \cdots , \\ &\text{etc.} \end{aligned} \tag{4.78}$$

3. If we substitute (4.61) for the covariant derivative $a_i|_k$, we see, because of the antimetry of the ε-tensor and the symmetry of the Christoffel symbols with respect to the two lower indices, that we can write for the curl of a vector

$$\operatorname{curl} \underline{a} = a_{i,k} \, e^{ijk} \, \underline{g}_j = a_{i,k} \, e^{i\ k}_{\ j} \, \underline{g}^j . \tag{4.79}$$

Although these equations contain partial derivatives, which are not tensor coordinates, they are very convenient for computing the curl.

Problem 4.14.
The following equations are given either in symbolic notation, in Cartesian coordinates, or in holonomic curvilinear coordinates. Translate each of them into the other two notations.

A. $\underline{a} \times \operatorname{curl} \underline{a} = \underline{b}$, B. $\dfrac{\partial \hat{a}_i}{\partial x_j} + \dfrac{\partial \hat{a}_j}{\partial x_i} = \hat{b}_{ij}$,

C. $a^i \, b_j|_i = c_j$, D. $e^i{}_{jk} \, e_{imn} \, a^j \, b^k \, c^m \, d^n = f$.

Simplify the last expression, using the Grassmann identity, before translating.

4.4.9 Physical Coordinates of Tensor-Analytical Operations

In the course of our presentation we arrived naturally at the holonomic coordinates of the tensors grad a, div $\underline{a}$, curl $\underline{a}$, etc. When the coordinates of these tensors are mentioned in physics books, e. g. in cylindrical or spherical coordinates, what is usually meant are the physical coordinates. It is not very difficult to compute the physical coordinates from arbitrary holonomic coordinates, using the transformation equations between holonomic and physical tensor coordinates.

Because these conversions are often used, we compute them here step-by-step for one example, i. e. for the divergence of a second-order tensor in cylindrical coordinates. In holonomic coordinates, we have

$$(\operatorname{div} \underline{\underline{a}})^i = a^{im}|_m .$$

Substituting successively 1, 2, and 3 for the free index and summing over the summation index, we obtain

$$(\operatorname{div} \underline{\underline{a}})^1 = a^{11}|_1 + a^{12}|_2 + a^{13}|_3,$$
$$(\operatorname{div} \underline{\underline{a}})^2 = a^{21}|_1 + a^{22}|_2 + a^{23}|_3,$$
$$(\operatorname{div} \underline{\underline{a}})^3 = a^{31}|_1 + a^{32}|_2 + a^{33}|_3.$$

Now we compute the covariant derivatives of the tensor coordinates in these equations. Using (4.61), we have

$$a^{ij}|_{\underline{j}} = a^{ij}{}_{,\underline{j}} + \Gamma^i_{m\underline{j}}\, a^{m\underline{j}} + \Gamma^{\underline{j}}_{m\underline{j}}\, a^{im}.$$

Taking into account that for cylindrical coordinates only Γ^1_{22}, Γ^2_{12}, and Γ^2_{21} are nonzero, we have

$$a^{11}|_1 = a^{11}{}_{,1},$$
$$a^{12}|_2 = a^{12}{}_{,2} + \Gamma^1_{22}\, a^{22} + \Gamma^2_{12}\, a^{11},$$
$$a^{13}|_3 = a^{13}{}_{,3},$$
$$a^{21}|_1 = a^{21}{}_{,1} + \Gamma^2_{21}\, a^{21},$$
$$a^{22}|_2 = a^{22}{}_{,2} + \Gamma^2_{12}\, a^{12} + \Gamma^2_{12}\, a^{21},$$
$$a^{23}|_3 = a^{23}{}_{,3},$$
$$a^{31}|_1 = a^{31}{}_{,1},$$
$$a^{32}|_2 = a^{32}{}_{,2} + \Gamma^2_{12}\, a^{31},$$
$$a^{33}|_3 = a^{33}{}_{,3}.$$

Next we substitute these covariant derivatives into the equations for the contravariant coordinates of the divergence; we write out the partial derivatives and we substitute for the Christoffel symbols their values $\Gamma^1_{22} = -R$, $\Gamma^2_{12} = \Gamma^2_{21} = 1/R$, so we obtain

$$(\operatorname{div} \underline{\underline{a}})^1 = \frac{\partial a^{11}}{\partial R} + \frac{\partial a^{12}}{\partial \varphi} - R\, a^{22} + \frac{1}{R}\, a^{11} + \frac{\partial a^{13}}{\partial z},$$

$$(\operatorname{div} \underline{\underline{a}})^2 = \frac{\partial a^{21}}{\partial R} + \frac{1}{R}\, a^{21} + \frac{\partial a^{22}}{\partial \varphi} + \frac{1}{R}\, a^{12} + \frac{1}{R}\, a^{21} + \frac{\partial a^{23}}{\partial z},$$

$$(\operatorname{div} \underline{\underline{a}})^3 = \frac{\partial a^{31}}{\partial R} + \frac{\partial a^{32}}{\partial \varphi} + \frac{1}{R}\, a^{31} + \frac{\partial a^{33}}{\partial z}.$$

We thus computed the divergence of a tensor in (*one* type of) holonomic cylindrical coordinates, and it only remains to convert the left and the right side, using (4.48), into physical cylindrical coordinates. We compute

$$g_{11} = g^{11} = 1, \quad g_{22} = R^2, \quad g^{22} = \frac{1}{R^2}, \quad g_{33} = g^{33} = 1.$$

Using (4.48), we find that $a_R = a^1, \quad a_\varphi = R\,a^2, \quad a_z = a^3$, and

$$a^{11} = a_{RR}, \quad a^{12} = \frac{1}{R}\,a_{R\varphi}, \quad a^{13} = a_{Rz}, \quad a^{21} = \frac{1}{R}\,a_{\varphi R},$$

$$a^{22} = \frac{1}{R^2}\,a_{\varphi\varphi}, \quad a^{23} = \frac{1}{R}\,a_{\varphi z}, \quad a^{31} = a_{zR}, \quad a^{32} = \frac{1}{R}\,a_{z\varphi}, \quad a^{33} = a_{zz},$$

and thus we have

$$(\operatorname{div} \underline{\underline{a}})_R = (\operatorname{div} \underline{\underline{a}})^1 = \frac{\partial a_{RR}}{\partial R} + \frac{\partial}{\partial \varphi}\frac{a_{R\varphi}}{R} - R\,\frac{a_{\varphi\varphi}}{R^2} + \frac{1}{R}\,a_{RR} + \frac{\partial a_{Rz}}{\partial z},$$

$$(\operatorname{div} \underline{\underline{a}})_\varphi = R(\operatorname{div} \underline{\underline{a}})^2 = R\Big(\frac{\partial}{\partial R}\frac{a_{\varphi R}}{R} + \frac{1}{R}\frac{a_{\varphi R}}{R} + \frac{\partial}{\partial \varphi}\frac{a_{\varphi\varphi}}{R^2}$$
$$+ \frac{1}{R}\frac{a_{R\varphi}}{R} + \frac{1}{R}\frac{a_{\varphi R}}{R} + \frac{\partial}{\partial z}\frac{a_{\varphi z}}{R}\Big),$$

$$(\operatorname{div} \underline{\underline{a}})_z = (\operatorname{div} \underline{\underline{a}})^3 = \frac{\partial a_{zR}}{\partial R} + \frac{\partial}{\partial \varphi}\frac{a_{z\varphi}}{R} + \frac{a_{zR}}{R} + \frac{\partial a_{zz}}{\partial z},$$

or, reordered,

$$(\operatorname{div} \underline{\underline{a}})_R = \frac{\partial a_{RR}}{\partial R} + \frac{1}{R}\frac{\partial a_{R\varphi}}{\partial \varphi} + \frac{\partial a_{Rz}}{\partial z} + \frac{a_{RR} - a_{\varphi\varphi}}{R},$$

$$(\operatorname{div} \underline{\underline{a}})_\varphi = \frac{\partial a_{\varphi R}}{\partial R} + \frac{1}{R}\frac{\partial a_{\varphi\varphi}}{\partial \varphi} + \frac{\partial a_{\varphi z}}{\partial z} + \frac{a_{R\varphi} + a_{\varphi R}}{R},$$

$$(\operatorname{div} \underline{\underline{a}})_z = \frac{\partial a_{zR}}{\partial R} + \frac{1}{R}\frac{\partial a_{z\varphi}}{\partial \varphi} + \frac{\partial a_{zz}}{\partial z} + \frac{a_{zR}}{R}.$$

Problem 4.15.
Compute in physical cylindrical coordinates:
A. grad a, B. div $\underline{a}$, C. rot $\underline{a}$, D. grad $\underline{a}$, E. (grad $\underline{a}$) $\cdot\, \underline{b}$.
The equations (B.1) to (B.13) from the appendix can be assumed to be known.

4.4.10 Second Covariant Derivative of a Tensor Coordinate. Laplace Operator

1. According to (4.54), together with (4.18), and also according to (4.63), the index of the partial derivative of the position vector or of a tensor transforms like a covariant index. This enabled us to introduce the covariant derivative (4.61) of a tensor coordinate in terms of the partial derivative of the tensor. For the second derivative this is not the case: the partial derivative of (4.54) is $\underline{x}_{,ij} = \underline{g}_{i,j} = \Gamma_{ij}^{m}\,\underline{g}_m$, and the partial derivative of (4.63) is

$$\frac{\partial^2 \mathscr{A}}{\partial \bar{u}^j\, \partial \bar{u}^i} = \frac{\partial u^m}{\partial \bar{u}^i}\frac{\partial u^n}{\partial \bar{u}^j}\frac{\partial^2 \mathscr{A}}{\partial u^n\, \partial u^m} + \frac{\partial^2 u^m}{\partial \bar{u}^j\, \partial \bar{u}^i}\frac{\partial \mathscr{A}}{\partial u^m}, \tag{4.80}$$

i. e. while the index i in $\mathscr{A}_{,i}$ transforms covariantly, the two indices in $\mathscr{A}_{,ij}$ do not transform like tensor indices. This is the reason why we do not use the second partial derivative of the position vector or of a tensor. Instead, we introduce the second covariant derivative. Since the covariant derivative of a tensor coordinate is again a tensor coordinate, the covariant derivative of the covariant derivative of a tensor coordinate is defined and is again a tensor coordinate. We call it the second covariant derivative of the original tensor coordinate and write

$$a^{i...j}_{m...n}|_{pq} := a^{i...j}_{m...n}|_p|_q. \tag{4.81}$$

2. The second covariant derivative of a tensor coordinate is clearly a coordinate of the gradient of the gradient of the tensor under consideration. From $\mathscr{B} = \operatorname{grad} \mathscr{A}$, $\mathscr{C} = \operatorname{grad} \mathscr{B} = \operatorname{grad} \operatorname{grad} \mathscr{A}$, we obtain $b_{i...jp} = a_{i...j}|_p$, $c_{i...jpq} = b_{i...jp}|_q = a_{i...j}|_{pq}$ or

$$\operatorname{grad} \operatorname{grad} \mathscr{A} = a_{i...j}|_{pq}\, \underline{g}^i \cdots \underline{g}^j\, \underline{g}^p\, \underline{g}^q \tag{4.82}$$

and other versions in other holonomic coordinates.

3. We compute the tensor curl grad $\mathscr{A}$ as follows:

$$\operatorname{grad} \mathscr{A} = a_{i...j}|_k\, \underline{g}^i \cdots \underline{g}^j\, \underline{g}^k = b_{i...jk}\, \underline{g}^i \cdots \underline{g}^j\, \underline{g}^k = \mathscr{B},$$
$$\operatorname{curl} \operatorname{grad} \mathscr{A} = \operatorname{curl} \mathscr{B} = b_{i...jk}|_n\, e^{k}{}_{m}{}^{n}\, \underline{g}^i \cdots \underline{g}^j\, \underline{g}^m = a_{i...j}|_{kn}\, e^{k}{}_{m}{}^{n}\, \underline{g}^i \cdots \underline{g}^j\, \underline{g}^m.$$

As we know, all Cartesian coordinates of this tensor are zero, which follows immediately from the fact that the second partial derivative of a quantity does not depend on the order of differentiation. However, if all Cartesian coordinates of a tensor are zero, according to the transformation law (4.19) for tensor coordinates, also all holonomic coordinates of this tensor are zero in any curvilinear coordinate system. Thus we have $a_{i...j}|_{kn}\, e^{k}{}_{m}{}^{n} = 0$. For $m = 1$, for example, this gives

$$a_{i...j}|_{32}\, e^{3}{}_{1}{}^{2} + a_{i...j}|_{23}\, e^{2}{}_{1}{}^{3} = (a_{i...j}|_{32} - a_{i...j}|_{23})\, e^{3}{}_{1}{}^{2} = 0.$$

Since $e^{3}{}_{1}{}^{2} \neq 0$, it follows that $a_{i...j}|_{32} = a_{i...j}|_{23}$. For $m = 2$ and $m = 3$ we have $a_{i...j}|_{13} = a_{i...j}|_{31}$, $a_{i...j}|_{12} = a_{i...j}|_{21}$. Thus, in general we obtain the following formula.

$$a^{i...j}_{m...n}|_{pq} = a^{i...j}_{m...n}|_{qp}. \tag{4.83}$$

The second covariant derivative of a tensor coordinate also does not depend on the order of differentiation.

4. We finally compute $\Delta \mathscr{A} := \operatorname{div} \operatorname{grad} \mathscr{A}$. For example, we have

$$\operatorname{grad} \underline{a} = a^i|_k\, \underline{g}_i\, \underline{g}^k = a^i|_m\, g^{mk}\, \underline{g}_i\, \underline{g}_k,$$
$$\operatorname{div} \operatorname{grad} \underline{a} = (a^i|_m\, g^{mk})|_k\, \underline{g}_i = g^{mk}\, a^i|_{mk}\, \underline{g}_i + \underbrace{g^{mk}|_k}_{\overset{(4.74)}{=}\,0}\, a^i|_m\, \underline{g}_i,$$

or, in general,

$$
\begin{aligned}
\Delta a &= g^{mn}\, a|_{mn}, \\
\Delta \underline{a} &= g^{mn}\, a^i|_{mn}\, \underline{g}_i = g^{mn}\, a_i|_{mn}\, \underline{g}^i, \\
\Delta \underline{\underline{a}} &= g^{mn}\, a^{ij}|_{mn}\, \underline{g}_i\, \underline{g}_j = g^{mn}\, a^i{}_j|_{mn}\, \underline{g}_i\, \underline{g}^j \\
&= g^{mn}\, a_i{}^j|_{mn}\, \underline{g}^i\, \underline{g}_j = g^{mn}\, a_{ij}|_{mn}\, \underline{g}^i\, \underline{g}^j,
\end{aligned}
\tag{4.84}
$$

etc.

Problem 4.16.
Translate into the other two notations:

A. $\dfrac{\partial^2 \hat{a}_m}{\partial x_k^2} = \hat{b}_m$, B. $\dfrac{\partial^2 \hat{a}_k}{\partial x_m\, \partial x_k} = \hat{b}_m$, C. $\mathrm{d}\,\underline{\underline{a}} = (\mathrm{grad}\ \underline{\underline{a}}) \cdot \mathrm{d}\,\underline{x}$.

Problem 4.17.
Compute in physical cylindrical coordinates: A. $\Delta\, a$, B. $\Delta\, \underline{a}$.

Problem 4.18.
A. Show, using the definition of the covariant derivative, that the following holds:

$$a_i|_{kl} - a_i|_{lk} = (\Gamma^m_{il,k} - \Gamma^m_{ik,l} + \Gamma^n_{il}\,\Gamma^m_{nk} - \Gamma^n_{ik}\,\Gamma^m_{nl})\, a_m.$$

B. Prove that the bracket represents the coordinates of the zero tensor of fourth order.

The tensor

$$R^m{}_{ijk} := \Gamma^m_{ik,j} - \Gamma^m_{ij,k} + \Gamma^n_{ik}\,\Gamma^m_{nj} - \Gamma^n_{ij}\,\Gamma^m_{nk}$$

is called (mixed) Riemann curvature tensor. It is characteristic for Euclidean spaces that this tensor is zero.

4.4.11 Integrals of Tensor Fields

4.4.11.1 Line, Surface, and Volume Elements

Let us first derive expressions for a line element, a surface element, and a volume element in terms of the curvilinear coordinates.

1. A line element is, according to (4.55), given as

$$\mathrm{d}\,\underline{x} = \mathrm{d}\,u^i\, \underline{g}_i. \tag{4.85}$$

We denote the line elements of the coordinate curves by $\mathrm{d}\underline{x}_1$,$\mathrm{d}\underline{x}_2$, $\mathrm{d}\underline{x}_3$; so here the index is not a covariant index. Since the basis vectors $\underline{g}_i$ are by construction tangent to the coordinate curves, it follows from (4.85) that

$$\mathrm{d}\underline{x}_1 = \mathrm{d}u^1\,\underline{g}_1, \qquad \mathrm{d}\underline{x}_2 = \mathrm{d}u^2\,\underline{g}_2, \qquad \mathrm{d}\underline{x}_3 = \mathrm{d}u^3\,\underline{g}_3. \tag{4.86}$$

This explains why we usually use the contravariant coordinates of the vector $\mathrm{d}\underline{x}$. From (4.85), together with (4.86), we obtain in component form

$$\mathrm{d}\underline{x} = \underbrace{\mathrm{d}u^1\,\underline{g}_1}_{\mathrm{d}\underline{x}_1} + \underbrace{\mathrm{d}u^2\,\underline{g}_2}_{\mathrm{d}\underline{x}_2} + \underbrace{\mathrm{d}u^3\,\underline{g}_3}_{\mathrm{d}\underline{x}_3}. \tag{4.87}$$

2. A surface element $\mathrm{d}\underline{A}$ is analogously given by

$$\mathrm{d}\underline{A} = \mathrm{d}A_i\,\underline{g}^i. \tag{4.88}$$

We denote the surface elements of the coordinate planes by $\mathrm{d}\underline{A}_1$, $\mathrm{d}\underline{A}_2$, $\mathrm{d}\underline{A}_3$; i. e. here again the index is not a covariant index. Since the basis vectors $\underline{g}^i$ are by construction perpendicular to the coordinate surfaces at each point, it follows from (4.88) that

$$\mathrm{d}\underline{A}_1 = \mathrm{d}A_1\,\underline{g}^1, \qquad \mathrm{d}\underline{A}_2 = \mathrm{d}A_2\,\underline{g}^2, \qquad \mathrm{d}\underline{A}_3 = \mathrm{d}A_3\,\underline{g}^3. \tag{4.89}$$

This again explains why we commonly use the covariant coordinates of the vector $\mathrm{d}\underline{A}$. From (4.88), together with (4.89), analogously to (4.87), we obtain

$$\mathrm{d}\underline{A} = \underbrace{\mathrm{d}A_1\,\underline{g}^1}_{\mathrm{d}\underline{A}_1} + \underbrace{\mathrm{d}A_2\,\underline{g}^2}_{\mathrm{d}\underline{A}_2} + \underbrace{\mathrm{d}A_3\,\underline{g}^3}_{\mathrm{d}\underline{A}_3}. \tag{4.90}$$

3. The three vectors $\mathrm{d}\underline{A}_i$ and the three vectors $\mathrm{d}\underline{x}_i$ are related as follows. For example, the surface element with the surface vector $\mathrm{d}\underline{A}_1$ is spanned by the line elements $\mathrm{d}\underline{x}_2$ and $\mathrm{d}\underline{x}_3$, i. e. we have

$$\mathrm{d}\underline{A}_1 = \pm\mathrm{d}\underline{x}_2\times\mathrm{d}\underline{x}_3, \qquad \mathrm{d}\underline{A}_2 = \pm\mathrm{d}\underline{x}_3\times\mathrm{d}\underline{x}_1, \qquad \mathrm{d}\underline{A}_3 = \pm\mathrm{d}\underline{x}_1\times\mathrm{d}\underline{x}_2. \tag{4.91}$$

The sign is determined by the orientation of the three vectors $\mathrm{d}\underline{A}_i$.

Substituting the line elements into (4.91) gives

$$\begin{aligned}\mathrm{d}\underline{A}_1 &\overset{(4.89)}{=} \mathrm{d}A_1\,\underline{g}^1 \overset{(4.91)}{=} \pm\mathrm{d}\underline{x}_2\times\mathrm{d}\underline{x}_3 \overset{(4.86)}{=} \pm\mathrm{d}u^2\,\underline{g}_2\times\mathrm{d}u^3\,\underline{g}_3 \\ &\overset{(3.21)}{=} \pm[\underline{g}_1, \underline{g}_2, \underline{g}_3]\,\mathrm{d}u^2\,\mathrm{d}u^3\,\underline{g}^1 \overset{(4.28)}{=} \pm\sqrt{g}\,\mathrm{d}u^2\,\mathrm{d}u^3\,\underline{g}^1,\ \text{i. e.}\end{aligned}$$

$$\begin{aligned} \mathrm{d}A_1 &= \pm\sqrt{g}\,\mathrm{d}u^2\,\mathrm{d}u^3, \\ \mathrm{d}A_2 &= \pm\sqrt{g}\,\mathrm{d}u^3\,\mathrm{d}u^1, \\ \mathrm{d}A_3 &= \pm\sqrt{g}\,\mathrm{d}u^1\,\mathrm{d}u^2. \end{aligned} \tag{4.92}$$

Here again the sign is determined by the orientation of the $\mathrm{d}\underline{A}_i$.

4. The volume element spanned by the line elements $\mathrm{d}\underline{x}_1$, $\mathrm{d}\underline{x}_2$, and $\mathrm{d}\underline{x}_3$ is given by (assuming right-handed systems we dropped the absolute-value signs)

$$\mathrm{d}V = [\mathrm{d}\underline{x}_1, \mathrm{d}\underline{x}_2, \mathrm{d}\underline{x}_3]. \tag{4.93}$$

Substituting the line elements, using (4.86), into (4.93), we obtain

$$\mathrm{d}V = [\mathrm{d}u^1\,\underline{g}_1, \mathrm{d}u^2\,\underline{g}_2, \mathrm{d}u^3\,\underline{g}_3] = [\underline{g}_1, \underline{g}_2, \underline{g}_3]\,\mathrm{d}u^1\,\mathrm{d}u^2\,\mathrm{d}u^3$$

or, using (4.28),

$$\mathrm{d}V = \sqrt{g}\,\mathrm{d}u^1\,\mathrm{d}u^2\,\mathrm{d}u^3. \tag{4.94}$$

4.4.11.2 Integrals in Curvilinear Coordinates

1. In (2.94) and (2.95) we defined various types of integrals of tensor fields, both in symbolic notation and in Cartesian coordinates. If, in the equations in symbolic notation, we write all tensors as the sum of their components with respect to a Cartesian coordinate system, we can pull the bases in front of the integral and cancel them (they are independent of position) and thus obtain equations in Cartesian coordinates. We have, e. g. for

$$\int \underline{\underline{a}}\,\mathrm{d}\underline{x} = \underline{\underline{\underline{B}}}, \tag{a}$$

$$\int \hat{a}_{ij}\,\underline{e}_i\,\underline{e}_j\,\mathrm{d}x_k\,\underline{e}_k = \underline{e}_i\,\underline{e}_j\,\underline{e}_k \int \hat{a}_{ij}\,\mathrm{d}x_k = \widehat{B}_{ijk}\,\underline{e}_i\,\underline{e}_j\,\underline{e}_k,$$

$$\int \hat{a}_{ij}\,\mathrm{d}x_k = \widehat{B}_{ijk}. \tag{b}$$

A version in holonomic curvilinear coordinates, i. e. a translation of (a) and (b) into holonomic curvilinear coodinates, does not exist in general. Clearly, we can write the tensors, e. g. in (a), in component form with respect to any curvilinear coordinate system, but only in the case of rectilinear coordinates we can pull the bases in front of the integrals and cancel them out. Curvilinear bases depend on position and they cannot be pulled in front of the integrals. Thus, in the case of curvilinear coordinates, integrals over spatial domains and hence integral theorems, including the theorems of Gauss and Stokes, cannot be written in “true” coordinate notation. The best we can do is to write the bases, using (4.5) or (4.7), with respect to a Cartesian basis and then pull this Cartesian basis in front of the integrals and cancel them out, e. g. we write

$$\int \underline{a}\,\mathrm{d}\,V = \int a_i\,\underline{g}^i\,\mathrm{d}\,V \overset{(4.7)}{=} \int a_i\,\frac{\partial u^i}{\partial x_j}\,\underline{e}_j\,\mathrm{d}\,V = \underline{e}_j \int a_i\,\frac{\partial u^i}{\partial x_j}\,\mathrm{d}\,V.$$

In this way, we obtain an expression for the integrand which contains curvilinear coordinates. However, according to the transformation law (4.19) for tensor coordinates, the integrand consists of the Cartesian coordinates of the respective tensor as a function of the curvilinear coordinates. Such expressions are sometimes useful, but let us keep in mind that ultimately they are expressions in Cartesian tensor coordinates.

2. However, there are two special cases where the integrand can be written in "true" curvilinear coordinates. First, if the integral is a scalar combination of tensor coordinates, then the integral does not contain a basis. An example is the Gauss theorem $(2.97)_1$ in the form

$$\int a^i|_i\,\mathrm{d}\,V = \oint a^i\,\mathrm{d}\,A_i \tag{4.95}$$

and the Stokes theorem $(2.100)_1$ in the form

$$\int a_i|_k\,e^{ijk}\,\mathrm{d}\,A_j = \oint a_i\,\mathrm{d}\,u^i. \tag{4.96}$$

Second, we obtain "true" expressions in curvilinear coordinates for coordinate systems, where one basis vector is independent of position. This is the case, e. g. for cylindrical coordinates, where the third basis vector (if we list the indices in the usual order) is independent of position and moreover is a unit vector, i. e. $\underline{g}_3 = \underline{g}^3 = \underline{e}_3$. Then all tensor coordinates whose free indices have the value three are also Cartesian coordinates, and we can integrate over them. This gives, for example, the Gauss theorem $(2.96)_1$ in the form

$$\int a|_3\,\mathrm{d}\,V = \oint a\,\mathrm{d}\,A_3 \tag{4.97}$$

or, using $(2.98)_1$, in the form

$$\int a_i|_k\,e^{i3k}\,\mathrm{d}\,V = \oint a_i\,e^{i3k}\,\mathrm{d}\,A_k. \tag{4.98}$$

4.5 Fundamentals of the Theory of Surfaces

The theory of surfaces is the part of differential geometry which investigates the properties of curved surfaces. This topic is not only important for many technical applications, but also historically relevant, as it was the starting point for the development of tensor analysis in the 19th century. From today's perspective, however, it is easier to derive the fundamental equations of the theory of surfaces as a special case of tensor analysis in three-dimensional curvilinear coordinate systems. We will take this path here.

4.5.1 Surface Coordinates and Moving Frames

1. Let u^1 and u^2 be two coordinates defined on a surface, which uniquely determine the position of any point P on the surface. Then the position vector of the point P is given in a Cartesian coordinate system as $\underline{x} = \overrightarrow{OP} = x_i\, \underline{e}_i$,

$$\underline{x} = \underline{x}(u^1, u^2), \qquad x_i = x_i(u^1, u^2). \tag{4.99}$$

We further assume that the surface is sufficiently smooth and that the functions x_i are sufficiently often differentiable with respect to u^1 and u^2. The coordinates u^1 and u^2 are also called surface parameters or surface coordinates.

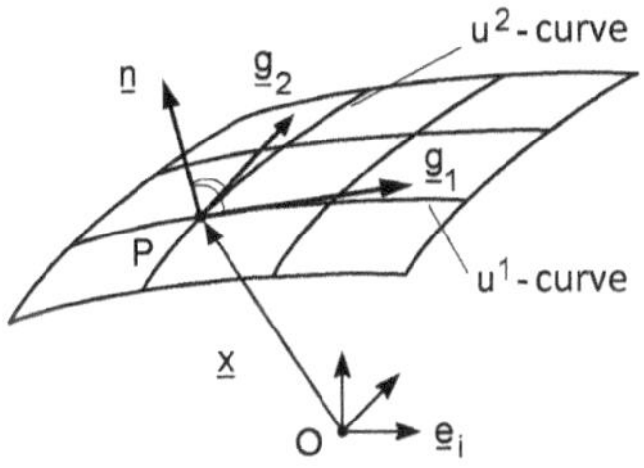

2. From the perspective of the surrounding space, the surface can be interpreted as a coordinate surface in a three-dimensional curvilinear coordinate system. The third coordinate u^3 has the same value for all points on the surface $u^3 = c$. Points with $u^3 \neq c$ are not included in the surface. With this interpretation we can assign two basis vectors to the surface, i. e. the covariant basis vectors $\underline{g}_1$ and $\underline{g}_2$, which are, according to (4.5), tangent to the coordinate curves of the u^1- and u^2-coordinates, respectively, i. e.

$$\underline{g}_1 = \frac{\partial x_i}{\partial u^1}\, \underline{e}_i = \underline{x}_{,1}, \qquad \underline{g}_2 = \frac{\partial x_i}{\partial u^2}\, \underline{e}_i = \underline{x}_{,2}. \tag{4.100}$$

Together, $\underline{g}_1$ and $\underline{g}_2$ span the tangent plane of the surface at the point P.

If we need a basis in the three-dimensional space, we add to $\underline{g}_1$ and $\underline{g}_2$ the normal vector $\underline{n}$ of the surface at the point P, so we have

$$\underline{n} = \frac{\underline{g}_1 \times \underline{g}_2}{|\underline{g}_1 \times \underline{g}_2|}. \tag{4.101}$$

The basis $\underline{g}_1, \underline{g}_2, \underline{n}$ is called a moving frame of the surface and, in general, it is varying from point to point. The normal vector $\underline{n}$ has length 1 and is perpendicular to both $\underline{g}_1$ and $\underline{g}_2$. The lengths of $\underline{g}_1$ and $\underline{g}_2$ and the angle enclosed by them depend on the definition of the coordinates u^1 and u^2.

The normal vector can be interpreted as a covariant basis vector $\underline{g}_3 = (\partial x_i/\partial u^3)\,\underline{e}_i$, if the third coordinate u^3 is at every point normal to the surface and scaled such that $\underline{g}_3 = \underline{n}$. This interpretation turns out to be useful, when we specialize the equations of tensor analysis from the three-dimensional space to the surface. However, according to definition (4.101), $\underline{g}_3 = \underline{n}$ does not depend on u^3; it is via $\underline{g}_1$ and $\underline{g}_2$ only determined by the surface coordinates u^1 and u^2.

3. For our further investigations we find it useful to extend the summation convention. As before, we denote coordinate indices in the three-dimensional space by small Latin letters $i, j, k, \ldots$ with the value set 1, 2, 3; and as before we sum over Latin indices, which appear twice, from 1 to 3. However, to analyze a surface, we need only two coordinates. We denote indices which only take the values 1, 2 by small Greek letters $\alpha, \beta, \ldots$, and we sum correspondingly over Greek indices appearing twice from 1 to 2. Thus, we write $(4.99)_2$ and (4.100) compactly as

$$x_i = x_i(u^\alpha), \tag{4.102}$$

$$\underline{g}_\alpha = \frac{\partial x_i}{\partial u^\alpha}\,\underline{e}_i = \underline{x}_{,\alpha}. \tag{4.103}$$

4. Consider a surface where a second set of coordinates $\tilde{u}^\beta$ is defined, which is related one-to-one (except at singular points) to the first set of coordinates u^α, by the transformation equations

$$u^\alpha = u^\alpha(\tilde{u}^\beta), \qquad \tilde{u}^\alpha = \tilde{u}^\alpha(u^\beta). \tag{4.104}$$

Then, analogously to (4.103), we have for the covariant basis vectors $\underline{\tilde{g}}_\alpha$

$$\underline{\tilde{g}}_\alpha = \frac{\partial x_i}{\partial \tilde{u}^\alpha}\,\underline{e}_i.$$

Using the chain rule, we get

$$\underline{\tilde{g}}_\alpha = \frac{\partial x_i}{\partial \tilde{u}^\alpha}\,\underline{e}_i = \frac{\partial x_i}{\partial u^\beta}\frac{\partial u^\beta}{\partial \tilde{u}^\alpha}\,\underline{e}_i = \frac{\partial u^\beta}{\partial \tilde{u}^\alpha}\underbrace{\frac{\partial x_i}{\partial u^\beta}\,\underline{e}_i}_{\underline{g}_\beta},$$

$$\underline{\tilde{g}}_\alpha = \frac{\partial u^\beta}{\partial \tilde{u}^\alpha}\,\underline{g}_\beta. \tag{4.105}$$

This equation corresponds to the transformation law (4.18) for covariant basis vectors, with the only difference that here summation is only from 1 to 2.

The new basis vectors $\underline{\tilde{g}}_\alpha$ can be represented as a linear combination of the old basis vectors $\underline{g}_\alpha$ (as we can expect from the transformation equations (4.104)), and thus, they also lie in the tangent plane spanned by $\underline{g}_\alpha$. In other words, only one tangent plane exists at a point P and it contains the tangent vectors of all curves through the point P.

5. The basis vectors $\underline{g}^i$, reciprocal to the moving frame, are given by (3.21) and with $\underline{g}_3 = \underline{n}$ as

$$\underline{g}^1 = \frac{\underline{g}_2 \times \underline{n}}{[\underline{g}_1, \underline{g}_2, \underline{n}]}, \qquad \underline{g}^2 = \frac{\underline{n} \times \underline{g}_1}{[\underline{g}_1, \underline{g}_2, \underline{n}]}, \qquad \underline{g}^3 = \frac{\underline{g}_1 \times \underline{g}_2}{[\underline{g}_1, \underline{g}_2, \underline{n}]}.$$

From the properties of the vector product, we know that $\underline{g}^1$ and $\underline{g}^2$ are perpendicular to $\underline{n}$ and that $\underline{n}$ is perpendicular to $\underline{g}_1$ and $\underline{g}_2$. Thus $\underline{g}^1$ and $\underline{g}^2$ lie in the plane spanned by $\underline{g}_1$ and $\underline{g}_2$, while $\underline{g}^3$ is perpendicular to $\underline{g}_1$ and $\underline{g}_2$. Hence $\underline{g}^3$ and $\underline{n}$ are collinear and since $[\underline{g}_1, \underline{g}_2, \underline{n}] = \underline{g}_1 \times \underline{g}_2 \cdot \underline{n} = |\underline{g}_1 \times \underline{g}_2|$, we have $\underline{g}^3 = \underline{n}$. The $\underline{g}^\alpha$ are again called contravariant basis vectors. According to Section 3.9.4, they satisfy, analogously to (3.24), the orthogonality relations for reciprocal bases in a plane. We have

$$\underline{g}_\alpha \cdot \underline{g}^\beta = \delta^\beta_\alpha. \tag{4.106}$$

For the second orthogonality relation (3.25) we obtain with $\underline{g}^3 = \underline{g}_3 = \underline{n}$

$$\underline{g}_\alpha \underline{g}^\alpha + \underline{n}\,\underline{n} = \underline{\underline{\delta}}. \tag{4.107}$$

4.5.2 Surface Vectors and Surface Tensors

1. The differential of the position vector on the surface is given, in terms of the coordinates u^α and $\tilde{u}^\alpha$, as

$$\begin{aligned} \mathrm{d}\underline{x} = \mathrm{d}x_i\,\underline{e}_i &= \underbrace{\frac{\partial x_i}{\partial u^\alpha}\,\underline{e}_i}_{\underline{g}_\alpha}\,\mathrm{d}u^\alpha = \mathrm{d}u^\alpha\,\underline{g}_\alpha \\ &= \underbrace{\frac{\partial x_i}{\partial \tilde{u}^\alpha}\,\underline{e}_i}_{\underline{\tilde{g}}_\alpha}\,\mathrm{d}\tilde{u}^\alpha = \mathrm{d}\tilde{u}^\alpha\,\underline{\tilde{g}}_\alpha. \end{aligned}$$

Thus, the differential of the position vector $\mathrm{d}\underline{x}$ also lies in the tangent plane. Substitution of the transformation law (4.105) for the basis vectors gives

$$\mathrm{d}\tilde{u}^\alpha\,\underline{\tilde{g}}_\alpha = \mathrm{d}u^\beta\,\underline{g}_\beta = \mathrm{d}u^\beta\,\frac{\partial \tilde{u}^\alpha}{\partial u^\beta}\,\underline{\tilde{g}}_\alpha.$$

So the coordinate differentials satisfy the transformation law

$$\mathrm{d}\tilde{u}^{\alpha} = \frac{\partial \tilde{u}^{\alpha}}{\partial u^{\beta}}\,\mathrm{d}u^{\beta}, \tag{4.108}$$

which is, according to (4.19), the transformation law for contravariant vector coordinates.

2. A quantity $\underline{a}$, which can be represented in the form $\underline{a} = a^{\alpha}\,\underline{g}_{\alpha} = \tilde{a}^{\alpha}\,\underline{\tilde{g}}_{\alpha}$ and whose coefficients a^{α} and $\tilde{a}^{\alpha}$ are related by the transformation equations (4.108), is called a surface vector, and the a^{α} and $\tilde{a}^{\alpha}$ are called its contravariant coordinates. Using the contravariant basis vectors, we can also write a surface vector in the form $\underline{a} = a_{\alpha}\,\underline{g}^{\alpha} = \tilde{a}_{\alpha}\,\underline{\tilde{g}}^{\alpha}$, i. e. in terms of its covariant coordinates.

Problem 4.19.
Derive the transformation laws for the contravariant basis vectors $\underline{g}^{\alpha}$ and for the covariant coordinates a_{α} of a surface vector $\underline{a}$.

Surface tensors of higher order are defined similarly, and their coordinates must analogously satisfy the transformation laws (4.19). Thus, computations with surface tensors are performed in the same way as with tensors in a three-dimensional curvilinear coordinate system. The only difference is that the values of the indices are limited to 1, 2.

Problem 4.20.
Metric coefficients on a curved surface can be defined by the scalar product $\underline{g}_{\alpha} \cdot \underline{g}_{\beta} = g_{\alpha\beta}$, just like in a three-dimensional curvilinear coordinate system. Show that the metric coefficients $g_{\alpha\beta}$ are the entirely covariant coordinates of a second-order surface tensor.

4.5.3 Metric Properties

1. Now we ask how we can measure lengths, angles, and surface areas on a curved surface; these quantities are also called metric properties of a surface. The fundamental idea is to replace the curved surface in the infinitesimal neighborhood of a point P by its tangent plane; there we can compute lengths, angles, and surface areas, using the tools of plane geometry. The transition to surface patches of finite size then follows directly by integration.

2. The differential of the position vector was introduced in Section 4.5.2 as

$$\mathrm{d}\underline{x} = \underline{g}_{\alpha}\,\mathrm{d}u^{\alpha}\,. \tag{4.109}$$

The scalar product of the differential with itself is

$$\mathrm{d}\underline{x} \cdot \mathrm{d}\underline{x} = (\underline{g}_{\alpha}\,\mathrm{d}u^{\alpha}) \cdot (\underline{g}_{\beta}\,\mathrm{d}u^{\beta}) = \underline{g}_{\alpha} \cdot \underline{g}_{\beta}\,\mathrm{d}u^{\alpha}\,\mathrm{d}u^{\beta}.$$

The scalar products of the (covariant) basis vectors of the surface form, analogously to (4.21), the (covariant) metric coefficients of the surface, i. e.

$$g_{\alpha\beta} := \underline{g}_{\alpha} \cdot \underline{g}_{\beta}, \tag{4.110}$$

and then

$$\mathrm{d}\,\underline{x} \cdot \mathrm{d}\,\underline{x} = g_{\alpha\beta}\,\mathrm{d}\,u^{\alpha}\,\mathrm{d}\,u^{\beta}. \tag{4.111}$$

This equation is called the first fundamental form of the surface. It offers another way to show, different from Problem 4.20, that the metric coefficients are tensor coordinates; since the left side is a scalar $\mathrm{d}\,\underline{x} \cdot \mathrm{d}\,\underline{x}$ and the differentials $\mathrm{d}\,u^{\alpha}$ and $\mathrm{d}\,u^{\beta}$ of the coordinates are, according to (4.108), the contravariant coordinates of a surface vector, the $g_{\alpha\beta}$ are, according to the quotient rule, the covariant coordinates of a second-order surface tensor. The metric coefficients are symmetric, since they are defined as scalar products,

$$g_{\alpha\beta} = g_{\beta\alpha}, \tag{4.112}$$

and furthermore, because $\mathrm{d}\,\underline{x} \cdot \mathrm{d}\,\underline{x} > 0$, their matrix is positive definite. The following notations are commonly used:

$$g_{11} =: E, \qquad g_{22} =: G, \qquad g_{12} = g_{21} =: F. \tag{4.113}$$

Substituting the definitions (4.110) and (4.103) gives

$$E = \underline{g}_1 \cdot \underline{g}_1 = \left(\frac{\partial x_i}{\partial u^1}\right)^2, \quad G = \underline{g}_2 \cdot \underline{g}_2 = \left(\frac{\partial x_i}{\partial u^2}\right)^2, \quad F = \underline{g}_1 \cdot \underline{g}_2 = \frac{\partial x_i}{\partial u^1}\,\frac{\partial x_i}{\partial u^2},$$

and with $u^1 = u$ and $u^2 = v$ these are the relations (2.87). Analogously to (4.21), we compute the contravariant metric coefficients as

$$g^{\alpha\beta} := \underline{g}^{\alpha} \cdot \underline{g}^{\beta}. \tag{4.114}$$

Using the orthogonality relations (4.106) and (4.107) and since $\underline{g}_{\alpha} \cdot \underline{n} = \underline{n} \cdot \underline{g}^{\alpha} = 0$, we obtain

$$g_{\alpha\gamma}\,g^{\gamma\beta} = \underline{g}_{\alpha} \cdot \underline{g}_{\gamma}\,\underline{g}^{\gamma} \cdot \underline{g}^{\beta} = \underline{g}_{\alpha} \cdot \left(\underline{\underline{\delta}} - \underline{n}\,\underline{n}\right) \cdot \underline{g}^{\beta} = \underline{g}_{\alpha} \cdot \underline{g}^{\beta} - \underline{g}_{\alpha} \cdot \underline{n}\,\underline{n} \cdot \underline{g}^{\beta} = \delta_{\alpha}^{\beta},$$

i. e. the matrices of the two types of metric coefficients are inverses of each other, so we have

$$g_{\alpha\gamma}\,g^{\gamma\beta} = \delta_{\alpha}^{\beta}. \tag{4.115}$$

Because the metric coefficients on a surface are only 2,2-matrices, we can easily compute the inverse and we obtain

$$g^{11} = \frac{g_{22}}{g}, \qquad g^{22} = \frac{g_{11}}{g}, \qquad g^{12} = g^{21} = -\frac{g_{12}}{g}, \tag{4.116}$$

where g is the determinant of the covariant metric coefficients,

$$g = g_{11}\, g_{22} - g_{12}\, g_{12}. \tag{4.117}$$

Problem 4.21.
Show, using a surface vector as an example, that similarly as in three dimensions, we can again use the metric coefficients $g_{\alpha\beta}$ and $g^{\alpha\beta}$ to raise and lower indices.

Using the definition (4.110) and the Grassmann identity (compare the solution to Problem 2.13 A), we can also write the determinant as

$$g = \underbrace{\underline{g}_1 \cdot \underline{g}_1}_{g_{11}}\ \underbrace{\underline{g}_2 \cdot \underline{g}_2}_{g_{22}} - \underbrace{\underline{g}_1 \cdot \underline{g}_2}_{g_{12}}\ \underbrace{\underline{g}_1 \cdot \underline{g}_2}_{g_{12}} = (\underline{g}_1 \times \underline{g}_2) \cdot (\underline{g}_1 \times \underline{g}_2) = |\underline{g}_1 \times \underline{g}_2|^2$$

or

$$\sqrt{g} = |\underline{g}_1 \times \underline{g}_2|. \tag{4.118}$$

Furthermore, g is also the determinant of the covariant metric coefficients of the moving frame. Since $\underline{g}_1 \times \underline{g}_2$ and $\underline{n}$ are, according to (4.101), collinear and $\underline{n}$ has length 1, it follows with $(4.25)_1$ that

$$\det g_{ij} = [\underline{g}_1, \underline{g}_2, \underline{n}]^2 = (\underline{g}_1 \times \underline{g}_2 \cdot \underline{n})^2 = |\underline{g}_1 \times \underline{g}_2|^2\, |\underline{n}|^2 = |\underline{g}_1 \times \underline{g}_2|^2 = g,$$

$$\det g_{ij} = g > 0. \tag{4.119}$$

3. For a curve defined on a surface, the surface coordinates u^α are functions of a curve parameter s, i. e. we have for the position vector to a point on the curve

$$\underline{x} = \underline{x}\,(u^\alpha(s)) = \underline{x}(s).$$

Using the chain rule and (4.103), we obtain for the vector element tangent to the curve

$$\mathrm{d}\underline{x} = \frac{\mathrm{d}\underline{x}}{\mathrm{d}s}\,\mathrm{d}s = \underline{x}_{,\alpha}\,\frac{\mathrm{d}u^\alpha}{\mathrm{d}s}\,\mathrm{d}s = \underline{g}_\alpha\,\frac{\mathrm{d}u^\alpha}{\mathrm{d}s}\,\mathrm{d}s = \underline{t}\,\mathrm{d}s.$$

Here

$$\underline{t} = \frac{\mathrm{d}\underline{x}}{\mathrm{d}s}, \qquad t^\alpha = \frac{\mathrm{d}u^\alpha}{\mathrm{d}s} \tag{4.120}$$

is the vector tangent to the curve.

To find the length L of a finite curve element between the point A (with the parameter value s_A) and the point B (with the parameter value s_B) we integrate over $|\mathrm{d}\underline{x}| = \sqrt{\mathrm{d}\underline{x} \cdot \mathrm{d}\underline{x}}$. Using (4.111) and (4.120), we get

$$L = \int_{s_A}^{s_B} \sqrt{g_{\alpha\beta}\, t^\alpha\, t^\beta}\, \mathrm{d}s. \tag{4.121}$$

4. For two surface curves intersecting at a point P on the surface, we can compute the angle φ between the two curves from the two tangent vectors $\underline{t}_1$ and $\underline{t}_2$. From

$$\underline{t}_1 \cdot \underline{t}_2 = |\underline{t}_1|\, |\underline{t}_2| \cos\varphi$$

it follows with (4.120) that

$$\cos\varphi = \frac{\underline{t}_1 \cdot \underline{t}_2}{|\underline{t}_1|\, |\underline{t}_2|} = \frac{g_{\alpha\beta}\, \underset{1}{t^\alpha}\, \underset{2}{t^\beta}}{\sqrt{g_{\mu\nu}\, \underset{1}{t^\mu}\, \underset{1}{t^\nu}}\, \sqrt{g_{\rho\sigma}\, \underset{2}{t^\rho}\, \underset{2}{t^\sigma}}}. \tag{4.122}$$

5. We consider two curve elements $\mathrm{d}\underline{x}_1$ and $\mathrm{d}\underline{x}_2$, which are tangent to the u^1- and u^2-coordinate curves,

$$\mathrm{d}\underline{x}_1 = \underline{g}_1\, \mathrm{d}u^1, \quad \mathrm{d}\underline{x}_2 = \underline{g}_2\, \mathrm{d}u^2.$$

The surface area of the parallelogram spanned by the two vectors is given by the length of the vector product

$$\mathrm{d}\underline{A} = \mathrm{d}\underline{x}_1 \times \mathrm{d}\underline{x}_2 = (\underline{g}_1 \times \underline{g}_2)\, \mathrm{d}u^1\, \mathrm{d}u^2,$$

or, using (4.118),

$$\mathrm{d}A = |\underline{g}_1 \times \underline{g}_2|\, \mathrm{d}u^1\, \mathrm{d}u^2 = \sqrt{g}\, \mathrm{d}u^1\, \mathrm{d}u^2.$$

Hence, we compute the surface area A of a finite surface area patch bounded by $u_A^1 \le u^1 \le u_B^1$ and $u_A^2(u^1) \le u^2 \le u_B^2(u^1)$ with the following integral:

$$A = \int_{u_A^1}^{u_B^1} \int_{u_A^2(u^1)}^{u_B^2(u^1)} \sqrt{g}\, \mathrm{d}u^1\, \mathrm{d}u^2. \tag{4.123}$$

We thus derived once more the equation for a surface integral of the first kind from Section 2.14.3; to see this, we compare the last equation with (2.88) and write g using the notations E, F, G from (4.113), so we have $g = E\,G - F^2$.

4.5.4 Curvature Properties

1. Planes are defined by the fact that their normal vector $\underline{n}$ points everywhere in the same direction, i. e. its differential always satisfies $\mathrm{d}\,\underline{n} = \underline{0}$. On a curved surface, the direction of $\underline{n}$ varies from point to point, i. e. we have $\mathrm{d}\,\underline{n} \neq \underline{0}$. This suggests to investigate $\mathrm{d}\,\underline{n}$ to gain further insight into the nature of curvature.

2. According to Section 4.5.1, the normal vector $\underline{n}$ is a function of the surface coordinates u^α, i. e. its differential is given by

$$\mathrm{d}\,\underline{n} = \underline{n}_{,\beta}\,\mathrm{d}\,u^\beta\,. \tag{4.124}$$

We could compute $\underline{n}_{,\beta}$ directly from the defining equation (4.101). However, this is complicated and messy, because we have to use the product rule and the quotient rule several times. It turns out to be more efficient to use some of the properties of the moving frame.

The normal vector $\underline{n}$ is by definition a unit vector, so its scalar product with itself is

$$\underline{n} \cdot \underline{n} = 1.$$

Partial differentiation of this equation with respect to the surface coordinates u^β gives

$$\underline{n}_{,\beta} \cdot \underline{n} = 0\,.$$

Hence the vectors $\underline{n}_{,\beta}$ are perpendicular to $\underline{n}$, i. e. they lie in the tangent plane of the surface at the point P. We can thus represent the $\underline{n}_{,\beta}$ as a linear combination of the basis vectors $\underline{g}_\alpha$ which span this tangent plane. Denoting the coefficients of this linear combination by $-b^\alpha{}_\beta$, we first obtain

$$\underline{n}_{,\beta} = -b^\alpha{}_\beta\,\underline{g}_\alpha. \tag{4.125}$$

We will show that the $b^\alpha{}_\beta$ are the contravariant–covariant coordinates of a second-order surface tensor, which is called the curvature tensor. To do this we compute the scalar product of the differential of the position vector with the differential of the normal vector at the point P, and we obtain, using (4.109), (4.124), and (4.125),

$$-\,\mathrm{d}\,\underline{n}\cdot\mathrm{d}\,\underline{x} = -(\underline{n}_{,\beta}\,\mathrm{d}\,u^\beta)\cdot(\underline{g}_\alpha\,\mathrm{d}\,u^\alpha) = b^\gamma{}_\beta\,\underbrace{\underline{g}_\gamma\cdot\underline{g}_\alpha}_{g_{\gamma\alpha}}\,\mathrm{d}\,u^\alpha\,\mathrm{d}\,u^\beta = \underbrace{b^\gamma{}_\beta\,g_{\gamma\alpha}}_{b_{\alpha\beta}}\,\mathrm{d}\,u^\alpha\,\mathrm{d}\,u^\beta.$$

The left side is the scalar $-\mathrm{d}\,\underline{n}\cdot\mathrm{d}\,\underline{x}$, the coordinate differentials $\mathrm{d}\,u^\alpha$ and $\mathrm{d}\,u^\beta$ are according to (4.108) the contravariant coordinates of a surface vector, and we already know from (4.111) that the metric coefficients $g_{\gamma\alpha}$ are the covariant coordinates of a surface tensor of second order. Hence it follows from the quotient rule that the $b^\gamma{}_\beta$ are also tensor coordinates, or more precisely, the coordinates of a surface tensor of second order. Thus, we can lower the first index using the metric coefficients, and we obtain the covariant coordinates $b_{\alpha\beta}$ of the curvature tensor, i. e. we obtain

$$-\mathrm{d}\underline{n}\cdot\mathrm{d}\underline{x} = b_{\alpha\beta}\,\mathrm{d}u^{\alpha}\,\mathrm{d}u^{\beta}. \tag{4.126}$$

This equation is called the second fundamental form of a surface.

3. To compute the $b_{\alpha\beta}$, we use the fact that $\underline{n}$ is perpendicular to the $\underline{g}_{\alpha}$, i. e. that their scalar product satisfies

$$\underline{n}\cdot\underline{g}_{\alpha} = 0.$$

Taking the partial derivative of this equation with respect to the surface coordinates u^{β} gives

$$\underline{n}_{,\beta}\cdot\underline{g}_{\alpha} = -\underline{n}\cdot\underline{g}_{\alpha,\beta},$$

from which we get by substituting (4.125) for $\underline{n}_{,\beta}$

$$-\,b^{\gamma}{}_{\beta}\,\underbrace{\underline{g}_{\gamma}\cdot\underline{g}_{\alpha}}_{g_{\gamma\alpha}} = -\underline{n}\cdot\underline{g}_{\alpha,\beta},$$

i. e. we obtain

$$b_{\alpha\beta} = \underline{n}\cdot\underline{g}_{\alpha,\beta}. \tag{4.127}$$

If we substitute the definition (4.101) for $\underline{n}$ and use $|\underline{g}_1\times\underline{g}_2| = \sqrt{g}$, we can write with (4.118) the covariant coordinates of the curvature tensor as a triple product, i. e.

$$b_{\alpha\beta} = \frac{1}{\sqrt{g}}\,[\underline{g}_1, \underline{g}_2, \underline{g}_{\alpha,\beta}]. \tag{4.128}$$

According to (2.45), this equation reads in coordinate notation, using the definition (4.103) of the basis vectors $\underline{g}_{\alpha}$,

$$b_{\alpha\beta} = \frac{1}{\sqrt{g}}\,\varepsilon_{ijk}\,\frac{\partial x_i}{\partial u^1}\,\frac{\partial x_j}{\partial u^2}\,\frac{\partial^2 x_k}{\partial u^{\alpha}\,\partial u^{\beta}}. \tag{4.129}$$

The curvature tensor is clearly symmetric, i. e.

$$b_{\alpha\beta} = b_{\beta\alpha}, \tag{4.130}$$

because the order of the partial derivatives with respect to u^{α} and u^{β} does not matter. The covariant coordinates of the curvature tensor are often denoted by

$$b_{11} =: L, \qquad b_{22} =: N, \qquad b_{12} = b_{21} =: M. \tag{4.131}$$

Problem 4.22.
The equation $z = f(x, y) = x\,y$ defines a surface over the Cartesian x, y-plane. Compute the coordinates $b^{\alpha}{}_{\beta}$ and $b_{\alpha\beta}$ of the curvature tensor for this surface.

Hint: Use the Cartesian coordinates x and y as surface parameters.

4. In order to derive a coordinate-independent expression for the curvature, it is reasonable to use the invariants of the curvature tensor. Since the coordinates of the curvature tensor form a 2,2-matrix, we have, according to (3.65), only two main invariants: the trace and the determinant. Both invariants have a name in surface theory; the trace is called mean curvature (after division by 2),

$$H := \frac{1}{2}\, b^{\alpha}{}_{\alpha} = \frac{1}{2}\left(b^{1}{}_{1} + b^{2}{}_{2}\right), \tag{4.132}$$

and the determinant is called Gaussian curvature,

$$K := \det b^{\alpha}{}_{\beta} = \left(b^{1}{}_{1}\, b^{2}{}_{2} - b^{1}{}_{2}\, b^{2}{}_{1}\right). \tag{4.133}$$

Problem 4.23.
Write the mean curvature and the Gaussian curvature in terms of E, F, G, L, M, N.

5. We extend our considerations about the curvature of a surface by investigating curves which result from the intersections of the curved surface and a plane in normal direction, i. e. a plane at the point P on the curved surface, which contains the normal vector; such a plane is also called a normal section of the surface. We again ask for the curvature of such a curve. To answer this question, we find it reasonable to compute how the tangent vector changes along the curve. To obtain comparable results, we request that the tangent vector $\underline{t}$ has always the same length, i. e. that the curve parameter s is scaled so that

$$|\underline{t}| = \left|\frac{d\underline{x}}{ds}\right| = 1.$$

Then we can define the absolute value of the change of the tangent vector as a measure for the curvature κ of the curve. In addition, it makes sense to assign a sign to the curvature, depending on which side of the curve the center of the curvature lies (i. e. the center of the circle, which approximates the curve at the point P under consideration), i. e.

$$\kappa := \pm\left|\frac{d\underline{t}}{ds}\right| = \pm\left|\frac{d^2\underline{x}}{ds^2}\right|.$$

From the scaling it follows that $\underline{t}\cdot\underline{t} = 1$, and hence by differentiating with respect to the curve parameter,

$$\frac{d\underline{t}}{ds}\cdot\underline{t} = 0,$$

i. e. $\mathrm{d}\underline{t}/\mathrm{d}s$ points in the direction of the normal vector $\underline{n}$. Since in addition $|\underline{n}| = 1$, we can write for the curvature

$$\kappa = \frac{\mathrm{d}\underline{t}}{\mathrm{d}s} \cdot \underline{n},$$

which also determines the sign; if $\mathrm{d}\underline{t}/\mathrm{d}s$ and $\underline{n}$ have the same direction, κ is positive; if they have opposite directions, it is negative. Furthermore, it follows from $\underline{t} \cdot \underline{n} = 0$ that

$$\frac{\mathrm{d}\underline{t}}{\mathrm{d}s} \cdot \underline{n} = -\underline{t} \cdot \frac{\mathrm{d}\underline{n}}{\mathrm{d}s},$$

and so we finally obtain with (4.120) the curvature of a normal section as

$$\kappa = -\frac{\mathrm{d}\underline{x}}{\mathrm{d}s} \cdot \frac{\mathrm{d}\underline{n}}{\mathrm{d}s}. \tag{4.134}$$

The plane of a normal section can be rotated arbitrarily around the normal vector. So it makes sense to ask for which planes the curvature attains its maximum and minimum values. The answer to this question is the solution to the extremum problem

$$\kappa = -\frac{\mathrm{d}\underline{x}}{\mathrm{d}s} \cdot \frac{\mathrm{d}\underline{n}}{\mathrm{d}s} \qquad \longrightarrow \qquad \text{extremum}$$

for the tangent vector $\underline{t}$, under the constraint $\underline{t} \cdot \underline{t} = 1$, i. e.

$$1 - \frac{\mathrm{d}\underline{x}}{\mathrm{d}s} \cdot \frac{\mathrm{d}\underline{x}}{\mathrm{d}s} = 0.$$

We introduce a Lagrange multiplier k and the corresponding Lagrange function

$$h = -\frac{\mathrm{d}\underline{x}}{\mathrm{d}s} \cdot \frac{\mathrm{d}\underline{n}}{\mathrm{d}s} + k\left(1 - \frac{\mathrm{d}\underline{x}}{\mathrm{d}s} \cdot \frac{\mathrm{d}\underline{x}}{\mathrm{d}s}\right),$$

substitute the first fundamental form (4.111) and the second fundamental form (4.126), and use (4.120) to obtain

$$h = b_{\alpha\beta} \frac{\mathrm{d}u^\alpha}{\mathrm{d}s} \frac{\mathrm{d}u^\beta}{\mathrm{d}s} + k\left(1 - g_{\alpha\beta} \frac{\mathrm{d}u^\alpha}{\mathrm{d}s} \frac{\mathrm{d}u^\beta}{\mathrm{d}s}\right) = b_{\alpha\beta}\, t^\alpha\, t^\beta + k\,(1 - g_{\alpha\beta}\, t^\alpha\, t^\beta).$$

Thus, the necessary condition $\partial h/\partial t^\gamma = 0$ for the extrema to exist leads, analogously to Section 3.12.3, to the generalized eigenvalue problem

$$(b_{\alpha\beta} - k\, g_{\alpha\beta})\, t^\beta = 0. \tag{4.135}$$

In the theory of surfaces these eigenvalues are called principal curvatures. Problem 3.14 shows that they are always real, because the $b_{\alpha\beta}$ are symmetric and the $g_{\alpha\beta}$ are symmetric and positive definite. The corresponding eigendirections are called principal curvature directions. The principal curvatures k_1 and k_2 are as usual computed by solving the characteristic equation $\det(b_{\alpha\beta} - k\, g_{\alpha\beta}) = 0$. If we use the mixed coordinates $b^\alpha{}_\beta$, we can write the characteristic equation, according to the multiplication theorem (1.13) for determinants, as

$$\det(b_{\alpha\beta} - k\, g_{\alpha\beta}) = \det\left(g_{\alpha\mu}\,(b^\mu{}_\beta - k\,\delta^\mu_\beta)\right) = (\det g_{\alpha\mu})\left(\det(b^\mu{}_\beta - k\,\delta^\mu_\beta)\right) = 0,$$

or, because $\det g_{\alpha\mu} > 0$, also as

$$\det(b^\alpha{}_\beta - k\,\delta^\alpha_\beta)$$

$$= \begin{vmatrix} b^1{}_1 - k & b^1{}_2 \\ b^2{}_1 & b^2{}_2 - k \end{vmatrix} = \underbrace{\left(b^1{}_1\, b^2{}_2 - b^1{}_2\, b^2{}_1\right)}_{K} - \underbrace{\left(b^1{}_1 + b^2{}_2\right)}_{2H} k + k^2 = 0.$$

Using (4.132) for the mean curvature H and (4.133) for the Gaussian curvature K, we get

$$k^2 - 2H\,k + K = 0\,. \tag{4.136}$$

Conversely, Vieta's formulas give

$$H = \frac{1}{2}(k_1 + k_2)\,, \qquad K = k_1\, k_2. \tag{4.137}$$

The principal curvature directions corresponding to the principal curvatures k_1 and k_2 are orthogonal to each other. We see this from the eigenvalue equations

$$b_{\alpha\beta}\, \underset{1}{t}^\beta = k_1\, g_{\alpha\beta}\, \underset{1}{t}^\beta, \quad b_{\alpha\beta}\, \underset{2}{t}^\beta = k_2\, g_{\alpha\beta}\, \underset{2}{t}^\beta,$$

if we multiply each equation by the other eigenvector and subtract the two equations; then we have

$$\underbrace{b_{\alpha\beta}\, \underset{2}{t}^\alpha \underset{1}{t}^\beta - b_{\alpha\beta}\, \underset{1}{t}^\alpha \underset{2}{t}^\beta}_{0} = \underbrace{k_1\, g_{\alpha\beta}\, \underset{2}{t}^\alpha \underset{1}{t}^\beta - k_2\, g_{\alpha\beta}\, \underset{1}{t}^\alpha \underset{2}{t}^\beta}_{(k_1 - k_2)\, g_{\alpha\beta}\, \underset{1}{t}^\alpha \underset{2}{t}^\beta}.$$

The left side is zero due to the symmetry of $b_{\alpha\beta}$; on the right side we can use the symmetry of $g_{\alpha\beta}$. Thus we get for $k_1 \neq k_2$

$$g_{\alpha\beta}\, \underset{1}{t}^\alpha \underset{2}{t}^\beta = 0.$$

With (4.122), however, this is equivalent to $\cos\varphi = 0$, i. e. the angle between the eigendirections $\underset{1}{t}^\alpha$ and $\underset{2}{t}^\beta$ is $\varphi = 90°$.

Problem 4.24.
Find the principal curvatures and the principal curvature directions for the surface $z = f(x, y) = x\,y$ of Problem 4.22.

4.5.5 Derivative Equations, Integrability Conditions

1. In the last section we used the partial derivative $\underline{n}_{,\beta}$ of the normal vector to gather more information about the curvature of a surface. In this section we generalize our considerations to the covariant basis vectors $\underline{g}_1$ and $\underline{g}_2$, i. e. we ask, in all generality, how the moving frame varies along the surface.

2. We can write the partial derivative of a covariant basis using (4.58) in terms of the Christoffel symbols as follows:

$$\underline{g}_{i,j} = \Gamma^m_{ij}\,\underline{g}_m\,.$$

This gives with $\underline{g}_3 = \underline{n}$ for the moving frame of a surface

$$\underline{g}_{\alpha,\beta} = \Gamma^\gamma_{\alpha\beta}\,\underline{g}_\gamma + \Gamma^3_{\alpha\beta}\,\underline{n}, \qquad \underline{n}_{,\beta} = \Gamma^\gamma_{3\beta}\,\underline{g}_\gamma + \Gamma^3_{3\beta}\,\underline{n}. \tag{a}$$

From Section 4.5.4 we already know that $\underline{n}_{,\beta}$ can be represented as a linear combination of the $\underline{g}_\alpha$, so comparing (a) with (4.125) gives

$$\Gamma^\gamma_{3\beta} = -b^\gamma{}_\beta = -g^{\gamma\alpha}\,b_{\alpha\beta}, \qquad \Gamma^3_{3\beta} = 0. \tag{4.138}$$

In the next step we use $\underline{g}_\alpha \cdot \underline{n} = 0$ and after taking the partial derivatives, we have

$$\underline{n}_{,\beta}\cdot\underline{g}_\alpha = -\underline{n}\cdot\underline{g}_{\alpha,\beta}\,.$$

Substituting $\underline{n}_{,\beta}$ and $\underline{g}_{\alpha,\beta}$ and using (a) together with (4.138), we obtain

$$-b^\gamma{}_\beta\,\underbrace{\underline{g}_\gamma\cdot\underline{g}_\alpha}_{g_{\gamma\alpha}} = -\underline{n}\cdot\left(\Gamma^\gamma_{\alpha\beta}\,\underline{g}_\gamma + \Gamma^3_{\alpha\beta}\,\underline{n}\right) = -\Gamma^\gamma_{\alpha\beta}\,\underbrace{\underline{g}_\gamma\cdot\underline{n}}_{0} - \Gamma^3_{\alpha\beta}\,\underbrace{\underline{n}\cdot\underline{n}}_{1},$$

$$b_{\alpha\beta} = \Gamma^3_{\alpha\beta}. \tag{4.139}$$

Thus, we finally obtain for the partial derivative of the moving frame

$$\underline{g}_{\alpha,\beta} = \Gamma^\gamma_{\alpha\beta}\,\underline{g}_\gamma + b_{\alpha\beta}\,\underline{n}, \tag{4.140}$$

$$\underline{n}_{,\beta} = -b^\gamma{}_\beta\,\underline{g}_\gamma. \tag{4.141}$$

These equations are known in the theory of surfaces as the derivative equations of Gauss and Weingarten. The $b_{\alpha\beta}$ are, according to (4.128), the covariant coordinates of the curvature tensor. The Christoffel symbols $\Gamma^\gamma_{\alpha\beta}$ can be computed from (4.60).

3. A by-product of obtaining the derivative equations is that, using (4.139), we can write the coordinates of the curvature tensor in terms of the Christoffel symbols. The

Christoffel symbols themselves are related to the partial derivatives of the metric coefficients according to (4.60), and thus the curvature properties of a surface are already determined by the metric coefficients.

Thus, we can, for a given set of metric coefficients, interpret the derivative equations (4.140) and (4.141) as the determining partial differential equations of the moving frame. However, the metric coefficients cannot be chosen arbitrarily. Since (4.140) and (4.141) consist of a total of eighteen equations for only nine vector coordinates, certain integrability conditions must be satisfied; otherwise the moving frame cannot be consistently determined.

The integrability conditions follow from the interchangeability of the mixed second partial derivatives. From (4.140) we have

$$\underline{g}_{\alpha,\beta\delta} = \left(\Gamma^{\gamma}_{\alpha\beta}\,\underline{g}_{\gamma} + b_{\alpha\beta}\,\underline{n}\right)_{,\delta} = \Gamma^{\gamma}_{\alpha\beta,\delta}\,\underline{g}_{\gamma} + \Gamma^{\gamma}_{\alpha\beta}\,\underline{g}_{\gamma,\delta} + b_{\alpha\beta,\delta}\,\underline{n} + b_{\alpha\beta}\,\underline{n}_{,\delta}.$$

The second term on the right side can be written with the help of renaming indices and equation (4.140) as

$$\Gamma^{\gamma}_{\alpha\beta}\,\underline{g}_{\gamma,\delta} = \Gamma^{\nu}_{\alpha\beta}\,\underline{g}_{\nu,\delta} = \Gamma^{\nu}_{\alpha\beta}\left(\Gamma^{\gamma}_{\nu\delta}\,\underline{g}_{\gamma} + b_{\nu\delta}\,\underline{n}\right),$$

and from (4.141) it further follows that the fourth term equals

$$b_{\alpha\beta}\,\underline{n}_{,\delta} = -b_{\alpha\beta}\,b^{\gamma}{}_{\delta}\,\underline{g}_{\gamma}.$$

Thus we obtain for $\underline{g}_{\alpha,\beta\delta}$, after sorting the coefficients of $\underline{g}_{\gamma}$ and $\underline{n}$,

$$\underline{g}_{\alpha,\beta\delta} = \left(\Gamma^{\gamma}_{\alpha\beta,\delta} + \Gamma^{\nu}_{\alpha\beta}\,\Gamma^{\gamma}_{\nu\delta} - b_{\alpha\beta}\,b^{\gamma}{}_{\delta}\right)\underline{g}_{\gamma} + \left(b_{\alpha\beta,\delta} + \Gamma^{\nu}_{\alpha\beta}\,b_{\nu\delta}\right)\underline{n}.$$

We now exchange β and δ, so

$$\underline{g}_{\alpha,\delta\beta} = \left(\Gamma^{\gamma}_{\alpha\delta,\beta} + \Gamma^{\nu}_{\alpha\delta}\,\Gamma^{\gamma}_{\nu\beta} - b_{\alpha\delta}\,b^{\gamma}{}_{\beta}\right)\underline{g}_{\gamma} + \left(b_{\alpha\delta,\beta} + \Gamma^{\nu}_{\alpha\delta}\,b_{\nu\beta}\right)\underline{n},$$

and subtract the two expressions. Since $\underline{g}_{\alpha,\beta\delta} - \underline{g}_{\alpha,\delta\beta} = \underline{0}$, it follows from the right sides that

$$\underline{0} = \left(\Gamma^{\gamma}_{\alpha\beta,\delta} - \Gamma^{\gamma}_{\alpha\delta,\beta} + \Gamma^{\nu}_{\alpha\beta}\,\Gamma^{\gamma}_{\nu\delta} - \Gamma^{\nu}_{\alpha\delta}\,\Gamma^{\gamma}_{\nu\beta} - \left(b_{\alpha\beta}\,b^{\gamma}{}_{\delta} - b_{\alpha\delta}\,b^{\gamma}{}_{\beta}\right)\right)\underline{g}_{\gamma}$$
$$+ \left(b_{\alpha\beta,\delta} - b_{\alpha\delta,\beta} + \Gamma^{\nu}_{\alpha\beta}\,b_{\nu\delta} - \Gamma^{\nu}_{\alpha\delta}\,b_{\nu\beta}\right)\underline{n}.$$

This equation is satisfied, if the parentheses in front of $\underline{g}_{\gamma}$ and $\underline{n}$ are zero, and this gives the following integrability conditions:

$$\Gamma^{\gamma}_{\alpha\beta,\delta} - \Gamma^{\gamma}_{\alpha\delta,\beta} + \Gamma^{\nu}_{\alpha\beta}\,\Gamma^{\gamma}_{\nu\delta} - \Gamma^{\nu}_{\alpha\delta}\,\Gamma^{\gamma}_{\nu\beta} = b_{\alpha\beta}\,b^{\gamma}{}_{\delta} - b_{\alpha\delta}\,b^{\gamma}{}_{\beta}, \qquad \text{(a)}$$

$$b_{\alpha\beta,\delta} - \Gamma^{\nu}_{\alpha\delta}\,b_{\nu\beta} = b_{\alpha\delta,\beta} - \Gamma^{\nu}_{\alpha\beta}\,b_{\nu\delta}.$$

The expression on the left side of (a) is called (mixed) Riemann curvature tensor of the surface

$$R^{\mu}{}_{\alpha\beta\gamma} := \Gamma^{\mu}_{\alpha\gamma,\beta} - \Gamma^{\mu}_{\alpha\beta,\gamma} + \Gamma^{\nu}_{\alpha\gamma}\,\Gamma^{\mu}_{\nu\beta} - \Gamma^{\nu}_{\alpha\beta}\,\Gamma^{\mu}_{\nu\gamma}, \tag{4.142}$$

i. e. we can write the integrability conditions as

$$R^{\mu}{}_{\alpha\beta\gamma} = b^{\mu}{}_{\beta}\,b_{\alpha\gamma} - b^{\mu}{}_{\gamma}\,b_{\alpha\beta}, \tag{4.143}$$

$$b_{\alpha\beta,\delta} - \Gamma^{\nu}_{\alpha\delta}\,b_{\nu\beta} = b_{\alpha\delta,\beta} - \Gamma^{\nu}_{\alpha\beta}\,b_{\nu\delta}. \tag{4.144}$$

In a three-dimensional Cartesian coordinate system all Christoffel symbols are zero by (4.60), because the metric coefficients $g_{ij} = \delta_{ij}$ are constant. Hence the three-dimensional Riemann curvature tensor $R^{i}{}_{jkl}$ is the zero tensor of fourth order (see Problem 4.18). This statement is also true for all three-dimensional curvilinear coordinate systems, which are defined with respect to a Cartesian coordinate system. However, for a curved surface, i. e. a two-dimensional structure embedded in the three-dimensional space, this statement is not true. If the coordinates of the curvature tensor are nonzero, then it follows from (4.143) that the two-dimensional Riemann curvature tensor $R^{\mu}{}_{\alpha\beta\gamma}$ is a nonzero tensor.

We can bring the integrability condition (4.144) into a more familiar form, if we add the term $\Gamma^{\nu}_{\beta\delta}\,b_{\alpha\nu}$ on both sides and use the symmetry of the Christoffel symbols $\Gamma^{\nu}_{\beta\delta} = \Gamma^{\nu}_{\delta\beta}$ with respect to the lower indices. Then we have

$$\underbrace{b_{\alpha\beta,\delta} - \Gamma^{\nu}_{\alpha\delta}\,b_{\nu\beta} - \Gamma^{\nu}_{\beta\delta}\,b_{\alpha\nu}}_{b_{\alpha\beta}|_{\delta}} = \underbrace{b_{\alpha\delta,\beta} - \Gamma^{\nu}_{\alpha\beta}\,b_{\nu\delta} - \Gamma^{\nu}_{\delta\beta}\,b_{\alpha\nu}}_{b_{\alpha\delta}|_{\beta}}.$$

Comparing this equation with (4.61) shows that it represents the covariant derivatives of the entirely covariant coordinates of the curvature tensor, i. e. instead of (4.144) we can also write

$$b_{\alpha\beta}|_{\gamma} - b_{\alpha\gamma}|_{\beta} = 0. \tag{4.145}$$

In the theory of surfaces (4.145) is called the Mainardi–Codazzi equation.

Problem 4.25.
Obtain the integrability conditions (4.143) and (4.144) from the relation $R^{i}{}_{jkl} = 0$ for the three-dimensional Riemann curvature tensor.

4. If we consider the entirely covariant coordinates of the Riemann curvature tensor, we can make further statements. From $R^{\mu}{}_{\beta\gamma\delta} = b^{\mu}{}_{\gamma}\,b_{\beta\delta} - b^{\mu}{}_{\delta}\,b_{\beta\gamma}$ it follows with (4.143), after multiplication by $g_{\alpha\mu}$, that

$$R_{\alpha\beta\gamma\delta} = b_{\alpha\gamma}\,b_{\beta\delta} - b_{\alpha\delta}\,b_{\beta\gamma}.$$

If we change the order of the first two indices, we get

$$R_{\beta\alpha\gamma\delta} = b_{\beta\gamma}\, b_{\alpha\delta} - b_{\beta\delta}\, b_{\alpha\gamma} = -\left(b_{\alpha\gamma}\, b_{\beta\delta} - b_{\alpha\delta}\, b_{\beta\gamma}\right) = -R_{\alpha\beta\gamma\delta},$$

and correspondingly, if we change the order of the last two indices, we obtain

$$R_{\alpha\beta\delta\gamma} = b_{\alpha\delta}\, b_{\beta\gamma} - b_{\alpha\gamma}\, b_{\beta\delta} = -\left(b_{\alpha\gamma}\, b_{\beta\delta} - b_{\alpha\delta}\, b_{\beta\gamma}\right) = -R_{\alpha\beta\gamma\delta}.$$

Thus, the entirely covariant coordinates of the Riemann curvature tensor are antimetric with respect to the first two indices and also with respect to the last two indices. In other words, 12 of the 16 coordinates are zero, namely the coordinates where the first and second index or the third and fourth index are equal. For the remaining four coordinates we have $R_{1212} = -R_{1221} = R_{2121} = -R_{2112}$, i. e. the entirely covariant coordinates of the Riemann curvature tensor are symmetric also with respect to the first and second index pair. It remains to compute only one coordinate, and we choose R_{1212}, so we write

$$R_{1212} = b_{11}\, b_{22} - b_{12}\, b_{21} = \det b_{\mu\nu}.$$

Using the multiplication theorem (1.13) and with (4.133), we can write this determinant in terms of the Gaussian curvature K, i. e.

$$R_{1212} = \det b_{\mu\nu} = \det\left(b^{\lambda}{}_{\nu}\, g_{\lambda\mu}\right) = \left(\det b^{\lambda}{}_{\nu}\right)\left(\det g_{\lambda\mu}\right) = K\, g.$$

In the last step we used properties of symmetry and antimetry that are similar to those of permutation symbols, so we write

$$R_{\alpha\beta\gamma\delta} = K\, g\, \varepsilon_{\alpha\beta}\, \varepsilon_{\gamma\delta}.$$

However, because of the permutation symbols this is not an equation between tensor coordinates. To make this an equation for tensor coordinates, we first have to introduce the holonomic coordinates of a two-dimensional ε-tensor. For the shared set of index values 1,2 we set $\mu = i,\ \nu = j$; then we can write $\varepsilon_{\mu\nu}$ as $\varepsilon_{\mu\nu} = \varepsilon_{ij3}$. We define the covariant coordinates analogously as

$$e_{\mu\nu} := e_{ij3}. \tag{4.146}$$

Then it follows from $(4.34)_1$ with (4.119) that

$$e_{\mu\nu} = e_{ij3} = \sqrt{g}\, \varepsilon_{ij3} = \sqrt{g}\, \varepsilon_{\mu\nu},$$

i. e. we finally obtain for the Riemann curvature tensor

$$R_{\alpha\beta\gamma\delta} = K\, e_{\alpha\beta}\, e_{\gamma\delta}. \tag{4.147}$$

Problem 4.26.
Show that the $e_{\mu\nu}$ satisfy the transformation law for the covariant coordinates of a second-order surface tensor.

Problem 4.27.
Compute, using the Christoffel symbols, the Gaussian curvature of a spherical surface and of a cylindrical surface.

5. For completeness it remains to check if further integrability conditions can be obtained from (4.141). We proceed similarly to No. 3 and compute

$$\underline{n}_{,\beta\delta} = \left(-b^{\gamma}{}_{\beta}\,\underline{g}_{\gamma}\right)_{,\delta} = -b^{\gamma}{}_{\beta,\delta}\,\underline{g}_{\gamma} - b^{\gamma}{}_{\beta}\,\underline{g}_{\gamma,\delta}.$$

Substituting (4.140) and sorting with respect to the $\underline{g}_{\gamma}$ and $\underline{n}$ gives

$$\underline{n}_{,\beta\delta} = -\left(b^{\gamma}{}_{\beta,\delta} + b^{\nu}{}_{\beta}\,\Gamma^{\gamma}_{\nu\delta}\right)\underline{g}_{\gamma} - b^{\nu}{}_{\beta}\,b_{\nu\delta}\,\underline{n}.$$

We again exchange β and δ, so

$$\underline{n}_{,\delta\beta} = -\left(b^{\gamma}{}_{\delta,\beta} + b^{\nu}{}_{\delta}\,\Gamma^{\gamma}_{\nu\beta}\right)\underline{g}_{\gamma} - b^{\nu}{}_{\delta}\,b_{\nu\beta}\,\underline{n},$$

and subtract the two expressions. Because $\underline{n}_{,\beta\delta} - \underline{n}_{,\delta\beta} = \underline{0}$ it follows from the right sides that

$$\underline{0} = -\left(b^{\gamma}{}_{\beta,\delta} - b^{\gamma}{}_{\delta,\beta} + b^{\nu}{}_{\beta}\,\Gamma^{\gamma}_{\nu\delta} - b^{\nu}{}_{\delta}\,\Gamma^{\gamma}_{\nu\beta}\right)\underline{g}_{\gamma} - \left(b^{\nu}{}_{\beta}\,b_{\nu\delta} - b^{\nu}{}_{\delta}\,b_{\nu\beta}\right)\underline{n},$$

i. e.

$$b^{\gamma}{}_{\beta,\delta} + \Gamma^{\gamma}_{\nu\delta}\,b^{\nu}{}_{\beta} = b^{\gamma}{}_{\delta,\beta} + \Gamma^{\gamma}_{\nu\beta}\,b^{\nu}{}_{\delta}, \tag{a}$$

and

$$b_{\nu\delta}\,b^{\nu}{}_{\beta} - b_{\nu\beta}\,b^{\nu}{}_{\delta} = 0 \tag{b}$$

must hold. Again we can write (a) in terms of covariant derivatives, if we subtract $\Gamma^{\nu}_{\beta\delta}\,b^{\gamma}{}_{\nu}$ on both sides and take $\Gamma^{\nu}_{\beta\delta} = \Gamma^{\nu}_{\delta\beta}$ into account. Then we have

$$\underbrace{b^{\gamma}{}_{\beta,\delta} + \Gamma^{\gamma}_{\nu\delta}\,b^{\nu}{}_{\beta} - \Gamma^{\nu}_{\beta\delta}\,b^{\gamma}{}_{\nu}}_{b^{\gamma}{}_{\beta}|_{\delta}} = \underbrace{b^{\gamma}{}_{\delta,\beta} + \Gamma^{\gamma}_{\nu\beta}\,b^{\nu}{}_{\delta} - \Gamma^{\nu}_{\delta\beta}\,b^{\gamma}{}_{\nu}}_{b^{\gamma}{}_{\delta}|_{\beta}}.$$

This equation is equivalent to (4.145), with the only difference that here we used the mixed coordinates of the curvature tensor.

We can write (b), by comparing it with (4.143), in terms of the Riemann curvature tensor. Thus we get with (4.147)

$$0 = R^{\nu}{}_{\nu\beta\delta} = g^{\mu\nu} R_{\mu\nu\beta\delta} = K \underbrace{g^{\mu\nu} e_{\mu\nu}}_{0} e_{\beta\delta}.$$

This equation is satisfied since $g^{\mu\nu}$ is symmetric and $e_{\mu\nu}$ is antimetric.

Thus, no further integrability conditions can be obtained from (4.141).

5 Representation of Tensor Functions

5.1 The Basic Idea of the Representation of Tensor Functions

1. Modeling a physical process often means to find a function relating two tensors. From the context we can assume that such a function exists, but it is not known and it can often be determined only from experiments. Typical examples of such functions are the stress–displacement relation $\underline{\underline{T}} = \underline{\underline{f}}(\underline{\underline{D}})$ for an elastic solid, where the (symmetric) stress tensor $\underline{\underline{T}}$ depends on an (also symmetric) displacement tensor $\underline{\underline{D}}$; or the yield-stress condition $\sigma = f(\underline{\underline{T}})$ in plasticity theory, which relates the experimentally determined yield-stress σ to the three-dimensional stress state described by the stress tensor $\underline{\underline{T}}$.

The functions f or $\underline{\underline{f}}$ in the examples are called, according to the order of the tensors, scalar-valued or tensor-valued tensor functions. Contrary to what one may expect, these functions cannot have an arbitrary form, because the transformation properties of a tensor impose certain restrictions. For example, if in a given Cartesian coordinate system $\sigma = f(T_{ij})$ and $T_{ij} = f_{ij}(D_{kl})$, then, under a change of the coordinate system, substituting the transformed coordinates $\tilde{T}_{ij}$ into the function f must result in the same scalar $\sigma = f(\tilde{T}_{ij})$, and similarly, substituting the transformed coordinates $\tilde{D}_{kl}$ into the function $\underline{\underline{f}}$ must give the transformed coordinates $\tilde{T}_{ij} = f_{ij}(\widetilde{D}_{kl})$. If we assume that the tensors are polar, then, according to the transformation equations (2.17), a scalar-valued function f must satisfy

$$f(T_{mn}) = f(\alpha_{im}\,\alpha_{jn}\,T_{ij}), \tag{a}$$

and a tensor-valued function $\underline{\underline{f}}$ is restricted by the condition

$$\alpha_{mi}\,\alpha_{nj}\,f_{mn}(D_{kl}) = f_{ij}(\alpha_{pk}\,\alpha_{ql}\,D_{pq}). \tag{b}$$

2. How to evaluate conditions of the type (a) or (b) and how to use them to construct functions relating tensors is the subject of a theory called the theory of the representation of tensor functions. The basic idea is that a scalar-valued function cannot depend on single tensor coordinates, but only on scalar combinations of tensor coordinates, which are clearly invariant under a coordinate transformation. This explains why, instead of the theory of the representation of tensor functions, we also speak of the theory of invariants. We will show that for any symmetric tensor, such as the stress tensor $\underline{\underline{T}}$, the three basic invariants tr $\underline{\underline{T}}$, tr $\underline{\underline{T}}^2$, and tr $\underline{\underline{T}}^3$ form a complete set of invariants, so we can write the scalar-valued function f in our example as a function of the three basic invariants (or, according to Section 3.15, of another equivalent set of invariants, such as the eigenvalues or the main invariants), so we have

$$\sigma = f(\operatorname{tr}\underline{\underline{T}}, \operatorname{tr}\underline{\underline{T}}^2, \operatorname{tr}\underline{\underline{T}}^3).$$

https://doi.org/10.1515/9783110404265-005

These considerations also apply to tensor-valued functions, if we make them scalar-valued by introducing an auxiliary tensor. In our example, then $\underline{\underline{T}} = \underline{\underline{f}}(\underline{\underline{D}})$ gives temporarily

$$\underline{\underline{T}} \cdot\cdot \underline{\underline{H}} = f(\underline{\underline{D}}, \underline{\underline{H}}),$$

where the auxiliary tensor $\underline{\underline{H}}$ must appear in such a way that it can be canceled out in the end.

Before we turn to the details, we have to clarify how many tensor invariants are needed for a (yet to be defined) representation to be complete. For this we first need to generalize the Cayley–Hamilton theorem, which in its basic version relates only powers of the same tensor, while in the theory of the representation of tensor functions, products of different tensors often appear as well. Since the tensors in physical applications are often polar, we restrict ourselves to polar tensors for the rest of this chapter (this includes scalars and vectors as tensors of order zero and one, respectively), without mentioning this every time. In the few cases where we consider axial tensors, or if the distinction between polar and axial tensors is important for the argument, we will mention this explicitly.

3. In physics, representation of tensor functions is as important as dimensional analysis. Physical quantities are the product of a numerical value and a unit, but the quantity itself is independent of the chosen unit, i. e. if we change the unit, the numerical value changes correspondingly. This property carries over to equations between physical quantities, where we say that an equation must be dimensionally homogeneous. Specifically, it follows that physical quantities cannot be related by an arbitrary function, but only by functions which are invariant under a change of units. This condition of invariance under a change of units has its analog for tensors in the relations (a) and (b), which result from the requirement that an equation must be invariant under a change of the coordinate system.

In dimensional analysis we use the condition of invariance under a change of units to replace relations between quantities by relations between dimensionless combinations of quantities (which do not need units). In many cases, this leads to restrictions on the permissible functions. Analogously, in the theory of the representation of tensor functions, we change to scalar combinations of tensor coordinates (i. e. tensor invariants), which again leads to restrictions on the permissible functions.

5.2 Generalized Cayley–Hamilton Theorem

1. According to (1.20), $\delta^{ijkl}_{pqrs} = 0$, and then (1.19) gives

$$\begin{vmatrix} \delta_{ip} & \delta_{iq} & \delta_{ir} & \delta_{is} \\ \delta_{jp} & \delta_{jq} & \delta_{jr} & \delta_{js} \\ \delta_{kp} & \delta_{kq} & \delta_{kr} & \delta_{ks} \\ \delta_{lp} & \delta_{lq} & \delta_{lr} & \delta_{ls} \end{vmatrix} = 0.$$

Contraction with $a_{pi}\, b_{qj}\, c_{rk}$ gives, if we multiply the first row by a_{pi}, the second row by b_{qj}, and the third row by c_{rk},

$$\begin{vmatrix} a_{pp} & a_{pq} & a_{pr} & a_{ps} \\ b_{qp} & b_{qq} & b_{qr} & b_{qs} \\ c_{rp} & c_{rq} & c_{rr} & c_{rs} \\ \delta_{lp} & \delta_{lq} & \delta_{lr} & \delta_{ls} \end{vmatrix} = 0.$$

We expand this determinant along the last row, so we obtain

$$-\delta_{lp} \begin{vmatrix} a_{pq} & a_{pr} & a_{ps} \\ b_{qq} & b_{qr} & b_{qs} \\ c_{rq} & c_{rr} & c_{rs} \end{vmatrix} + \delta_{lq} \begin{vmatrix} a_{pp} & a_{pr} & a_{ps} \\ b_{qp} & b_{qr} & b_{qs} \\ c_{rp} & c_{rr} & c_{rs} \end{vmatrix} \\ - \delta_{lr} \begin{vmatrix} a_{pp} & a_{pq} & a_{ps} \\ b_{qp} & b_{qq} & b_{qs} \\ c_{rp} & c_{rq} & c_{rs} \end{vmatrix} + \delta_{ls} \begin{vmatrix} a_{pp} & a_{pq} & a_{pr} \\ b_{qp} & b_{qq} & b_{qr} \\ c_{rp} & c_{rq} & c_{rr} \end{vmatrix} = 0.$$

Multiplying δ_{lp}, δ_{lq}, and δ_{lr} into the first, second, and third row of the first three determinants, respectively, gives

$$-\begin{vmatrix} a_{lq} & a_{lr} & a_{ls} \\ b_{qq} & b_{qr} & b_{qs} \\ c_{rq} & c_{rr} & c_{rs} \end{vmatrix} + \begin{vmatrix} a_{pp} & a_{pr} & a_{ps} \\ b_{lp} & b_{lr} & b_{ls} \\ c_{rp} & c_{rr} & c_{rs} \end{vmatrix} - \begin{vmatrix} a_{pp} & a_{pq} & a_{ps} \\ b_{qp} & b_{qq} & b_{qs} \\ c_{lp} & c_{lq} & c_{ls} \end{vmatrix} \\ + \delta_{ls} \begin{vmatrix} a_{pp} & a_{pq} & a_{pr} \\ b_{qp} & b_{qq} & b_{qr} \\ c_{rp} & c_{rq} & c_{rr} \end{vmatrix} = 0.$$

We expand each of the determinants and obtain

$$\begin{aligned} &- a_{lq}\, b_{qr}\, c_{rs} - a_{lr}\, b_{qs}\, c_{rq} - a_{ls}\, b_{qq}\, c_{rr} + a_{ls}\, b_{qr}\, c_{rq} + a_{lr}\, b_{qq}\, c_{rs} + a_{lq}\, b_{qs}\, c_{rr} \\ &+ a_{pp}\, b_{lr}\, c_{rs} + a_{pr}\, b_{ls}\, c_{rp} + a_{ps}\, b_{lp}\, c_{rr} - a_{ps}\, b_{lr}\, c_{rp} - a_{pr}\, b_{lp}\, c_{rs} - a_{pp}\, b_{ls}\, c_{rr} \\ &- a_{pp}\, b_{qq}\, c_{ls} - a_{pq}\, b_{qs}\, c_{lp} - a_{ps}\, b_{qp}\, c_{lq} + a_{ps}\, b_{qq}\, c_{lp} + a_{pq}\, b_{qp}\, c_{ls} + a_{pp}\, b_{qs}\, c_{lq} \\ &+ \delta_{ls}\, (a_{pp}\, b_{qq}\, c_{rr} + a_{pq}\, b_{qr}\, c_{rp} + a_{pr}\, b_{qp}\, c_{rq} - a_{pr}\, b_{qq}\, c_{rp} - a_{pq}\, b_{qp}\, c_{rr} \\ &\qquad - a_{pp}\, b_{qr}\, c_{rq}) = 0. \end{aligned}$$

Translating this expression into symbolic notation gives

$$\begin{aligned}
&-\underline{\underline{a}}\cdot\underline{\underline{b}}\cdot\underline{\underline{c}}-\underline{\underline{a}}\cdot\underline{\underline{c}}\cdot\underline{\underline{b}}-\underline{\underline{a}}\operatorname{tr}\underline{\underline{b}}\operatorname{tr}\underline{\underline{c}}+\underline{\underline{a}}\operatorname{tr}(\underline{\underline{b}}\cdot\underline{\underline{c}})+\underline{\underline{a}}\cdot\underline{\underline{c}}\operatorname{tr}\underline{\underline{b}}+\underline{\underline{a}}\cdot\underline{\underline{b}}\operatorname{tr}\underline{\underline{c}}\\
&+\underline{\underline{b}}\cdot\underline{\underline{c}}\operatorname{tr}\underline{\underline{a}}+\underline{\underline{b}}\operatorname{tr}(\underline{\underline{a}}\cdot\underline{\underline{c}})+\underline{\underline{b}}\cdot\underline{\underline{a}}\operatorname{tr}\underline{\underline{c}}-\underline{\underline{b}}\cdot\underline{\underline{c}}\cdot\underline{\underline{a}}-\underline{\underline{b}}\cdot\underline{\underline{a}}\cdot\underline{\underline{c}}-\underline{\underline{b}}\operatorname{tr}\underline{\underline{a}}\operatorname{tr}\underline{\underline{c}}\\
&-\underline{\underline{c}}\operatorname{tr}\underline{\underline{a}}\operatorname{tr}\underline{\underline{b}}-\underline{\underline{c}}\cdot\underline{\underline{a}}\cdot\underline{\underline{b}}-\underline{\underline{c}}\cdot\underline{\underline{b}}\cdot\underline{\underline{a}}+\underline{\underline{c}}\cdot\underline{\underline{a}}\operatorname{tr}\underline{\underline{b}}+\underline{\underline{c}}\operatorname{tr}(\underline{\underline{a}}\cdot\underline{\underline{b}})+\underline{\underline{c}}\cdot\underline{\underline{b}}\operatorname{tr}\underline{\underline{a}}\\
&+\underline{\underline{\delta}}\big(\operatorname{tr}\underline{\underline{a}}\operatorname{tr}\underline{\underline{b}}\operatorname{tr}\underline{\underline{c}}+\operatorname{tr}(\underline{\underline{a}}\cdot\underline{\underline{b}}\cdot\underline{\underline{c}})+\operatorname{tr}(\underline{\underline{a}}\cdot\underline{\underline{c}}\cdot\underline{\underline{b}})-\operatorname{tr}(\underline{\underline{a}}\cdot\underline{\underline{c}})\operatorname{tr}\underline{\underline{b}}-\operatorname{tr}(\underline{\underline{a}}\cdot\underline{\underline{b}})\operatorname{tr}\underline{\underline{c}}\\
&\qquad\qquad\qquad\qquad\qquad\qquad\qquad\qquad\qquad-\operatorname{tr}(\underline{\underline{b}}\cdot\underline{\underline{c}})\operatorname{tr}\underline{\underline{a}}\big)=\underline{\underline{0}}.
\end{aligned}$$

Finally, after sorting these terms and reversing the signs, we obtain

$$\begin{aligned}
&\underline{\underline{a}}\cdot\left(\underline{\underline{b}}\cdot\underline{\underline{c}}+\underline{\underline{c}}\cdot\underline{\underline{b}}\right)+\underline{\underline{b}}\cdot\left(\underline{\underline{a}}\cdot\underline{\underline{c}}+\underline{\underline{c}}\cdot\underline{\underline{a}}\right)+\underline{\underline{c}}\cdot\left(\underline{\underline{a}}\cdot\underline{\underline{b}}+\underline{\underline{b}}\cdot\underline{\underline{a}}\right)\\
&-\left(\underline{\underline{b}}\cdot\underline{\underline{c}}+\underline{\underline{c}}\cdot\underline{\underline{b}}\right)\operatorname{tr}\underline{\underline{a}}-\left(\underline{\underline{a}}\cdot\underline{\underline{c}}+\underline{\underline{c}}\cdot\underline{\underline{a}}\right)\operatorname{tr}\underline{\underline{b}}-\left(\underline{\underline{a}}\cdot\underline{\underline{b}}+\underline{\underline{b}}\cdot\underline{\underline{a}}\right)\operatorname{tr}\underline{\underline{c}}\\
&+\left[\operatorname{tr}\underline{\underline{b}}\operatorname{tr}\underline{\underline{c}}-\operatorname{tr}(\underline{\underline{b}}\cdot\underline{\underline{c}})\right]\underline{\underline{a}}+\left[\operatorname{tr}\underline{\underline{a}}\operatorname{tr}\underline{\underline{c}}-\operatorname{tr}(\underline{\underline{a}}\cdot\underline{\underline{c}})\right]\underline{\underline{b}}\\
&\qquad\qquad\qquad\qquad\qquad+\left[\operatorname{tr}\underline{\underline{a}}\operatorname{tr}\underline{\underline{b}}-\operatorname{tr}(\underline{\underline{a}}\cdot\underline{\underline{b}})\right]\underline{\underline{c}}\\
&-\big[\operatorname{tr}\underline{\underline{a}}\operatorname{tr}\underline{\underline{b}}\operatorname{tr}\underline{\underline{c}}+\operatorname{tr}(\underline{\underline{a}}\cdot\underline{\underline{b}}\cdot\underline{\underline{c}})+\operatorname{tr}(\underline{\underline{c}}\cdot\underline{\underline{b}}\cdot\underline{\underline{a}})\\
&\qquad-\operatorname{tr}(\underline{\underline{b}}\cdot\underline{\underline{c}})\operatorname{tr}\underline{\underline{a}}-\operatorname{tr}(\underline{\underline{a}}\cdot\underline{\underline{c}})\operatorname{tr}\underline{\underline{b}}-\operatorname{tr}(\underline{\underline{a}}\cdot\underline{\underline{b}})\operatorname{tr}\underline{\underline{c}}\big]\underline{\underline{\delta}}=\underline{\underline{0}}.
\end{aligned}\tag{5.1}$$

2. If we set $\underline{\underline{a}}=\underline{\underline{b}}=\underline{\underline{c}}$, then (5.1) simplifies to

$$6\underline{\underline{a}}^3-6\operatorname{tr}\underline{\underline{a}}\,\underline{\underline{a}}^2+3\left[\operatorname{tr}^2\underline{\underline{a}}-\operatorname{tr}\underline{\underline{a}}^2\right]\underline{\underline{a}}-\left[\operatorname{tr}^3\underline{\underline{a}}-3\operatorname{tr}\underline{\underline{a}}\operatorname{tr}\underline{\underline{a}}^2+2\operatorname{tr}\underline{\underline{a}}^3\right]\underline{\underline{\delta}}=0.$$

After dividing by −6 and using (3.98) we see that this is the Cayley–Hamilton theorem (3.94), so we call (5.1) the generalized Cayley–Hamilton theorem.

3. For the representation of tensor functions, the first row of (5.1) is of particular importance, so we introduce the following symbol:

$$\underline{\underline{\Sigma}}:=\underline{\underline{a}}\cdot\left(\underline{\underline{b}}\cdot\underline{\underline{c}}+\underline{\underline{c}}\cdot\underline{\underline{b}}\right)+\underline{\underline{b}}\cdot\left(\underline{\underline{a}}\cdot\underline{\underline{c}}+\underline{\underline{c}}\cdot\underline{\underline{a}}\right)+\underline{\underline{c}}\cdot\left(\underline{\underline{a}}\cdot\underline{\underline{b}}+\underline{\underline{b}}\cdot\underline{\underline{a}}\right).\tag{5.2}$$

Then we obtain from (5.1), after solving for the first row, i. e.

$$\begin{aligned}
\underline{\underline{\Sigma}}=&\left(\underline{\underline{b}}\cdot\underline{\underline{c}}+\underline{\underline{c}}\cdot\underline{\underline{b}}\right)\operatorname{tr}\underline{\underline{a}}+\left(\underline{\underline{a}}\cdot\underline{\underline{c}}+\underline{\underline{c}}\cdot\underline{\underline{a}}\right)\operatorname{tr}\underline{\underline{b}}+\left(\underline{\underline{a}}\cdot\underline{\underline{b}}+\underline{\underline{b}}\cdot\underline{\underline{a}}\right)\operatorname{tr}\underline{\underline{c}}\\
&-\left[\operatorname{tr}\underline{\underline{b}}\operatorname{tr}\underline{\underline{c}}-\operatorname{tr}(\underline{\underline{b}}\cdot\underline{\underline{c}})\right]\underline{\underline{a}}-\left[\operatorname{tr}\underline{\underline{a}}\operatorname{tr}\underline{\underline{c}}-\operatorname{tr}(\underline{\underline{a}}\cdot\underline{\underline{c}})\right]\underline{\underline{b}}\\
&\qquad\qquad\qquad\qquad\qquad-\left[\operatorname{tr}\underline{\underline{a}}\operatorname{tr}\underline{\underline{b}}-\operatorname{tr}(\underline{\underline{a}}\cdot\underline{\underline{b}})\right]\underline{\underline{c}}\\
&+\big[\operatorname{tr}\underline{\underline{a}}\operatorname{tr}\underline{\underline{b}}\operatorname{tr}\underline{\underline{c}}+\operatorname{tr}(\underline{\underline{a}}\cdot\underline{\underline{b}}\cdot\underline{\underline{c}})+\operatorname{tr}(\underline{\underline{c}}\cdot\underline{\underline{b}}\cdot\underline{\underline{a}})\\
&\qquad-\operatorname{tr}(\underline{\underline{b}}\cdot\underline{\underline{c}})\operatorname{tr}\underline{\underline{a}}-\operatorname{tr}(\underline{\underline{a}}\cdot\underline{\underline{c}})\operatorname{tr}\underline{\underline{b}}-\operatorname{tr}(\underline{\underline{a}}\cdot\underline{\underline{b}})\operatorname{tr}\underline{\underline{c}}\big]\underline{\underline{\delta}}.
\end{aligned}\tag{5.3}$$

5.3 Invariants of Vectors and Second-Order Tensors

Let us recall our definition of an invariant. An invariant is a scalar combination of tensor coordinates which (for polar scalars) does not change under a change of the coordinate system, or (for axial scalars) at most its sign changes. Following our definition from Section 5.1, however, we use the term invariant in this chapter only for polar scalars.

Problem 5.1.
Show, using the transformation equations (2.17) or (2.18), that tr $\underline{\underline{T}}$ and tr $\underline{\underline{T}}^2$ are invariants of a (polar or axial) second-order tensor.

In Section 3.15, we saw that the eigenvalues, the main invariants, and the basic invariants are three different sets of three invariants of a second-order tensor. The elements of each of these three sets are independent of each other, i. e. no element of a set can be computed from the other two elements of the same set; but the three sets are not independent of each other, i. e. if we know the invariants of one set, then we can compute the invariants of the other sets. However, it is not clear if the three invariants of a set form a complete set, i. e. if further invariants exist, which cannot be computed from these invariants. A complete set of invariants is called a basis (of invariants).

The easiest way to compute invariants is from appropriate contractions of tensor coordinates. An example are the main invariants of a second-order tensor. Such invariants are polynomials of tensor coordinates and we call them polynomial invariants. In the theory of the representation of tensor functions we usually consider only polynomial invariants and we use the following terms. A polynomial invariant which can be written as a polynomial of other polynomial invariants is called reducible (with respect to these invariants); otherwise it is called irreducible. A complete set of irreducible invariants, i. e. a set such that all other polynomial invariants can be written in terms of these irreducible invariants, is called an integrity basis. However, it is also possible that polynomial invariants are related by a polynomial and yet none of them can be written *as a polynomial* of the other invariants. Such polynomials are called syzygies;[1] an example is equation (2.46) for the scalar products of the six vectors $\underline{a}$, $\underline{b}$, $\underline{c}$, $\underline{d}$, $\underline{e}$, and $\underline{f}$.

An example of nonpolynomial invariants are the eigenvalues of a second-order tensor. We computed them in Section 3.11.2 from a cubic equation whose coefficients are polynomial invariants, but an eigenvalue cannot be written as a polynomial of tensor coordinates.

---

1 Greek: pair (from *syn* [together] and *zygon* [yoke]), i. e. literally, a pair coupled together by a yoke, e. g. in astronomy a generic term for conjunction and opposition of two planets, and in poetry the juxtaposition of two equal metrical feet.

The polynomial invariants are a subset of all the invariants of a tensor. Since reducible invariants are polynomial combinations of tensor coordinates, it may still happen that there are more irreducible invariants than independent invariants. In other words, an integrity basis can contain more elements than a basis.

5.3.1 Invariants of Vectors

A single vector $\underline{u}$ has only one irreducible invariant, i. e. its square $u_i\, u_i$.

From two vectors $\underline{u}$ and $\underline{v}$ we can form a total of three irreducible invariants:

- the squares $u_i\, u_i$ and $v_i\, v_i$ of each of the vectors;
- the scalar product $u_i\, v_i$.

The scalar product is an example of a so-called simultaneous invariant, i. e. an invariant which consists of coordinates of different tensors.

As the number of vectors increases, the number of irreducible invariants quickly increases. For three vectors $\underline{u}$, $\underline{v}$, $\underline{w}$, there are already seven independent scalar combinations:

- the squares $u_i u_i$, $v_i v_i$, $w_i\, w_i$, of each of the vectors;
- the scalar products $u_i v_i$, $u_i w_i$, $v_i w_i$, of each of the vector pairs;
- the triple product $\varepsilon_{ijk}\, u_i\, v_j\, w_k$ of all three vectors.

The triple product plays a particular role. If the vectors $\underline{u}$, $\underline{v}$, $\underline{w}$ are polar, then the first six invariants are also polar scalars, whereas the triple product is, due to the ε-tensor, an axial scalar. However, according to our definition, an axial scalar is no invariant, i. e. from three (polar) vectors $\underline{u}$, $\underline{v}$, $\underline{w}$, we can form six irreducible invariants.

5.3.2 Independent Invariants of a Second-Order Tensor

1. Let us first determine the number of independent invariants of an arbitrary second-order tensor $\underline{\underline{T}}$. To this end, we decompose the tensor into its symmetric part $\underline{\underline{S}}$ and its antimetric part $\underline{\underline{A}}$. We can write the antimetric part $\underline{\underline{A}}$, according to Section 3.3, in terms of its corresponding vector $\underline{b}$, so we have

$$T_{ij} = S_{ij} + A_{ij} = S_{ij} + \varepsilon_{ijk}\, b_k .$$

If we assume that $\underline{\underline{T}}$ is polar, then $\underline{\underline{S}}$ and $\underline{\underline{A}}$ are also polar and $\underline{b}$ is axial.

The decomposition into a symmetric part and an antimetric part is, according to Section 2.6, No. 7, independent of the coordinate system. So we can also use the principal axis system of $\underline{\underline{S}}$ to represent the coordinates of the full tensor $\underline{\underline{T}}$, and thus we have for its coordinate matrix

$$\tilde{T}_{ij} = \begin{pmatrix} \sigma_1 & 0 & 0 \\ 0 & \sigma_2 & 0 \\ 0 & 0 & \sigma_3 \end{pmatrix} + \begin{pmatrix} 0 & \beta_3 & -\beta_2 \\ -\beta_3 & 0 & \beta_1 \\ \beta_2 & -\beta_1 & 0 \end{pmatrix},$$

$$\tilde{T}_{ij} = \begin{pmatrix} \sigma_1 & \beta_3 & -\beta_2 \\ -\beta_3 & \sigma_2 & \beta_1 \\ \beta_2 & -\beta_1 & \sigma_3 \end{pmatrix}. \tag{5.4}$$

Here the σ_i are the (not necessarily different) eigenvalues of $\underline{\underline{S}}$ and the β_i are the coordinates of $\underline{b}$ in the principal axis system of $\underline{\underline{S}}$. We already know from Section 3.11.2 that the eigenvalues σ_i are invariants. The coordinates of $\underline{b}$ can be interpreted as the projections of the vector $\underline{b}$ onto the principal axes of $\underline{\underline{S}}$; hence the β_i are also invariants, because the direction of the principal axes of $\underline{\underline{S}}$ and the direction of $\underline{b}$ are independent of the coordinate system.

Thus the principal axis system of the symmetric part $\underline{\underline{S}}$ is a distinguished coordinate system of the tensor $\underline{\underline{T}}$; the coordinate matrix of $\underline{\underline{T}}$ is then (and only then) antimetric, except on the main diagonal.

We conclude that, in general, a second-order tensor has six independent invariants; more than six independent invariants are not possible, because the tensor is uniquely determined by the σ_i and β_i. However, these invariants are not polynomial invariants, i. e. they cannot be written as polynomials of tensor coordinates in an arbitrary Cartesian coordinate system.

2. If the symmetric part $\underline{\underline{S}}$ has multiple eigenvalues, the number of independent invariants is reduced.

For an eigenvalue of multiplicity two, we choose the x, y-plane as the eigenplane of the principal axis system and we put the x-axis such that $\underline{b}$ lies in the x, z-plane. Then the tensor $\underline{\underline{T}}$ has the coordinate matrix

$$\tilde{T}_{ij} = \begin{pmatrix} \sigma_1 & 0 & 0 \\ 0 & \sigma_1 & 0 \\ 0 & 0 & \sigma_3 \end{pmatrix} + \begin{pmatrix} 0 & \beta_3 & 0 \\ -\beta_3 & 0 & \beta_1 \\ 0 & -\beta_1 & 0 \end{pmatrix} = \begin{pmatrix} \sigma_1 & \beta_3 & 0 \\ -\beta_3 & \sigma_1 & \beta_1 \\ 0 & -\beta_1 & \sigma_3 \end{pmatrix},$$

i. e. only four independent invariants exist.

For an eigenvalue of multiplicity three, every coordinate system is also a principal axis system of $\underline{\underline{S}}$. So we can put the x-axis in the direction of $\underline{b}$ and we get for the coordinate matrix of $\underline{\underline{T}}$

$$\tilde{T}_{ij} = \begin{pmatrix} \sigma & 0 & 0 \\ 0 & \sigma & 0 \\ 0 & 0 & \sigma \end{pmatrix} + \begin{pmatrix} 0 & 0 & 0 \\ 0 & 0 & \beta \\ 0 & -\beta & 0 \end{pmatrix} = \begin{pmatrix} \sigma & 0 & 0 \\ 0 & \sigma & \beta \\ 0 & -\beta & \sigma \end{pmatrix},$$

i. e. $\underline{\underline{T}}$ has only two independent invariants.

3. If the tensor is symmetric, we have $\underline{\underline{A}} = \underline{\underline{0}}$. A symmetric tensor has at most three independent invariants, namely if all its eigenvalues are different. If the tensor has an eigenvalue of multiplicity two, it has only two invariants; and if it has an eigenvalue of multiplicity three, only one invariant exists.

If the tensor is antimetric, we have $\underline{\underline{S}} = \underline{\underline{0}}$. An antimetric tensor has only one independent invariant, i. e. the square of the corresponding vector.

In addition, we consider orthogonal tensors. In a coordinate system in which the z-axis is the axis of rotation or the axis of rotary reflection, an orthogonal tensor has, according to (3.75) or (3.76), the coordinate matrix

$$\widetilde{R}_{ij} = \begin{pmatrix} \cos\vartheta & -\sin\vartheta & 0 \\ \sin\vartheta & \cos\vartheta & 0 \\ 0 & 0 & \pm 1 \end{pmatrix}.$$

This is a representation of a tensor with respect to the principal axis system of its symmetric part, so we note that with the angle ϑ only one independent invariant exists.

5.3.3 Irreducible Invariants of Second-Order Tensors

1. Irreducible invariants are by definition polynomial invariants, which are obtained from appropriate contractions of tensor coordinates. For second-order tensors, we can write the polynomial invariants also as the trace of a sequence of scalar products between these tensors. We begin with a few statements, which we need later:

I. The trace of a sequence of scalar products of second-order tensors does not change if we cyclically permute the order of the factors in the scalar products, i. e.

$$\operatorname{tr}(\underline{\underline{a}} \cdot \underline{\underline{b}} \cdot \ldots \cdot \underline{\underline{c}} \cdot \underline{\underline{d}}) = a_{ij}\, b_{jk} \cdots c_{mn}\, d_{ni} = b_{jk} \cdots c_{mn}\, d_{ni}\, a_{ij} = \operatorname{tr}(\underline{\underline{b}} \cdot \ldots \cdot \underline{\underline{c}} \cdot \underline{\underline{d}} \cdot \underline{\underline{a}}),$$

so we have

$$\operatorname{tr}(\underline{\underline{a}} \cdot \underline{\underline{b}} \cdot \ldots \cdot \underline{\underline{c}} \cdot \underline{\underline{d}}) = \operatorname{tr}(\underline{\underline{b}} \cdot \ldots \cdot \underline{\underline{c}} \cdot \underline{\underline{d}} \cdot \underline{\underline{a}}) = \cdots = \operatorname{tr}(\underline{\underline{d}} \cdot \underline{\underline{a}} \cdot \underline{\underline{b}} \cdot \ldots \cdot \underline{\underline{c}}). \tag{5.5}$$

II. The trace of a sequence of scalar products of second-order tensors does not change if we transpose each factor, while reversing the order of the factors, i. e.

$$\begin{aligned} &\operatorname{tr}(\underline{\underline{a}} \cdot \underline{\underline{b}} \cdot \ldots \cdot \underline{\underline{c}} \cdot \underline{\underline{d}}) = a_{ij}\, b_{jk} \cdots c_{mn}\, d_{ni} \\ &= d^{\mathrm{T}}_{in}\, c^{\mathrm{T}}_{nm} \cdots b^{\mathrm{T}}_{kj}\, a^{\mathrm{T}}_{ji} = \operatorname{tr}(\underline{\underline{d}}^{\mathrm{T}} \cdot \underline{\underline{c}}^{\mathrm{T}} \cdot \ldots \cdot \underline{\underline{b}}^{\mathrm{T}} \cdot \underline{\underline{a}}^{\mathrm{T}}), \end{aligned}$$

so we have

$$\operatorname{tr}(\underline{\underline{a}} \cdot \underline{\underline{b}} \cdot \ldots \cdot \underline{\underline{c}} \cdot \underline{\underline{d}}) = \operatorname{tr}(\underline{\underline{d}}^{\mathrm{T}} \cdot \underline{\underline{c}}^{\mathrm{T}} \cdot \ldots \cdot \underline{\underline{b}}^{\mathrm{T}} \cdot \underline{\underline{a}}^{\mathrm{T}}). \tag{5.6}$$

In particular, if all factors are equal, we have

$$\operatorname{tr} \underline{\underline{a}}^n = \operatorname{tr} (\underline{\underline{a}}^{\mathrm{T}})^n = \operatorname{tr} (\underline{\underline{a}}^n)^{\mathrm{T}}, \tag{5.7}$$

where n is a natural number.

III. The trace of a scalar product of a symmetric tensor $\underline{\underline{s}} = \underline{\underline{s}}^{\mathrm{T}}$ and an antimetric tensor $\underline{\underline{a}} = -\underline{\underline{a}}^{\mathrm{T}}$ is zero.

Since

$$(\underline{\underline{s}} \cdot \underline{\underline{a}})^{\mathrm{T}} \overset{(2.35)}{=} \underline{\underline{a}}^{\mathrm{T}} \cdot \underline{\underline{s}}^{\mathrm{T}} = -\underline{\underline{a}} \cdot \underline{\underline{s}} \tag{a}$$

and also

$$\operatorname{tr}(\underline{\underline{s}} \cdot \underline{\underline{a}}) \overset{(5.7)}{=} \operatorname{tr} (\underline{\underline{s}} \cdot \underline{\underline{a}})^{\mathrm{T}} \overset{(a)}{=} -\operatorname{tr}(\underline{\underline{a}} \cdot \underline{\underline{s}}) \overset{(5.5)}{=} -\operatorname{tr}(\underline{\underline{s}} \cdot \underline{\underline{a}}),$$

we obtain

$$\operatorname{tr}(\underline{\underline{s}} \cdot \underline{\underline{a}}) = 0. \tag{5.8}$$

2. In order to write the irreducible invariants of a second-order tensor as the trace of a sequence of scalar products, the main question is how many of these invariants form an integrity basis. Answering this question here in all generality would be beyond the scope of this book. We restrict ourselves to the case that the factors in the scalar product depend only on two different tensors $\underline{\underline{a}}$ and $\underline{\underline{b}}$. We will follow a systematical approach and increase the number of the factors in the scalar product step-by-step; the number of these factors is also called the degree of the invariant.

With only one factor, there are only two ways to form invariants in the described manner, namely the trace of each of the two tensors themselves, i. e.

$$I_{11} := \operatorname{tr} \underline{\underline{a}}, \qquad I_{12} := \operatorname{tr} \underline{\underline{b}}. \tag{5.9}$$

Scalar products of two factors can be formed in four different ways: $\underline{\underline{a}}^2$, $\underline{\underline{b}}^2$, $\underline{\underline{a}} \cdot \underline{\underline{b}}$, $\underline{\underline{b}} \cdot \underline{\underline{a}}$; after taking cyclical permutations (5.5) into account, only three invariants of degree two remain:

$$I_{21} := \operatorname{tr} \underline{\underline{a}}^2, \qquad I_{22} := \operatorname{tr} \underline{\underline{b}}^2, \qquad I_{23} := \operatorname{tr}(\underline{\underline{a}} \cdot \underline{\underline{b}}). \tag{5.10}$$

From three factors, there are altogether eight possible scalar products: $\underline{\underline{a}}^3$, $\underline{\underline{b}}^3$, $\underline{\underline{a}}^2 \cdot \underline{\underline{b}}$, $\underline{\underline{a}} \cdot \underline{\underline{b}} \cdot \underline{\underline{a}}$, $\underline{\underline{b}} \cdot \underline{\underline{a}}^2$, $\underline{\underline{b}}^2 \cdot \underline{\underline{a}}$, $\underline{\underline{b}} \cdot \underline{\underline{a}} \cdot \underline{\underline{b}}$, $\underline{\underline{a}} \cdot \underline{\underline{b}}^2$; after computing the traces and taking cyclical permutations according to (5.5) into account, only four invariants of degree three remain:

$$\begin{aligned} I_{31} &:= \operatorname{tr} \underline{\underline{a}}^3, & I_{32} &:= \operatorname{tr} \underline{\underline{b}}^3, \\ I_{33} &:= \operatorname{tr}(\underline{\underline{a}}^2 \cdot \underline{\underline{b}}), & I_{34} &:= \operatorname{tr}(\underline{\underline{a}} \cdot \underline{\underline{b}}^2). \end{aligned} \tag{5.11}$$

The invariants of degree one, two, and three, which we have considered so far, are all irreducible, because none can be written in terms of the others. However, for invariants of degree four, a fundamental restriction appears, since invariants such as $\operatorname{tr}(\underline{\underline{a}}^3 \cdot \underline{\underline{b}})$ and $\operatorname{tr}(\underline{\underline{b}}^3 \cdot \underline{\underline{a}})$ are reducible. To see this, we solve the Cayley–Hamilton theorem (3.94) for $\underline{\underline{a}}^3$ and scalar multiply it from the right by $\underline{\underline{b}}$, so we have

$$\underline{\underline{a}}^3 \cdot \underline{\underline{b}} = A''\, \underline{\underline{a}}^2 \cdot \underline{\underline{b}} - A'\, \underline{\underline{a}} \cdot \underline{\underline{b}} + A\, \underline{\underline{b}}.$$

Computing the trace gives

$$\operatorname{tr}(\underline{\underline{a}}^3 \cdot \underline{\underline{b}}) = A'' \operatorname{tr}(\underline{\underline{a}}^2 \cdot \underline{\underline{b}}) - A' \operatorname{tr}(\underline{\underline{a}} \cdot \underline{\underline{b}}) + A \operatorname{tr} \underline{\underline{b}},$$

i. e. $\operatorname{tr}(\underline{\underline{a}}^3 \cdot \underline{\underline{b}})$ can be written in terms of invariants of degree less than or equal to three, since A'', A', and A are, according to (3.99), also functions of $\operatorname{tr} \underline{\underline{a}}$, $\operatorname{tr} \underline{\underline{a}}^2$, and $\operatorname{tr} \underline{\underline{a}}^3$. The same is true for $\operatorname{tr}(\underline{\underline{b}}^3 \cdot \underline{\underline{a}})$, because $\underline{\underline{a}}$ and $\underline{\underline{b}}$ can be interchanged in the above derivation.

We can extend these considerations to invariants of the form $\operatorname{tr}(\underline{\underline{c}} \cdot \underline{\underline{a}}^3 \cdot \underline{\underline{d}})$ and $\operatorname{tr}(\underline{\underline{c}} \cdot \underline{\underline{b}}^3 \cdot \underline{\underline{d}})$, where $\underline{\underline{c}}$ and $\underline{\underline{d}}$ are arbitrary second-order tensors; in particular they can be scalar products of the factors $\underline{\underline{a}}$ and $\underline{\underline{b}}$. Such invariants can be reduced, using the Cayley–Hamilton theorem (3.94), to invariants of a smaller degree. In other words, invariants with a term of power three (or higher) are reducible. Hence we only need to investigate sequences of scalar products with the elements $\underline{\underline{a}}^2$, $\underline{\underline{b}}^2$, $\underline{\underline{a}}$, $\underline{\underline{b}}$.

For scalar products with at least four factors, we can derive another relation, using the generalized Cayley–Hamilton theorem, which will narrow down the search for irreducible invariants. For this we set in (5.1) $\underline{\underline{a}} = \underline{\underline{b}} = \underline{\underline{n}}$ and scalar multiply from the right by an arbitrary tensor $\underline{\underline{d}}$. For the first row we get with (5.2)

$$\begin{aligned}\underline{\underline{\Sigma}} \cdot \underline{\underline{d}} &= \left[\underline{\underline{n}} \cdot \left(\underline{\underline{n}} \cdot \underline{\underline{c}} + \underline{\underline{c}} \cdot \underline{\underline{n}}\right) + \underline{\underline{n}} \cdot \left(\underline{\underline{n}} \cdot \underline{\underline{c}} + \underline{\underline{c}} \cdot \underline{\underline{n}}\right) + \underline{\underline{c}} \cdot \left(\underline{\underline{n}} \cdot \underline{\underline{n}} + \underline{\underline{n}} \cdot \underline{\underline{n}}\right)\right] \cdot \underline{\underline{d}} \\ &= 2\left[\underline{\underline{n}} \cdot \underline{\underline{c}} \cdot \underline{\underline{n}} + \underline{\underline{n}}^2 \cdot \underline{\underline{c}} + \underline{\underline{c}} \cdot \underline{\underline{n}}^2\right] \cdot \underline{\underline{d}} = 2\left[\underline{\underline{n}} \cdot \underline{\underline{c}} \cdot \underline{\underline{n}} \cdot \underline{\underline{d}} + \underline{\underline{n}}^2 \cdot \underline{\underline{c}} \cdot \underline{\underline{d}} + \underline{\underline{c}} \cdot \underline{\underline{n}}^2 \cdot \underline{\underline{d}}\right].\end{aligned}$$

Computing the trace, taking cyclic permutations into account, yields

$$\begin{aligned}&\operatorname{tr}(\underline{\underline{\Sigma}} \cdot \underline{\underline{d}}) = 2\left[\operatorname{tr}(\underline{\underline{n}} \cdot \underline{\underline{c}} \cdot \underline{\underline{n}} \cdot \underline{\underline{d}}) + \operatorname{tr}(\underline{\underline{n}}^2 \cdot \underline{\underline{c}} \cdot \underline{\underline{d}}) + \operatorname{tr}(\underline{\underline{n}}^2 \cdot \underline{\underline{d}} \cdot \underline{\underline{c}})\right], \\ &\operatorname{tr}(\underline{\underline{n}} \cdot \underline{\underline{c}} \cdot \underline{\underline{n}} \cdot \underline{\underline{d}}) = -\left(\operatorname{tr}(\underline{\underline{n}}^2 \cdot \underline{\underline{c}} \cdot \underline{\underline{d}}) + \operatorname{tr}(\underline{\underline{n}}^2 \cdot \underline{\underline{d}} \cdot \underline{\underline{c}})\right) + \frac{1}{2} \operatorname{tr}(\underline{\underline{\Sigma}} \cdot \underline{\underline{d}}).\end{aligned}$$

Using (5.3), we can write $\operatorname{tr}(\underline{\underline{\Sigma}} \cdot \underline{\underline{d}})$ in terms of invariants of at least one degree less than the degree of $\operatorname{tr}(\underline{\underline{n}} \cdot \underline{\underline{c}} \cdot \underline{\underline{n}} \cdot \underline{\underline{d}})$, $\operatorname{tr}(\underline{\underline{n}}^2 \cdot \underline{\underline{c}} \cdot \underline{\underline{d}})$, and $\operatorname{tr}(\underline{\underline{n}}^2 \cdot \underline{\underline{d}} \cdot \underline{\underline{c}})$. In the theory of the representation of tensor functions we also say that $\operatorname{tr}(\underline{\underline{n}} \cdot \underline{\underline{c}} \cdot \underline{\underline{n}} \cdot \underline{\underline{d}})$ and $-\left(\operatorname{tr}(\underline{\underline{n}}^2 \cdot \underline{\underline{c}} \cdot \underline{\underline{d}}) + \operatorname{tr}(\underline{\underline{n}}^2 \cdot \underline{\underline{d}} \cdot \underline{\underline{c}})\right)$ are equivalent and we write

$$\operatorname{tr}(\underline{\underline{n}} \cdot \underline{\underline{c}} \cdot \underline{\underline{n}} \cdot \underline{\underline{d}}) \equiv -\left(\operatorname{tr}(\underline{\underline{n}}^2 \cdot \underline{\underline{c}} \cdot \underline{\underline{d}}) + \operatorname{tr}(\underline{\underline{n}}^2 \cdot \underline{\underline{d}} \cdot \underline{\underline{c}})\right). \tag{5.12}$$

Equivalent invariants have the same degree and differ only in their (reducible) invariants of lower degrees; in particular, we can replace an irreducible invariant in an integrity basis by an equivalent invariant. If a computation shows that an invariant is equivalent to zero, then this invariant is reducible. The important implication of the equivalence relation (5.12) is that invariants with two factors which are equal but appear separated from each other can be written as the sum of two other invariants in which this factor appears squared.

With these preparations we can now reduce our search for fourth-order irreducible invariants to the following scalar products: $\underline{\underline{a}}^2 \cdot \underline{\underline{b}} \cdot \underline{\underline{a}}$, $\underline{\underline{a}} \cdot \underline{\underline{b}} \cdot \underline{\underline{a}}^2$, $\underline{\underline{b}}^2 \cdot \underline{\underline{a}} \cdot \underline{\underline{b}}$, $\underline{\underline{b}} \cdot \underline{\underline{a}} \cdot \underline{\underline{b}}^2$, $\underline{\underline{a}} \cdot \underline{\underline{b}} \cdot \underline{\underline{a}} \cdot \underline{\underline{b}}$, $\underline{\underline{b}} \cdot \underline{\underline{a}} \cdot \underline{\underline{b}} \cdot \underline{\underline{a}}$, $\underline{\underline{a}} \cdot \underline{\underline{b}}^2 \cdot \underline{\underline{a}}$, $\underline{\underline{b}} \cdot \underline{\underline{a}}^2 \cdot \underline{\underline{b}}$, $\underline{\underline{a}}^2 \cdot \underline{\underline{b}}^2$, $\underline{\underline{b}}^2 \cdot \underline{\underline{a}}^2$. After computing the trace and using cyclic permutation we have $\mathrm{tr}(\underline{\underline{a}}^2 \cdot \underline{\underline{b}} \cdot \underline{\underline{a}}) = \mathrm{tr}(\underline{\underline{a}} \cdot \underline{\underline{b}} \cdot \underline{\underline{a}}^2) = \mathrm{tr}(\underline{\underline{a}}^3 \cdot \underline{\underline{b}})$ and $\mathrm{tr}(\underline{\underline{b}}^2 \cdot \underline{\underline{a}} \cdot \underline{\underline{b}}) = \mathrm{tr}(\underline{\underline{b}} \cdot \underline{\underline{a}} \cdot \underline{\underline{b}}^2) = \mathrm{tr}(\underline{\underline{b}}^3 \cdot \underline{\underline{a}})$, so these invariants are reducible. With the same reasoning we have $\mathrm{tr}(\underline{\underline{a}} \cdot \underline{\underline{b}} \cdot \underline{\underline{a}} \cdot \underline{\underline{b}}) = \mathrm{tr}(\underline{\underline{b}} \cdot \underline{\underline{a}} \cdot \underline{\underline{b}} \cdot \underline{\underline{a}})$ and $\mathrm{tr}(\underline{\underline{a}}^2 \cdot \underline{\underline{b}}^2) = \mathrm{tr}(\underline{\underline{b}}^2 \cdot \underline{\underline{a}}^2) = \mathrm{tr}(\underline{\underline{a}} \cdot \underline{\underline{b}}^2 \cdot \underline{\underline{a}}) = \mathrm{tr}(\underline{\underline{b}} \cdot \underline{\underline{a}}^2 \cdot \underline{\underline{b}})$, so only two invariants remain, i. e. $\mathrm{tr}(\underline{\underline{a}} \cdot \underline{\underline{b}} \cdot \underline{\underline{a}} \cdot \underline{\underline{b}})$ and $\mathrm{tr}(\underline{\underline{a}}^2 \cdot \underline{\underline{b}}^2)$, and we have to check if they are irreducible. Setting in (5.12) $\underline{\underline{n}} = \underline{\underline{a}}$, $\underline{\underline{c}} = \underline{\underline{d}} = \underline{\underline{b}}$ we obtain

$$\mathrm{tr}(\underline{\underline{a}} \cdot \underline{\underline{b}} \cdot \underline{\underline{a}} \cdot \underline{\underline{b}}) \equiv -2\,\mathrm{tr}(\underline{\underline{a}}^2 \cdot \underline{\underline{b}}^2),$$

i. e. both invariants are equivalent. Hence only one irreducible invariant of degree four exists and we choose

$$I_{41} := \mathrm{tr}(\underline{\underline{a}}^2 \cdot \underline{\underline{b}}^2). \tag{5.13}$$

When investigating scalar products with five factors, we consider cyclic permutations from the outset and then we see that there are only two ways of computing invariants of degree five, which are composed of the elements $\underline{\underline{a}}^2$, $\underline{\underline{b}}^2$, $\underline{\underline{a}}$, $\underline{\underline{b}}$, namely $\mathrm{tr}(\underline{\underline{a}} \cdot \underline{\underline{b}} \cdot \underline{\underline{a}} \cdot \underline{\underline{b}}^2)$ and $\mathrm{tr}(\underline{\underline{b}} \cdot \underline{\underline{a}} \cdot \underline{\underline{b}} \cdot \underline{\underline{a}}^2)$. From (5.12) it follows that

$$\mathrm{tr}(\underline{\underline{a}} \cdot \underline{\underline{b}} \cdot \underline{\underline{a}} \cdot \underline{\underline{b}}^2) \equiv -2\,\mathrm{tr}(\underline{\underline{a}}^2 \cdot \underline{\underline{b}}^3),$$
$$\mathrm{tr}(\underline{\underline{b}} \cdot \underline{\underline{a}} \cdot \underline{\underline{b}} \cdot \underline{\underline{a}}^2) \equiv -2\,\mathrm{tr}(\underline{\underline{b}}^2 \cdot \underline{\underline{a}}^3);$$

however, because of $\underline{\underline{a}}^3$ and $\underline{\underline{b}}^3$ both invariants are reducible. In other words, no irreducible invariants of degree five exist.

In order to find the irreducible invariants of degree six, which can be formed from the elements $\underline{\underline{a}}^2$, $\underline{\underline{b}}^2$, $\underline{\underline{a}}$, $\underline{\underline{b}}$, we start, after taking cyclic permutations into account, from the invariants $\mathrm{tr}(\underline{\underline{a}} \cdot \underline{\underline{b}}^2 \cdot \underline{\underline{a}} \cdot \underline{\underline{b}}^2)$, $\mathrm{tr}(\underline{\underline{b}} \cdot \underline{\underline{a}}^2 \cdot \underline{\underline{b}} \cdot \underline{\underline{a}}^2)$, $\mathrm{tr}(\underline{\underline{a}} \cdot \underline{\underline{b}} \cdot \underline{\underline{a}} \cdot \underline{\underline{b}} \cdot \underline{\underline{a}} \cdot \underline{\underline{b}})$, $\mathrm{tr}(\underline{\underline{a}}^2 \cdot \underline{\underline{b}}^2 \cdot \underline{\underline{a}} \cdot \underline{\underline{b}})$, $\mathrm{tr}(\underline{\underline{a}}^2 \cdot \underline{\underline{b}} \cdot \underline{\underline{a}} \cdot \underline{\underline{b}}^2)$. We then investigate, using (5.12), which of them are irreducible. For the first two invariants we have

$$\operatorname{tr}(\underline{\underline{a}} \cdot \underline{\underline{b}}^2 \cdot \underline{\underline{a}} \cdot \underline{\underline{b}}^2) \equiv -2 \operatorname{tr}(\underline{\underline{a}}^2 \cdot \underline{\underline{b}}^4),$$
$$\operatorname{tr}(\underline{\underline{b}} \cdot \underline{\underline{a}}^2 \cdot \underline{\underline{b}} \cdot \underline{\underline{a}}^2) \equiv -2 \operatorname{tr}(\underline{\underline{b}}^2 \cdot \underline{\underline{a}}^4);$$

but because of $\underline{\underline{a}}^4$ and $\underline{\underline{b}}^4$ these invariants are reducible. For the third invariant we have

$$\operatorname{tr}(\underline{\underline{a}} \cdot (\underline{\underline{b}} \cdot \underline{\underline{a}} \cdot \underline{\underline{b}}) \cdot \underline{\underline{a}} \cdot \underline{\underline{b}}) \equiv -\operatorname{tr}(\underline{\underline{a}}^2 \cdot \underline{\underline{b}} \cdot \underline{\underline{a}} \cdot \underline{\underline{b}}^2) - \operatorname{tr}(\underline{\underline{a}}^2 \cdot \underline{\underline{b}}^2 \cdot \underline{\underline{a}} \cdot \underline{\underline{b}}),$$

i. e. the third invariant is equivalent to the (negative) sum of the last two invariants. For the last two invariants we obtain

$$\operatorname{tr}(\underline{\underline{a}}^2 \cdot \underline{\underline{b}}^2 \cdot \underline{\underline{a}} \cdot \underline{\underline{b}}) = \operatorname{tr}(\underline{\underline{a}} \cdot (\underline{\underline{a}} \cdot \underline{\underline{b}}^2) \cdot \underline{\underline{a}} \cdot \underline{\underline{b}}) \equiv -\underbrace{\operatorname{tr}(\underline{\underline{a}}^3 \cdot \underline{\underline{b}}^3)}_{\equiv 0} - \operatorname{tr}(\underline{\underline{a}}^2 \cdot \underline{\underline{b}} \cdot \underline{\underline{a}} \cdot \underline{\underline{b}}^2),$$

i. e. they are equivalent and can only be reduced as a sum, but not individually, to invariants of lower degree. So we can regard one of them as irreducible and we choose

$$I_{61} := \operatorname{tr}(\underline{\underline{a}}^2 \cdot \underline{\underline{b}} \cdot \underline{\underline{a}} \cdot \underline{\underline{b}}^2). \tag{5.14}$$

It turns out that we do not have to look further for invariants of degree seven or higher, because they will always be reducible. This follows from two facts. First, a factor can appear in irreducible invariants only as itself or squared. Second, the equivalence relation (5.12) states that invariants appearing twice can be replaced by invariants with quadratic factors. From the elements $\underline{\underline{a}}^2$, $\underline{\underline{b}}^2$, $\underline{\underline{a}}$, $\underline{\underline{b}}$ we can form, without repetition, at most invariants of degree six, because for higher degrees at least one of the elements appears twice. If the element appearing twice is $\underline{\underline{a}}$ or $\underline{\underline{b}}$, we can replace the invariant under consideration, using (5.12), with equivalent invariants with $\underline{\underline{a}}^2$ or $\underline{\underline{b}}^2$; however, the elements $\underline{\underline{a}}^2$ and $\underline{\underline{b}}^2$ already appear, so that applying (5.12) again leads to invariants with $\underline{\underline{a}}^4$ or $\underline{\underline{b}}^4$, which are reducible. If the element appearing twice is $\underline{\underline{a}}^2$ or $\underline{\underline{b}}^2$, then applying (5.12) once is already sufficient to reduce the invariant under consideration to invariants of lower degree.

3. We summarize the results of our investigations.

We computed invariants as the trace of a sequence of scalar products of second-order tensors, and we restricted our search for invariants to the case that the factors in the sequence of scalar products are two different tensors $\underline{\underline{a}}$ and $\underline{\underline{b}}$. Using the Cayley–Hamilton theorem (3.94) and its generalization (5.1), we could show that invariants with more than six factors are always reducible. Finally, we were able to find with (5.9), (5.10), (5.11), (5.13), and (5.14) a total of 11 irreducible invariants (some of which are simultaneous invariants, since they are built from the coordinates of different tensors):

$$
\begin{aligned}
&I_{11} = \operatorname{tr} \underline{\underline{a}}, \qquad I_{12} = \operatorname{tr} \underline{\underline{b}},\\
&I_{21} = \operatorname{tr} \underline{\underline{a}}^2, \qquad I_{22} = \operatorname{tr} \underline{\underline{b}}^2, \qquad I_{23} = \operatorname{tr}(\underline{\underline{a}} \cdot \underline{\underline{b}}),\\
&I_{31} = \operatorname{tr} \underline{\underline{a}}^3, \qquad I_{32} = \operatorname{tr} \underline{\underline{b}}^3, \qquad I_{33} = \operatorname{tr}(\underline{\underline{a}}^2 \cdot \underline{\underline{b}}), \qquad I_{34} = \operatorname{tr}(\underline{\underline{a}} \cdot \underline{\underline{b}}^2),\\
&I_{41} = \operatorname{tr}(\underline{\underline{a}}^2 \cdot \underline{\underline{b}}^2),\\
&I_{61} = \operatorname{tr}(\underline{\underline{a}}^2 \cdot \underline{\underline{b}} \cdot \underline{\underline{a}} \cdot \underline{\underline{b}}^2).
\end{aligned}
\tag{5.15}
$$

It remains to answer the question if these irreducible invariants form a complete set. We will do this by means of examples.

4. As a first example, we consider a single tensor $\underline{\underline{T}}$ and set in (5.15) $\underline{\underline{a}} = \underline{\underline{T}}$, $\underline{\underline{b}} = \underline{\underline{T}}^{\mathrm{T}}$. Since the trace does not change if we transpose its argument and since, according to (5.7), transposition and exponentiation are interchangeable within a trace, we see immediately that some of the invariants in (5.15) are the same. We have

$$I_{11} = I_{12} = \operatorname{tr} \underline{\underline{T}}, \quad I_{21} = I_{22} = \operatorname{tr} \underline{\underline{T}}^2, \quad I_{31} = I_{32} = \operatorname{tr} \underline{\underline{T}}^3,$$

and further investigation of I_{33} and I_{34} shows that

$$I_{34} = \operatorname{tr}(\underline{\underline{T}} \cdot (\underline{\underline{T}}^{\mathrm{T}})^2) \overset{(5.6)}{=} \operatorname{tr}(\underline{\underline{T}}^2 \cdot \underline{\underline{T}}^{\mathrm{T}}) = I_{33}.$$

Thus a single second-order tensor has, in general, seven irreducible invariants:

$$
\begin{aligned}
I_1 &:= T_{ii} &&= \operatorname{tr} \underline{\underline{T}},\\
I_2 &:= T_{ij}\, T_{ji} &&= \operatorname{tr} \underline{\underline{T}}^2,\\
I_3 &:= T_{ij}\, T_{ij} &&= \operatorname{tr}(\underline{\underline{T}} \cdot \underline{\underline{T}}^{\mathrm{T}}),\\
I_4 &:= T_{ij}\, T_{jk}\, T_{ki} &&= \operatorname{tr} \underline{\underline{T}}^3,\\
I_5 &:= T_{ij}\, T_{jk}\, T_{ik} &&= \operatorname{tr}(\underline{\underline{T}}^2 \cdot \underline{\underline{T}}^{\mathrm{T}}),\\
I_6 &:= T_{ij}\, T_{jk}\, T_{lk}\, T_{il} &&= \operatorname{tr}(\underline{\underline{T}}^2 \cdot (\underline{\underline{T}}^{\mathrm{T}})^2),\\
I_7 &:= T_{ij}\, T_{jk}\, T_{lk}\, T_{lm}\, T_{nm}\, T_{in} &&= \operatorname{tr}(\underline{\underline{T}}^2 \cdot \underline{\underline{T}}^{\mathrm{T}} \cdot \underline{\underline{T}} \cdot (\underline{\underline{T}}^{\mathrm{T}})^2).
\end{aligned}
\tag{5.16}
$$

These invariants in fact form an integrity basis. Both $\underline{\underline{T}}$ and $\underline{\underline{T}}^{\mathrm{T}}$ are needed and it is not possible to write all invariants in terms of traces using only $\underline{\underline{T}}$. We can see this, for example, if we compare I_2 and I_3. In index notation we have $I_2 = T_{ij}\, T_{ji}$ and $I_3 = T_{ij}\, T_{ij}$; both are scalars obtained by contracting tensor coordinates, but without the transposed tensor, we cannot write I_3 as a trace.[2]

---

2 This explains why (5.15) is not a complete set of irreducible invariants of the tensors $\underline{\underline{a}}$ and $\underline{\underline{b}}$; to obtain an integrity basis, we have to include also the transposed tensors $\underline{\underline{a}}^{\mathrm{T}}$ and $\underline{\underline{b}}^{\mathrm{T}}$, i. e. we have to investigate four different tensors.

If we compare our results with Section 5.3.2, we see that, in general, a second-order tensor has six independent invariants, but seven irreducible invariants. We can clearly write all irreducible invariants in terms of the independent invariants from Section 5.3.2, No. 1, if we write the traces in (5.16) with respect to the principal axis system of the symmetric part of $\underline{\underline{T}}$.

Symmetric tensors obey $\underline{\underline{T}}^{\mathrm{T}} = \underline{\underline{T}}$. Then $I_2 = I_3$ and $I_4 = I_5$, and we can reduce $I_6 = \operatorname{tr} \underline{\underline{T}}^4$ and $I_7 = \operatorname{tr} \underline{\underline{T}}^6$, using the Cayley–Hamilton theorem (3.95), to the remaining invariants. So, from the irreducible invariants in (5.16), only the three basic invariants $I_1 = \operatorname{tr} \underline{\underline{T}}$, $I_2 = \operatorname{tr} \underline{\underline{T}}^2$, $I_4 = \operatorname{tr} \underline{\underline{T}}^3$ remain.

Antimetric tensors satisfy $\underline{\underline{T}}^{\mathrm{T}} = -\underline{\underline{T}}$; hence $I_1 = 0$. We know $\underline{\underline{T}}^2$ is symmetric because $T_{ij}\, T_{jk} = (-T_{ji})\,(-T_{kj}) = T_{kj}\, T_{ji}$, so I_2 is nonzero, but we have $I_3 = -I_2$. Here $\underline{\underline{T}}^3$ is again antimetric because $T_{ij}\, T_{jk}\, T_{kl} = (-T_{ji})\,(-T_{kj})\,(-T_{lk}) = -T_{lk}\, T_{kj}\, T_{ji}$, so we have $I_4 = 0$ and thus also $I_5 = -I_4 = 0$. It further follows that $I_6 = \operatorname{tr} \underline{\underline{T}}^4$ and $I_7 = -\operatorname{tr} \underline{\underline{T}}^6$, so the Cayley–Hamilton theorem (3.95) helps us reduce I_6 and I_7 to the only remaining invariant $I_2 = \operatorname{tr} \underline{\underline{T}}^2$.

For orthogonal tensors we have $\underline{\underline{T}}^{\mathrm{T}} = \underline{\underline{T}}^{-1}$. Then $I_3 = I_6 = I_7 = 3$. In other words, I_3, I_6, and I_7 do not contain any information about a particular orthogonal tensor and thus do not count as invariants. It further follows that $I_5 = I_1$, so we are left with the three basic invariants $I_1 = \operatorname{tr} \underline{\underline{T}}$, $I_2 = \operatorname{tr} \underline{\underline{T}}^2$, $I_4 = \operatorname{tr} \underline{\underline{T}}^3$. The determinant of an orthogonal tensor is, according to Section 3.13.2, equal to ± 1, and the cotensor is, according to (3.14), equal to the tensor itself, up to the sign. Thus, we have for the main invariants, according to (3.47)–(3.49), $A'' = \operatorname{tr} \underline{\underline{T}}$, $A' = \pm \operatorname{tr} \underline{\underline{T}}$, $A = \pm 1$; then it follows from (3.100) that $\operatorname{tr} \underline{\underline{T}}^2 = \operatorname{tr}^2 \underline{\underline{T}} \mp 2 \operatorname{tr} \underline{\underline{T}}$, $\operatorname{tr} \underline{\underline{T}}^3 = \operatorname{tr}^3 \underline{\underline{T}} \mp 3 \operatorname{tr}^2 \underline{\underline{T}} \pm 3$, so an orthogonal tensor has only one irreducible invariant $I_1 = \operatorname{tr}(\underline{\underline{T}})$.

5. As our next example we consider the case of two symmetric tensors $\underline{\underline{U}}$ and $\underline{\underline{V}}$, which is also important for physical applications. Now we do not need to distinguish between the tensors and their transposed tensors, so we certainly can obtain an integrity basis from (5.15). We set $\underline{\underline{a}} = \underline{\underline{U}}$, $\underline{\underline{b}} = \underline{\underline{V}}$ in (5.15) and keep the first 10 invariants. Only the invariant of degree six does not belong to the integrity basis because it turns out to be reducible. On the one hand, from the symmetry of the tensors and by taking cyclic permutations and transposition (5.5)–(5.7) into account, it follows that

$$\operatorname{tr}(\underline{\underline{U}}^2 \cdot \underline{\underline{V}} \cdot \underline{\underline{U}} \cdot \underline{\underline{V}}^2) = \operatorname{tr}((\underline{\underline{V}}^2)^{\mathrm{T}} \cdot \underline{\underline{U}}^{\mathrm{T}} \cdot \underline{\underline{V}}^{\mathrm{T}} \cdot (\underline{\underline{U}}^2)^{\mathrm{T}}) = \operatorname{tr}(\underline{\underline{U}}^2 \cdot \underline{\underline{V}}^2 \cdot \underline{\underline{U}} \cdot \underline{\underline{V}}),$$

and on the other hand, from (5.12), we get

$$\operatorname{tr}(\underline{\underline{U}}^2 \cdot \underline{\underline{V}} \cdot \underline{\underline{U}} \cdot \underline{\underline{V}}^2) = \operatorname{tr}(\underline{\underline{U}} \cdot (\underline{\underline{U}} \cdot \underline{\underline{V}}) \cdot \underline{\underline{U}} \cdot \underline{\underline{V}}^2) \equiv -\underbrace{\operatorname{tr}(\underline{\underline{U}}^3 \cdot \underline{\underline{V}}^3)}_{\equiv 0} - \operatorname{tr}(\underline{\underline{U}}^2 \cdot \underline{\underline{V}}^2 \cdot \underline{\underline{U}} \cdot \underline{\underline{V}}).$$

The comparison of these two relations yields

$$\operatorname{tr}(\underline{\underline{U}}^2 \cdot \underline{\underline{V}} \cdot \underline{\underline{U}} \cdot \underline{\underline{V}}^2) \equiv -\operatorname{tr}(\underline{\underline{U}}^2 \cdot \underline{\underline{V}} \cdot \underline{\underline{U}} \cdot \underline{\underline{V}}^2),$$

which is only possible if $\operatorname{tr}(\underline{\underline{U}}^2 \cdot \underline{\underline{V}} \cdot \underline{\underline{U}} \cdot \underline{\underline{V}}^2) \equiv 0$, and thus it is reducible.

So we find that the integrity basis of two symmetric tensors $\underline{\underline{U}}$ and $\underline{\underline{V}}$ consists of ten irreducible invariants:

- the basic invariants
  $\operatorname{tr} \underline{\underline{U}}$, $\operatorname{tr} \underline{\underline{U}}^2$, $\operatorname{tr} \underline{\underline{U}}^3$ and $\operatorname{tr} \underline{\underline{V}}$, $\operatorname{tr} \underline{\underline{V}}^2$, $\operatorname{tr} \underline{\underline{V}}^3$ of each of the tensors and
- the simultaneous invariants
  $\operatorname{tr}(\underline{\underline{U}} \cdot \underline{\underline{V}})$, $\operatorname{tr}(\underline{\underline{U}}^2 \cdot \underline{\underline{V}})$, $\operatorname{tr}(\underline{\underline{U}} \cdot \underline{\underline{V}}^2)$, $\operatorname{tr}(\underline{\underline{U}}^2 \cdot \underline{\underline{V}}^2)$.

Problem 5.2.

Find all irreducible invariants which can be computed from two antimetric tensors $\underline{\underline{A}}$ and $\underline{\underline{B}}$, and compare the result with the invariants of two vectors from Section 5.3.1.

Hint: To show that $\operatorname{tr}(\underline{\underline{A}}^2 \cdot \underline{\underline{B}}^2)$ is reducible, represent the two antimetric tensors $\underline{\underline{A}}$ and $\underline{\underline{B}}$ temporarily by their corresponding vectors.

6. As a last example, we investigate the invariants of a symmetric tensor $\underline{\underline{S}}$ and an antimetric tensor $\underline{\underline{A}}$. Also for this case we can obtain an integrity basis from (5.15), since the tensors and their transposed tensors differ at most by a sign. If we set $\underline{\underline{a}} = \underline{\underline{S}}$, $\underline{\underline{b}} = \underline{\underline{A}}$ in (5.15), some invariants are zero, because not only the trace of an antimetric tensor and its (also antimetric) third power is zero but also, according to (5.8), the trace of the scalar product of a symmetric tensor and an antimetric tensor is zero. Thus, we have $I_{12} = I_{23} = I_{32} = I_{33} = 0$. Contrary to the case of two symmetric tensors, here the invariant of degree six is not reducible. Using cyclic permutations and transposition (5.5)–(5.7), we initially have

$$\operatorname{tr}(\underline{\underline{S}}^2 \cdot \underline{\underline{A}} \cdot \underline{\underline{S}} \cdot \underline{\underline{A}}^2) = \operatorname{tr}((\underline{\underline{A}}^2)^{\mathrm{T}} \cdot \underline{\underline{S}}^{\mathrm{T}} \cdot \underline{\underline{A}}^{\mathrm{T}} \cdot (\underline{\underline{S}}^2)^{\mathrm{T}}) = -\operatorname{tr}(\underline{\underline{A}}^2 \cdot \underline{\underline{S}} \cdot \underline{\underline{A}} \cdot \underline{\underline{S}}^2),$$

and from (5.12) we further obtain that

$$\operatorname{tr}(\underline{\underline{S}}^2 \cdot \underline{\underline{A}} \cdot \underline{\underline{S}} \cdot \underline{\underline{A}}^2) = \operatorname{tr}(\underline{\underline{S}} \cdot (\underline{\underline{S}} \cdot \underline{\underline{A}}) \cdot \underline{\underline{S}} \cdot \underline{\underline{A}}^2) \equiv -\underbrace{\operatorname{tr}(\underline{\underline{S}}^3 \cdot \underline{\underline{A}}^3)}_{=0} - \operatorname{tr}(\underline{\underline{S}}^2 \cdot \underline{\underline{A}}^2 \cdot \underline{\underline{S}} \cdot \underline{\underline{A}}).$$

Comparing these two equations leads to the trivial statement $\operatorname{tr}(\underline{\underline{S}}^2 \cdot \underline{\underline{A}} \cdot \underline{\underline{S}} \cdot \underline{\underline{A}}^2) \equiv \operatorname{tr}(\underline{\underline{S}}^2 \cdot \underline{\underline{A}} \cdot \underline{\underline{S}} \cdot \underline{\underline{A}}^2)$, i. e. we obtain no additional information about this invariant.

We thus obtain for the integrity basis of a symmetric tensor $\underline{\underline{S}}$ and an antimetric tensor $\underline{\underline{A}}$ a total of seven irreducible invariants:

- the basic invariants
  $\operatorname{tr} \underline{\underline{S}}$, $\operatorname{tr} \underline{\underline{S}}^2$, $\operatorname{tr} \underline{\underline{S}}^3$ and $\operatorname{tr} \underline{\underline{A}}^2$ of each of the tensors,
- and also the simultaneous invariants
  $\operatorname{tr}(\underline{\underline{S}} \cdot \underline{\underline{A}}^2)$, $\operatorname{tr}(\underline{\underline{S}}^2 \cdot \underline{\underline{A}}^2)$, $\operatorname{tr}(\underline{\underline{S}}^2 \cdot \underline{\underline{A}} \cdot \underline{\underline{S}} \cdot \underline{\underline{A}}^2)$.

Problem 5.3.
Find all irreducible invariants which can be formed from a polar symmetric tensor $\underline{\underline{S}}$ and a vector $\underline{u}$, and compare the result with the invariants of a symmetric tensor and an antimetric tensor. Also, distinguish the cases where $\underline{u}$ is polar or axial.

5.3.4 Summary

We finally summarize the results of Section 5.3 in a table. Here $\underline{\underline{T}}$ denotes an arbitrary tensor, $\underline{\underline{R}}$ is an orthogonal tensor, $\underline{\underline{S}}$, $\underline{\underline{U}}$, $\underline{\underline{V}}$ are symmetric tensors, $\underline{\underline{A}}$, $\underline{\underline{B}}$ are antimetric tensors, and $\underline{u}$, $\underline{v}$, $\underline{w}$ are vectors. All vectors and tensors are polar.

Arguments	Integrity basis
$\underline{u}$	$\underline{u} \cdot \underline{u}$
$\underline{u}, \underline{v}$	$\underline{u} \cdot \underline{u},\ \underline{v} \cdot \underline{v},\ \underline{u} \cdot \underline{v}$
$\underline{u}, \underline{v}, \underline{w}$	$\underline{u} \cdot \underline{u},\ \underline{v} \cdot \underline{v},\ \underline{w} \cdot \underline{w},\ \underline{u} \cdot \underline{v},\ \underline{u} \cdot \underline{w},\ \underline{v} \cdot \underline{w}$
$\underline{\underline{T}}$	$\operatorname{tr} \underline{\underline{T}},\ \operatorname{tr} \underline{\underline{T}}^2,\ \operatorname{tr} \underline{\underline{T}}^3,\ \operatorname{tr}(\underline{\underline{T}} \cdot \underline{\underline{T}}^{\mathsf{T}}),\ \operatorname{tr}(\underline{\underline{T}}^2 \cdot \underline{\underline{T}}^{\mathsf{T}}),$ $\operatorname{tr}(\underline{\underline{T}}^2 \cdot (\underline{\underline{T}}^{\mathsf{T}})^2),\ \operatorname{tr}(\underline{\underline{T}}^2 \cdot \underline{\underline{T}}^{\mathsf{T}} \cdot \underline{\underline{T}} \cdot (\underline{\underline{T}}^{\mathsf{T}})^2)$
$\underline{\underline{S}}$	$\operatorname{tr} \underline{\underline{S}},\ \operatorname{tr} \underline{\underline{S}}^2,\ \operatorname{tr} \underline{\underline{S}}^3$
$\underline{\underline{A}}$	$\operatorname{tr} \underline{\underline{A}}^2$
$\underline{\underline{R}}$	$\operatorname{tr} \underline{\underline{R}}$
$\underline{\underline{U}}, \underline{\underline{V}}$	$\operatorname{tr} \underline{\underline{U}},\ \operatorname{tr} \underline{\underline{U}}^2,\ \operatorname{tr} \underline{\underline{U}}^3,\ \operatorname{tr} \underline{\underline{V}},\ \operatorname{tr} \underline{\underline{V}}^2,\ \operatorname{tr} \underline{\underline{V}}^3,$ $\operatorname{tr}(\underline{\underline{U}} \cdot \underline{\underline{V}}),\ \operatorname{tr}(\underline{\underline{U}}^2 \cdot \underline{\underline{V}}),\ \operatorname{tr}(\underline{\underline{U}} \cdot \underline{\underline{V}}^2),\ \operatorname{tr}(\underline{\underline{U}}^2 \cdot \underline{\underline{V}}^2)$
$\underline{\underline{A}}, \underline{\underline{B}}$	$\operatorname{tr} \underline{\underline{A}}^2,\ \operatorname{tr} \underline{\underline{B}}^2,\ \operatorname{tr}(\underline{\underline{A}} \cdot \underline{\underline{B}})$
$\underline{\underline{S}}, \underline{\underline{A}}$	$\operatorname{tr} \underline{\underline{S}},\ \operatorname{tr} \underline{\underline{S}}^2,\ \operatorname{tr} \underline{\underline{S}}^3,\ \operatorname{tr} \underline{\underline{A}}^2,\ \operatorname{tr}(\underline{\underline{S}} \cdot \underline{\underline{A}}^2),$ $\operatorname{tr}(\underline{\underline{S}}^2 \cdot \underline{\underline{A}}^2),\ \operatorname{tr}(\underline{\underline{S}}^2 \cdot \underline{\underline{A}} \cdot \underline{\underline{S}} \cdot \underline{\underline{A}}^2)$
$\underline{\underline{S}}, \underline{u}$	$\operatorname{tr} \underline{\underline{S}},\ \operatorname{tr} \underline{\underline{S}}^2,\ \operatorname{tr} \underline{\underline{S}}^3,\ \underline{u} \cdot \underline{u},\ \underline{u} \cdot \underline{\underline{S}} \cdot \underline{u},\ \underline{u} \cdot \underline{\underline{S}}^2 \cdot \underline{u}$

5.4 Isotropic Tensor Functions

5.4.1 Invariance Conditions

Tensors cannot be related by arbitrary functions, because the function value must satisfy the corresponding transformation law for tensor coordinates under a change to another Cartesian coordinate system. Let polar vectors $\underline{v}, \ldots, \underline{w}$ and polar second-order tensors $\underline{\underline{M}}, \ldots, \underline{\underline{N}}$ be arguments of tensor functions $\mathscr{F}$ of various orders. Under a change of the Cartesian coordinate system we have for the coordinates of the arguments, according to (2.17), the transformation equations

$$\widetilde{v}_i = \alpha_{pi}\, v_p, \;\ldots, \widetilde{w}_j = \alpha_{qj}\, w_q,$$
$$\widetilde{M}_{kl} = \alpha_{pk}\, \alpha_{ql}\, M_{pq}, \;\ldots, \widetilde{N}_{mn} = \alpha_{pm}\, \alpha_{qn}\, N_{pq}.$$

Depending on the order of the tensors, we obtain different conditions for the function $\mathscr{F}$:

- A polar scalar $s = f(\underline{v}, \ldots, \underline{w}, \underline{\underline{M}}, \ldots, \underline{\underline{N}})$ must remain unchanged under a transformation of the coordinate system,

  $$f(v_i, \ldots, w_j, M_{kl}, \ldots, N_{mn}) = f(\widetilde{v}_i, \ldots, \widetilde{w}_j, \widetilde{M}_{kl}, \ldots, \widetilde{N}_{mn}),$$

  and then it follows for the function f that

  $$\begin{aligned} &f(v_i, \ldots, w_j, M_{kl}, \ldots, N_{mn}) \\ &\quad = f(\alpha_{pi}\, v_p, \ldots, \alpha_{qj}\, w_q, \alpha_{pk}\, \alpha_{ql}\, M_{pq}, \ldots, \alpha_{pm}\, \alpha_{qn}\, N_{pq}). \end{aligned} \tag{5.17}$$

- For a polar vector $\underline{u} = \underline{f}(\underline{v}, \ldots, \underline{w}, \underline{\underline{M}}, \ldots, \underline{\underline{N}})$, substituting the transformed coordinates $\widetilde{v}_i, \ldots, \widetilde{w}_j, \widetilde{M}_{kl}, \ldots, \widetilde{N}_{mn}$ must give the transformed vector coordinates

  $$\widetilde{u}_r = \alpha_{ur}\, u_u = f_r(\widetilde{v}_i, \ldots, \widetilde{w}_j, \widetilde{M}_{kl}, \ldots, \widetilde{N}_{mn}).$$

  In the original coordinates, we have

  $$u_u = f_u(v_i, \ldots, w_j, M_{kl}, \ldots, N_{mn}),$$

  so it follows for the function $\underline{f}$ that

  $$\begin{aligned} &\alpha_{ur}\, f_u(v_i, \ldots, w_j, M_{kl}, \ldots, N_{mn}) \\ &\quad = f_r(\alpha_{pi}\, v_p, \ldots, \alpha_{qj}\, w_q, \alpha_{pk}\, \alpha_{ql}\, M_{pq}, \ldots, \alpha_{pm}\, \alpha_{qn}\, N_{pq}). \end{aligned} \tag{5.18}$$

- For a polar second-order tensor $\underline{\underline{T}} = \underline{\underline{f}}(\underline{v}, \ldots, \underline{w}, \underline{\underline{M}}, \ldots, \underline{\underline{N}})$, substituting the transformed coordinates $\widetilde{v}_i, \ldots, \widetilde{w}_j, \widetilde{M}_{kl}, \ldots, \widetilde{N}_{mn}$ must give the transformed tensor coordinates

  $$\widetilde{T}_{rs} = \alpha_{ur}\, \alpha_{vs}\, T_{uv} = f_{rs}(\widetilde{v}_i, \ldots, \widetilde{w}_j, \widetilde{M}_{kl}, \ldots, \widetilde{N}_{mn}).$$

  With

  $$T_{uv} = f_{uv}(v_i, \ldots, w_j, M_{kl}, \ldots, N_{mn}),$$

  we then obtain for the function $\underline{\underline{f}}$

  $$\begin{aligned} &\alpha_{ur}\, \alpha_{vs}\, f_{uv}(v_i, \ldots, w_j, M_{kl}, \ldots, N_{mn}) \\ &\quad = f_{rs}\,(\alpha_{pi}\, v_p, \ldots, \alpha_{qj}\, w_q, \alpha_{pk}\, \alpha_{ql}\, M_{pq}, \ldots, \alpha_{pm}\, \alpha_{qn}\, N_{pq}). \end{aligned} \tag{5.19}$$

If the transformation coefficients a_{ij} form an arbitrary orthogonal matrix (we also say that they encompass all orthogonal transformations), we call (5.17), (5.18) and (5.19) isotropic tensor functions.

The functions (5.17), (5.18), and (5.19) can in general further depend on polar scalars; since this does not impose any restrictions on the functions, we did not include scalars in the list of arguments.

We derived the invariance conditions here only for polar tensors, but they can be easily transferred to axial tensors by means of the transformation equations (2.18).

5.4.2 Scalar-Valued Functions

In order to satisfy the invariance condition (5.17), a scalar cannot depend on individual tensor coordinates, because, in general, coordinates change under a transformation of the coordinate system, so a scalar can only depend on combinations of coordinates which are invariants and thus also scalars. If it is possible to form an integrity basis with P irreducible invariants $I_1, \ldots, I_P$ from the coordinates $\underline{v}, \ldots, \underline{w}, \underline{\underline{M}}, \ldots, \underline{\underline{N}}$, then the scalar-valued function f satisfies

$$s = f(\underline{v}, \ldots, \underline{w}, \underline{\underline{M}}, \ldots, \underline{\underline{N}}) = f(I_1, I_2, \ldots, I_P). \tag{5.20}$$

The theory of the representation of tensor functions does not provide any more information about the function f, so for a particular physical process one has to rely on experiments. The type and the number P of the invariants $I_1, \ldots, I_P$ depend on the particular case and must be determined according to the rules from Section 5.3.

5.4.3 Vector-Valued Functions

1. We can obtain a scalar from a vector by means of the scalar product of the vector and another vector. This suggests that, with the help of the scalar product of the vector-valued function $\underline{f}$ and an arbitrary auxiliary vector $\underline{h}$ and by adding the auxiliary vector to the argument list, we can reduce the problem of finding a representation of a vector-valued function to the problem of finding a representation of a scalar-valued function:

$$\underline{u} \cdot \underline{h} = \underline{f}(\underline{v}, \ldots, \underline{w}, \underline{\underline{M}}, \ldots, \underline{\underline{N}}) \cdot \underline{h} = f(\underline{v}, \ldots, \underline{w}, \underline{\underline{M}}, \ldots, \underline{\underline{N}}, \underline{h}).$$

Then f is, according to Section 5.4.2, a function of the P irreducible invariants $I_1, \ldots, I_P$, which can be formed from the $\underline{v}, \ldots, \underline{w}, \underline{\underline{M}}, \ldots, \underline{\underline{N}}$ and the simultaneous invariants with a vector $\underline{h}$. However, since $\underline{h}$ is only an auxiliary vector, which must cancel out from the vector-valued function $\underline{f}$ at the end, we need only those simultaneous invariants which are linear in $\underline{h}$. These simultaneous invariants have the form $\underline{J}_i \cdot \underline{h}$, where $\underline{J}_i$ is a set of Q vectors, which are computed from the $\underline{v}, \ldots, \underline{w}, \underline{\underline{M}}, \ldots, \underline{\underline{N}}$ and which have to be determined for each case individually. The vectors $\underline{J}_i$ are also called the generators of the representation and a complete set of generators is called a function basis. We start by writing $\underline{u} \cdot \underline{h}$ as a linear combination of the $\underline{J}_i \cdot \underline{h}$, i. e.

$$\underline{u} \cdot \underline{h} = k_1 \underline{J}_1 \cdot \underline{h} + \cdots + k_Q \underline{J}_Q \cdot \underline{h}, \tag{a}$$

where the coefficients $k_1, \ldots, k_Q$ are scalars which can still depend on the invariants $I_1, \ldots, I_P$ of the integrity basis for $\underline{v}, \ldots, \underline{w}, \underline{\underline{M}}, \ldots, \underline{\underline{N}}$, i. e. we have

$$k_i = k_i(I_1, \ldots, I_P).$$

Since each term in (a) is a scalar product with the auxiliary vector $\underline{h}$, we can cancel $\underline{h}$ out, and thus we obtain a representation of the original vector-valued function $\underline{f}$, which inherently satisfies the invariance condition (5.18) due to its construction. We have

$$\underline{u} = k_1 \underline{J}_1 + \cdots + k_Q \underline{J}_Q. \tag{5.21}$$

We show by means of two examples how to determine the generators $\underline{J}_i$ of this representation for a particular case.

2. In the first example, we seek a representation of a vector $\underline{u}$ which depends only on another vector $\underline{v}$,

$$\underline{u} = \underline{f}(\underline{v}).$$

Scalar multiplication with an auxiliary vector $\underline{h}$ gives

$$\underline{u} \cdot \underline{h} = f(\underline{v}, \underline{h}).$$

According to Section 5.3.1, the vector $\underline{v}$ has only one irreducible invariant, i. e. its square $\underline{v} \cdot \underline{v}$, and from the vectors $\underline{v}$ and $\underline{h}$ we can form also only one irreducible simultaneous invariant, and this is linear in $\underline{h}$, namely the scalar product $\underline{v} \cdot \underline{h}$. So the representation of the vector $\underline{u}$ has only one generator, namely the vector $\underline{v}$ itself, and we have

$$\underline{u} = k(\underline{v} \cdot \underline{v})\, \underline{v}.$$

The function $k(\underline{v} \cdot \underline{v})$ cannot be determined further from the theory of the representation of tensor functions. For example, for a physical process, the missing information must be obtained from experiments.

3. In a second example, we extend the function $\underline{f}$ from No. 2 and include a symmetric tensor $\underline{\underline{S}}$, so we seek a representation for

$$\underline{u} = \underline{f}(\underline{v}, \underline{\underline{S}})$$

or, after scalar multiplication by an auxiliary vector $\underline{h}$,

$$\underline{u} \cdot \underline{h} = f(\underline{v}, \underline{\underline{S}}, \underline{h}).$$

As in No. 2, the scalar product $\underline{v} \cdot \underline{h}$ is a simultaneous invariant which is linear in $\underline{h}$. In addition, other linear simultaneous invariants can be formed with the help of the scalar product of $\underline{\underline{S}}$ and $\underline{v}$ on the left and $\underline{h}$ on the right; since $\underline{\underline{S}}$ is symmetric, the order does not matter. Doing the same with the integer powers of $\underline{\underline{S}}$ yields initially the invariants $\underline{v} \cdot \underline{\underline{S}} \cdot \underline{h}$, $\underline{v} \cdot \underline{\underline{S}}^2 \cdot \underline{h}$, $\underline{v} \cdot \underline{\underline{S}}^3 \cdot \underline{h}$, etc. Since $\underline{v} \cdot \underline{h}$ can also be written as $\underline{v} \cdot \underline{\underline{\delta}} \cdot \underline{h}$, we see that $\underline{v} \cdot \underline{\underline{S}}^3 \cdot \underline{h}$ (and correspondingly any expression with higher powers of $\underline{\underline{S}}$) is reducible, because we can write $\underline{\underline{S}}^3$, using the Cayley–Hamilton theorem (3.94), in terms of $\underline{\underline{S}}^2$, $\underline{\underline{S}}$, and $\underline{\underline{\delta}}$. Hence there are three generators for the representation of the vector $\underline{u}$, i. e.

$$\underline{u} = k_1\, \underline{v} + k_2\, \underline{v} \cdot \underline{\underline{S}} + k_3\, \underline{v} \cdot \underline{\underline{S}}^2.$$

The coefficients k_1, k_2, k_3 are scalar-valued functions of the invariants of $\underline{v}$ and $\underline{\underline{S}}$. Since here we consider only polar vectors and tensors, it follows from the result to Problem 5.3 that

$$k_i = f(\operatorname{tr} \underline{\underline{S}}, \operatorname{tr} \underline{\underline{S}}^2, \operatorname{tr} \underline{\underline{S}}^3, \underline{v} \cdot \underline{v}, \underline{v} \cdot \underline{\underline{S}} \cdot \underline{v}, \underline{v} \cdot \underline{\underline{S}}^2 \cdot \underline{v}).$$

Problem 5.4.
Find the representation of a vector $\underline{u}$ which depends on a polar tensor $\underline{\underline{T}}$ for the following cases:
- $\underline{\underline{T}}$ is symmetric or antimetric;
- $\underline{u}$ is polar or axial.

5.4.4 Tensor-Valued Functions

1. It is easy to generalize our work from Section 5.4.3 for vector-valued functions to tensor-valued functions; we simply introduce an auxiliary tensor $\underline{\underline{H}}$ and use the double scalar product; this again ensures that the result inherently satisfies the invariance condition (5.19) for tensor-valued functions. Then, from

$$\underline{\underline{T}} = \underline{\underline{f}}(\underline{v}, \ldots, \underline{w}, \underline{\underline{M}}, \ldots, \underline{\underline{N}})$$

we first get

$$\underline{\underline{T}} \cdot\cdot \underline{\underline{H}} = f(\underline{v}, \ldots, \underline{w}, \underline{\underline{M}}, \ldots, \underline{\underline{N}}, \underline{\underline{H}}).$$

The generators $\underline{\underline{J}}_i$ are now second-order tensors, because in order for $\underline{\underline{H}}$ to cancel out at the end, the simultaneous invariants with the auxiliary tensor must have the linear form $\underline{\underline{J}}_i \cdot\cdot \underline{\underline{H}}$. Then we can represent $\underline{\underline{T}} \cdot\cdot \underline{\underline{H}}$ as the linear combination of all Q simultaneous invariants, i. e. we have

$$\underline{\underline{T}} \cdot\cdot \underline{\underline{H}} = k_1 \underline{\underline{J}}_1 \cdot\cdot \underline{\underline{H}} + \cdots + k_Q \underline{\underline{J}}_Q \cdot\cdot \underline{\underline{H}},$$

and, after canceling $\underline{\underline{H}}$ out, we obtain for the original tensor-valued function $\underline{\underline{f}}$

$$\underline{\underline{T}} = k_1 \underline{\underline{J}}_1 + \cdots + k_Q \underline{\underline{J}}_Q. \tag{5.22}$$

As in Section 5.4.3, the coefficients $k_1, \ldots, k_Q$ are scalars which can depend on the invariants $I_1, \ldots, I_P$ of an integrity basis for $\underline{v}, \ldots, \underline{w}, \underline{\underline{M}}, \ldots, \underline{\underline{N}}$, i. e.

$$k_i = k_i(I_1, \ldots, I_P).$$

The type and number Q of the generators $\underline{\underline{J}}_1, \ldots, \underline{\underline{J}}_Q$ can only be determined for each particular case; we will again explain this with two examples.

2. We first consider a second-order tensor $\underline{\underline{T}}$ which depends only on a symmetric second-order tensor $\underline{\underline{S}}$, so we seek a representation for

$$\underline{\underline{T}} = \underline{\underline{f}}(\underline{\underline{S}}).$$

The double scalar product with an auxiliary tensor $\underline{\underline{H}}$ gives

$$\underline{\underline{T}} \cdot\cdot \underline{\underline{H}} = f(\underline{\underline{S}}, \underline{\underline{H}}).$$

Since $\underline{\underline{S}}$ is symmetric and because we only need the invariants which are linear in $\underline{\underline{H}}$, we can start from (5.15) and set there $\underline{\underline{a}} = \underline{\underline{S}}$ and $\underline{\underline{b}} = \underline{\underline{H}}$. From this we obtain the three basic invariants $\operatorname{tr}\underline{\underline{S}}$, $\operatorname{tr}\underline{\underline{S}}^2$, $\operatorname{tr}\underline{\underline{S}}^3$ of the symmetric tensor $\underline{\underline{S}}$ and, using the symmetry of $\underline{\underline{S}}$, the invariants which are linear in $\underline{\underline{H}}$,

$$\operatorname{tr}\underline{\underline{H}} = H_{ii} = \delta_{ij} H_{ij} = \underline{\underline{\delta}} \cdot\cdot \underline{\underline{H}},$$
$$\operatorname{tr}(\underline{\underline{S}} \cdot \underline{\underline{H}}) = S_{ij} H_{ji} = S_{ij} H_{ij} = \underline{\underline{S}} \cdot\cdot \underline{\underline{H}},$$
$$\operatorname{tr}(\underline{\underline{S}}^2 \cdot \underline{\underline{H}}) = S_{ij} S_{jk} H_{ki} = S_{kj} S_{ji} H_{ki} = \underline{\underline{S}}^2 \cdot\cdot \underline{\underline{H}}.$$

Hence we have three generators $\underline{\underline{\delta}}$, $\underline{\underline{S}}$, $\underline{\underline{S}}^2$, and the representation of the tensor $\underline{\underline{T}}$ is

$$\underline{\underline{T}} = k_1 \underline{\underline{\delta}} + k_2 \underline{\underline{S}} + k_3 \underline{\underline{S}}^2,$$

where the coefficients k_1, k_2, k_3 are scalar-valued functions of the three basic invariants of $\underline{\underline{S}}$, i. e.

$$k_i = f(\operatorname{tr}\underline{\underline{S}}, \operatorname{tr}\underline{\underline{S}}^2, \operatorname{tr}\underline{\underline{S}}^3).$$

Since we assumed that the tensor $\underline{\underline{S}}$ is symmetric, we immediately obtain from the theory of the representation of tensor functions that the tensor $\underline{\underline{T}}$ must also be symmetric.

3. As a second example, we consider a tensor $\underline{\underline{T}}$ which depends on a symmetric tensor $\underline{\underline{S}}$ and also on a vector $\underline{v}$, i. e. we are looking for a representation for

$$\underline{\underline{T}} = \underline{\underline{f}}(\underline{\underline{S}}, \underline{v}).$$

The double scalar product with the auxiliary tensor $\underline{\underline{H}}$ gives

$$\underline{\underline{T}} \cdot\cdot \underline{\underline{H}} = f(\underline{\underline{S}}, \underline{v}, \underline{\underline{H}}).$$

We can keep the invariants of $\underline{\underline{S}}$ and $\underline{v}$ from Section 5.4.3, No. 3, so we only need to compute the simultaneous invariants which are linear in $\underline{\underline{H}}$. As in No. 2 we initially find

$$\underline{\underline{\delta}} \cdot\cdot \underline{\underline{H}},\ \underline{\underline{S}} \cdot\cdot \underline{\underline{H}},\ \underline{\underline{S}}^2 \cdot\cdot \underline{\underline{H}}.$$

Another simultaneous invariant can be formed with the vector $\underline{v}$:

$$\underline{v}\,\underline{v} \cdot\cdot \underline{\underline{H}}.$$

In order to form the simultaneous invariants with both $\underline{\underline{S}}$ and $\underline{v}$, we start from $\underline{v}\,\underline{v} \cdot\cdot \underline{\underline{H}}$ and form scalar products with $\underline{v}\,\underline{v}$ and $\underline{\underline{S}}$ or $\underline{\underline{S}}^2$ on the right or on the left; we can neglect higher powers of $\underline{\underline{S}}$, because of the Cayley–Hamilton theorem (3.95). This gives

$$(\underline{\underline{S}} \cdot \underline{v}\,\underline{v}) \cdot\cdot \underline{\underline{H}},\ (\underline{v}\,\underline{v} \cdot \underline{\underline{S}}) \cdot\cdot \underline{\underline{H}},\ (\underline{\underline{S}}^2 \cdot \underline{v}\,\underline{v}) \cdot\cdot \underline{\underline{H}},\ (\underline{v}\,\underline{v} \cdot \underline{\underline{S}}^2) \cdot\cdot \underline{\underline{H}},$$
$$(\underline{\underline{S}} \cdot \underline{v}\,\underline{v} \cdot \underline{\underline{S}}) \cdot\cdot \underline{\underline{H}},\ (\underline{\underline{S}}^2 \cdot \underline{v}\,\underline{v} \cdot \underline{\underline{S}}) \cdot\cdot \underline{\underline{H}},\ (\underline{\underline{S}} \cdot \underline{v}\,\underline{v} \cdot \underline{\underline{S}}^2) \cdot\cdot \underline{\underline{H}},\ (\underline{\underline{S}}^2 \cdot \underline{v}\,\underline{v} \cdot \underline{\underline{S}}^2) \cdot\cdot \underline{\underline{H}}.$$

It remains to check if these simultaneous invariants are irreducible. Clearly, $(\underline{\underline{S}} \cdot \underline{v}\,\underline{v}) \cdot\cdot \underline{\underline{H}}$ and $(\underline{v}\,\underline{v} \cdot \underline{\underline{S}}) \cdot\cdot \underline{\underline{H}}$ are irreducible. However, the scalar products $\underline{\underline{S}} \cdot \underline{v}\,\underline{v} \cdot \underline{\underline{S}}$, $\underline{\underline{S}}^2 \cdot \underline{v}\,\underline{v}$, and $\underline{v}\,\underline{v} \cdot \underline{\underline{S}}^2$ are linked via the generalized Cayley–Hamilton theorem (5.1), for example, if we set $\underline{\underline{a}} = \underline{\underline{c}} = \underline{\underline{S}}$ and $\underline{\underline{b}} = \underline{v}\,\underline{v}$. Thus, from the three invariants $(\underline{\underline{S}}^2 \cdot \underline{v}\,\underline{v}) \cdot\cdot \underline{\underline{H}}$, $(\underline{v}\,\underline{v} \cdot \underline{\underline{S}}^2) \cdot\cdot \underline{\underline{H}}$, and $(\underline{\underline{S}} \cdot \underline{v}\,\underline{v} \cdot \underline{\underline{S}}) \cdot\cdot \underline{\underline{H}}$ only two are irreducible, and we choose $\underline{\underline{S}}^2 \cdot \underline{v}\,\underline{v}$ and $\underline{v}\,\underline{v} \cdot \underline{\underline{S}}^2$ for the function basis. The remaining invariants $(\underline{\underline{S}}^2 \cdot \underline{v}\,\underline{v} \cdot \underline{\underline{S}}) \cdot\cdot \underline{\underline{H}}$, $(\underline{\underline{S}} \cdot \underline{v}\,\underline{v} \cdot \underline{\underline{S}}^2) \cdot\cdot \underline{\underline{H}}$, and $(\underline{\underline{S}}^2 \cdot \underline{v}\,\underline{v} \cdot \underline{\underline{S}}^2) \cdot\cdot \underline{\underline{H}}$ are reducible, since, according to the generalized Cayley–Hamilton theorem (5.1), the scalar products $\underline{\underline{S}}^2 \cdot \underline{v}\,\underline{v} \cdot \underline{\underline{S}}$, $\underline{\underline{S}} \cdot \underline{v}\,\underline{v} \cdot \underline{\underline{S}}^2$, and $\underline{\underline{S}}^2 \cdot \underline{v}\,\underline{v} \cdot \underline{\underline{S}}^2$ can be written in terms of $\underline{\underline{\delta}}$, $\underline{\underline{S}}$, $\underline{\underline{S}}^2$, $\underline{v}\,\underline{v}$, $\underline{\underline{S}} \cdot \underline{v}\,\underline{v}$, $\underline{v}\,\underline{v} \cdot \underline{\underline{S}}$, $\underline{\underline{S}}^2 \cdot \underline{v}\,\underline{v}$, $\underline{v}\,\underline{v} \cdot \underline{\underline{S}}^2$.

Hence the function $\underline{\underline{f}}$ has eight generators, and we can represent the tensor $\underline{\underline{T}}$ as

$$\underline{\underline{T}} = k_1\,\underline{\underline{\delta}} + k_2\,\underline{\underline{S}} + k_3\,\underline{\underline{S}}^2 + k_4\,\underline{v}\,\underline{v} + k_5\,\underline{\underline{S}} \cdot \underline{v}\,\underline{v} + k_6\,\underline{v}\,\underline{v} \cdot \underline{\underline{S}} + k_7\,\underline{\underline{S}}^2 \cdot \underline{v}\,\underline{v} + k_8\,\underline{v}\,\underline{v} \cdot \underline{\underline{S}}^2.$$

The coefficients $k_1, \ldots, k_8$ are, as in Section 5.4.3, No. 3, scalar-valued functions of the invariants of $\underline{\underline{S}}$ and $\underline{v}$, i. e.

$$k_i = f(\operatorname{tr}\underline{\underline{S}}, \operatorname{tr}\underline{\underline{S}}^2, \operatorname{tr}\underline{\underline{S}}^3, \underline{v}\cdot\underline{v}, \underline{v}\cdot\underline{\underline{S}}\cdot\underline{v}, \underline{v}\cdot\underline{\underline{S}}^2\cdot\underline{v}).$$

In contrast to No. 2, here we cannot conclude from the symmetry of $\underline{\underline{S}}$ that $\underline{\underline{T}}$ is also symmetric. This is only true if $k_5 = k_6$ and $k_7 = k_8$. But if $\underline{\underline{T}}$ is symmetric, then we can represent $\underline{\underline{T}}$ also in the form

$$\underline{\underline{T}} = k_1^*\,\underline{\underline{\delta}} + k_2^*\,\underline{\underline{S}} + k_3^*\,\underline{\underline{S}}^2 + k_4^*\,\underline{v}\,\underline{v} + k_5^*\,\left(\underline{\underline{S}}\cdot\underline{v}\,\underline{v} + \underline{v}\,\underline{v}\cdot\underline{\underline{S}}\right) + k_6^*\,\underline{\underline{S}}\cdot\underline{v}\,\underline{v}\cdot\underline{\underline{S}}.$$

If $\underline{\underline{T}}$ is symmetric, then $\underline{\underline{S}}^2\cdot\underline{v}\,\underline{v}$ and $\underline{v}\,\underline{v}\cdot\underline{\underline{S}}^2$ can only appear as a sum, and according to the generalized Cayley–Hamilton theorem (5.1), we can replace this sum by the generator $\underline{\underline{S}}\cdot\underline{v}\,\underline{v}\cdot\underline{\underline{S}}$, which is symmetric from the outset. The coefficients $k_1^*, \ldots, k_6^*$ are again scalar-valued functions of the invariants of $\underline{\underline{S}}$ and $\underline{v}$.

Problem 5.5.
Find the representation of a polar second-order tensor $\underline{\underline{T}}$ which depends on a polar vector or on an axial vector $\underline{v}$.

5.4.5 Summary

We summarize the results of Section 5.4 in a table. Here $\underline{\underline{S}}$ denotes a polar symmetric tensor, $\underline{\underline{A}}$ a polar antimetric tensor, $\underline{v}$ a polar vector, and $\underline{u}$ an axial vector.

Argument	Function basis	
	vector-valued functions	
	polar	axial
$\underline{v}$	$\underline{v}$	
$\underline{u}$		$\underline{u}$
$\underline{\underline{S}}$	–	
$\underline{\underline{A}}$		$\underline{\underline{\epsilon}}\cdot\cdot\underline{\underline{A}}$
$\underline{\underline{S}}, \underline{v}$	$\underline{v}, \underline{\underline{S}}\cdot\underline{v}, \underline{\underline{S}}^2\cdot\underline{v}$	
	tensor-valued functions	
	polar	axial
$\underline{v}$	$\underline{\underline{\delta}}, \underline{v}\,\underline{v}$	$\underline{\underline{\epsilon}}\cdot\underline{v}$
$\underline{u}$	$\underline{\underline{\delta}}, \underline{u}\,\underline{u}, \underline{\underline{\epsilon}}\cdot\underline{u}$	
$\underline{\underline{S}}$	$\underline{\underline{\delta}}, \underline{\underline{S}}, \underline{\underline{S}}^2$	
$\underline{\underline{S}}, \underline{v}$	$\underline{\underline{\delta}}, \underline{\underline{S}}, \underline{\underline{S}}^2, \underline{v}\,\underline{v}, \underline{\underline{S}}\cdot\underline{v}\,\underline{v}, \underline{v}\,\underline{v}\cdot\underline{\underline{S}}, \underline{\underline{S}}^2\cdot\underline{v}\,\underline{v}, \underline{v}\,\underline{v}\cdot\underline{\underline{S}}^2$	

5.5 Considering Anisotropy

1. In the previous section, we called a tensor function isotropic if the transformation coefficients in the invariance conditions (5.17), (5.18) and (5.19) include all orthogonal transformations. We could also consider only a subset of certain permissible orthogonal transformations, e. g. only rotations about a fixed axis. Such considerations are important in physics and material science, if the problem is to describe the properties of a material in more detail; many materials, such as crystals and composite materials, are characterized by a directional dependency, i. e. they behave differently under rotations, depending on the direction of the axis of rotation. In such cases, often certain rotations about distinguished axes and angles of rotation exist, where the behavior of these materials remains the same. These rotations can be used to classify a material; we also say that the material possesses a certain symmetry or belongs to a certain symmetry group. Mathematically, such a symmetry group is defined as the set of orthogonal transformations under which the behavior of the material remains unchanged.

2. The invariance conditions from Section 5.4.1 can also be used if only a subset of all orthogonal transformations is allowed. Without further elaboration, we only mention here that, in general, the number of the invariants increases if the permissible transformations are restricted. An example is the restriction to only proper orthogonal transformations. Then we also speak of hemitropic invariants and hemitropic tensor functions. They are formed similarly as in Sections 5.3 and 5.4, but now we also have to take the simultaneous invariants with the ε-tensor into account, because for proper orthogonal transformations, we do not need to distinguish between polar and axial tensors. Another example are rotations only about a certain axis. If we choose this axis as the z-axis of a Cartesian coordinate system, then the matrix of the transformation coefficients has, according to (3.75) and with Section 3.13.4, the form

$$\alpha_{ij} = \begin{pmatrix} \cos\varphi & -\sin\varphi & 0 \\ \sin\varphi & \cos\varphi & 0 \\ 0 & 0 & 1 \end{pmatrix}.$$

Then it follows from the transformation equations (2.17) that the coordinate u_3 of a vector and the coordinate T_{33} of a second-order tensor remain unchanged under the transformation and thus they have to be counted as invariants.

3. All tensors which are invariant under the transformations of a certain symmetry group form an anisotropy class. We explain this by means of three examples with second-order tensors.

The anisotropy class of general anisotropy includes all tensors; its symmetry group consists of the identical transformation $\alpha_{ij} = \delta_{ij}$ and the inversion $\alpha_{ij} = -\delta_{ij}$, because only under these transformations the coordinates do not change.

We have

$$\tilde{T}_{ij} = \alpha_{mi}\,\alpha_{nj}\,T_{mn} = \delta_{mi}\,\delta_{nj}\,T_{mn} = T_{ij}.$$

On the other hand, the anisotropy class of isotropy includes all tensors whose coordinates are invariant under arbitrary orthogonal transformations. We already introduced such tensors as isotropic tensors; they have the form

$$\underline{\underline{T}} = k\,\underline{\underline{\delta}},$$

because for the transformed coordinates, we have, according to the orthogonality relation (2.6),

$$\tilde{T}_{ij} = \alpha_{mi}\,\alpha_{nj}\,(k\,\delta_{mn}) = k\,\alpha_{mi}\,\alpha_{mj} = k\,\delta_{ij}.$$

As a third example, we consider the anisotropy class of transverse isotropy. Its symmetry group includes all rotations and rotary reflections with a fixed axis. The general form of a transversely isotropic tensor of second order follows from Problem 5.5. A transversely isotropic tensor can be seen as a tensor which depends only on the direction $\underline{n}$ of the axis of rotation or rotary reflection: $\underline{\underline{T}} = \underline{\underline{f}}(\underline{n})$. Thus, it remains to ensure in the solution to Problem 5.5 that $\underline{n}$ is an (axial) unit vector, and we have

$$\underline{\underline{T}} = \alpha\,\underline{\underline{\delta}} + \beta\,\underline{n}\,\underline{n} + \gamma\,\underline{\underline{\underline{\varepsilon}}}\cdot\underline{n}. \tag{5.23}$$

Since $\underline{n}\cdot\underline{n} = 1$, here the α, β, γ are, unlike in Problem 5.5, not functions, but arbitrary constants.

If we choose the z-axis of the Cartesian coordinate system as the axis of rotation or rotary reflection, i. e. if $n_1 = n_2 = 0$, $n_3 = 1$, then $\underline{\underline{T}}$ has the coordinate matrix

$$\begin{aligned} T_{ij} &= \begin{pmatrix} \alpha & 0 & 0 \\ 0 & \alpha & 0 \\ 0 & 0 & \alpha \end{pmatrix} + \begin{pmatrix} 0 & 0 & 0 \\ 0 & 0 & 0 \\ 0 & 0 & \beta \end{pmatrix} + \begin{pmatrix} 0 & \gamma & 0 \\ -\gamma & 0 & 0 \\ 0 & 0 & 0 \end{pmatrix} \\ &= \begin{pmatrix} \alpha & \gamma & 0 \\ -\gamma & \alpha & 0 \\ 0 & 0 & \alpha+\beta \end{pmatrix}. \end{aligned}$$

It is easy to see that, if we evaluate the transformation equations, the coordinates of $\underline{\underline{T}}$ remain unchanged under a rotation of the coordinate system about the z-axis, i. e.

we have

$$\tilde{T}_{ij} = \alpha_{mi}\,\alpha_{nj}\,T_{mn} = \alpha^{\mathrm{T}}_{im}\,T_{mn}\,\alpha_{nj}$$

$$= \begin{pmatrix} \cos\varphi & \sin\varphi & 0 \\ -\sin\varphi & \cos\varphi & 0 \\ 0 & 0 & 1 \end{pmatrix} \begin{pmatrix} \alpha & \gamma & 0 \\ -\gamma & \alpha & 0 \\ 0 & 0 & \alpha+\beta \end{pmatrix} \begin{pmatrix} \cos\varphi & -\sin\varphi & 0 \\ \sin\varphi & \cos\varphi & 0 \\ 0 & 0 & 1 \end{pmatrix}$$

$$= \begin{pmatrix} \alpha\cos\varphi - \gamma\sin\varphi & \alpha\sin\varphi + \gamma\cos\varphi & 0 \\ -\alpha\sin\varphi - \gamma\cos\varphi & \alpha\cos\varphi - \gamma\sin\varphi & 0 \\ 0 & 0 & \alpha+\beta \end{pmatrix} \begin{pmatrix} \cos\varphi & -\sin\varphi & 0 \\ \sin\varphi & \cos\varphi & 0 \\ 0 & 0 & 1 \end{pmatrix}$$

$$= \begin{pmatrix} \alpha\cos^2\varphi - \gamma\sin\varphi\,\cos\varphi + \alpha\sin^2\varphi + \gamma\sin\varphi\,\cos\varphi & -\alpha\cos\varphi\,\sin\varphi + \gamma\sin^2\varphi + \alpha\cos\varphi\sin\varphi + \gamma\cos^2\varphi & 0 \\ -\alpha\cos\varphi\,\sin\varphi - \gamma\cos^2\varphi + \alpha\cos\varphi\,\sin\varphi - \gamma\sin^2\varphi & \alpha\sin^2\varphi + \gamma\sin\varphi\,\cos\varphi + \alpha\cos^2\varphi - \gamma\sin\varphi\,\cos\varphi & 0 \\ 0 & 0 & \alpha+\beta \end{pmatrix}$$

$$= \begin{pmatrix} \alpha & \gamma & 0 \\ -\gamma & \alpha & 0 \\ 0 & 0 & \alpha+\beta \end{pmatrix}.$$

The class of transversely isotropic tensors has some subclasses of tensors which we have already seen earlier. If we set $\alpha = \cos\vartheta, \beta = \pm 1 - \cos\vartheta, \gamma = -\sin\vartheta,$ then we obtain for $\underline{\underline{T}}$, according to (3.77) and (3.81), the general form of an orthogonal tensor, with the angle of rotation ϑ and the axis of rotation or rotary reflection $\underline{n}$. If we set $\alpha = \beta = 0$, then $\underline{\underline{T}}$ is antimetric and $\gamma\,\underline{n}$ is the corresponding vector of $\underline{\underline{T}}$. If we choose $\gamma = 0$, then the tensor is symmetric with an eigenvalue α of multiplicity 2 and another eigenvalue $\alpha + \beta$ of multiplicity 1; $\underline{n}$ is the eigendirection corresponding to the eigenvalue $\alpha + \beta$ of multiplicity 1.

We summarize the results of our discussion on anisotropy classes of second-order tensors in a table. For the anisotropy classes, the number of tensors $\underline{\underline{T}}$, which are contained in the class, increases from the first class to the last; the tensors of a particular anisotropy class are always included in the next anisotropy class. For the symmetry groups, the number of orthogonal matrices α_{ij} which are included in the class increases from the last class to the first; the matrices of a symmetry group are always contained in the previous symmetry group.

4. Anisotropies can also be considered by isotropic tensor functions. To explain this we consider the stress–strain relation for an elastic body. When we introduced the function $\underline{\underline{T}} = \underline{\underline{f}}(\underline{\underline{D}})$ between the symmetric stress tensor $\underline{\underline{T}}$ and the symmetric strain

Anisotropy class	General tensor $\underline{\underline{T}}$	Symmetry group α_{ij}
Isotropy	$k\,\underline{\underline{\delta}}$	arbitrary
Transverse isotropy	$\alpha\,\underline{\underline{\delta}} + \beta\,\underline{n}\,\underline{n} + \gamma\,\underline{\underline{\underline{\varepsilon}}}\cdot\underline{n}$	$\begin{pmatrix} \cos\varphi & -\sin\varphi & 0 \\ \sin\varphi & \cos\varphi & 0 \\ 0 & 0 & 1 \end{pmatrix}$, $\underline{n} = \underline{e}_z$
General anisotropy	arbitrary	$\pm\delta_{ij}$

tensor $\underline{\underline{D}}$ in Section 5.1, we silently assumed an isotropic solid, and, according to Section 5.4.4, No. 2, the function $\underline{\underline{f}}$ has the following representation:

$$\underline{\underline{T}} = k_1\,\underline{\underline{\delta}} + k_2\,\underline{\underline{D}} + k_3\,\underline{\underline{D}}^2, \quad k_i = f(\operatorname{tr}\underline{\underline{D}}, \operatorname{tr}\underline{\underline{D}}^2, \operatorname{tr}\underline{\underline{D}}^3).$$

If, on the other hand, the solid is transversely isotropic, then it has a distinguished direction $\underline{n}$, which we have to include in the argument list of the function $\underline{\underline{f}}$, i. e. we are looking for a representation for $\underline{\underline{T}} = \underline{\underline{f}}(\underline{\underline{D}}, \underline{n})$. We can use the result from Section 5.4.4, No. 3, if in addition we take into account that $\underline{\underline{T}}$ is symmetric and that $\underline{n}$ is a unit vector, i. e. that $\underline{n}\cdot\underline{n}$ is no invariant, i. e.

$$\begin{aligned}\underline{\underline{T}} &= k_1^*\,\underline{\underline{\delta}} + k_2^*\,\underline{\underline{D}} + k_3^*\,\underline{\underline{D}}^2 + k_4^*\,\underline{n}\,\underline{n} + k_5^*(\underline{\underline{D}}\cdot\underline{n}\,\underline{n} + \underline{n}\,\underline{n}\cdot\underline{\underline{D}}) + k_6^*\,\underline{\underline{D}}\cdot\underline{n}\,\underline{n}\cdot\underline{\underline{D}},\\ k_i^* &= f(\operatorname{tr}\underline{\underline{D}}, \operatorname{tr}\underline{\underline{D}}^2, \operatorname{tr}\underline{\underline{D}}^3, \underline{n}\cdot\underline{\underline{D}}\cdot\underline{n}, \underline{n}\cdot\underline{\underline{D}}^2\cdot\underline{n}).\end{aligned} \tag{a}$$

This expression includes the stress–strain relation for an isotropic solid as a special case, if we set $k_4^* = k_5^* = k_6^* = 0$ and if k_1^*, k_2^*, and k_3^* do not depend on $\underline{n}$.

For small strains we can approximate (a) by a linear relation between the coordinates of $\underline{\underline{T}}$ and $\underline{\underline{D}}$. Then $k_3^* = k_6^* = 0$, $k_2^* = \alpha$, and $k_5^* = \beta$ are constant, and k_1^* and k_4^* are linear functions of the invariants $\operatorname{tr}\underline{\underline{D}}$ and $\underline{n}\cdot\underline{\underline{D}}\cdot\underline{n}$ which are linear in $\underline{\underline{D}}$, so we have

$$k_1^* = \kappa\operatorname{tr}\underline{\underline{D}} + \lambda\,\underline{n}\cdot\underline{\underline{D}}\cdot\underline{n}, \quad k_4^* = \mu\operatorname{tr}\underline{\underline{D}} + \nu\,\underline{n}\cdot\underline{\underline{D}}\cdot\underline{n}.$$

Then (a) simplifies in index notation to

$$\begin{aligned}T_{ij} = {} & (\kappa\,D_{pp} + \lambda\,n_p\,D_{pq}\,n_q)\,\delta_{ij} + (\mu\,D_{pp} + \nu\,n_p\,D_{pq}\,n_q)\,n_i\,n_j\\ & + \alpha\,D_{ij} + \beta\,(D_{ip}\,n_p\,n_j + n_i\,n_p\,D_{pj})\,.\end{aligned}$$

If we introduce appropriate Kronecker symbols, we can factor out D_{kl} as follows:

$$\begin{aligned}T_{ij} &= (\kappa\,\delta_{pk}\,\delta_{pl}\,D_{kl} + \lambda\,n_p\,\delta_{pk}\,D_{kl}\,\delta_{lq}\,n_q)\,\delta_{ij}\\ &\quad + (\mu\,\delta_{pk}\,\delta_{pl}\,D_{kl} + \nu\,n_p\,\delta_{pk}\,D_{kl}\,\delta_{lq}\,n_q)\,n_i\,n_j + \alpha\,\delta_{ik}\,\delta_{jl}\,D_{kl}\\ &\quad + \beta\,(\delta_{ik}\,D_{kl}\,\delta_{lp}\,n_p\,n_j + n_i\,n_p\,\delta_{pk}\,D_{kl}\,\delta_{lj})\\ &= \{\kappa\,\delta_{ij}\,\delta_{kl} + \lambda\,\delta_{ij}\,n_k\,n_l + \mu\,n_i\,n_j\,\delta_{kl} + \nu\,n_i\,n_j\,n_k\,n_l + \alpha\,\delta_{ik}\,\delta_{jl}\\ &\quad + \beta\,(\delta_{ik}\,n_j\,n_l + n_i\,n_k\,\delta_{jl})\}D_{kl}.\end{aligned}$$

Since both T_{ij} and D_{kl} are symmetric, we can write

$$T_{ij} = \{\kappa\,\delta_{ij}\,\delta_{kl} + \lambda\,\delta_{ij}\,n_k\,n_l + \mu\,n_i\,n_j\,\delta_{kl} + \nu\,n_i\,n_j\,n_k\,n_l + \tfrac{1}{2}\,\alpha\,(\delta_{ik}\,\delta_{jl} + \delta_{il}\,\delta_{jk}) \\ + \tfrac{1}{2}\,\beta\,(\delta_{ik}\,n_j\,n_l + n_i\,n_k\,\delta_{jl} + \delta_{il}\,n_j\,n_k + n_i\,n_l\,\delta_{jk})\}D_{kl}.$$

The expression in the curly brackets can be interpreted as the coordinates of a fourth-order elasticity tensor $\underline{\underline{\underline{\underline{E}}}}$ with the symmetry properties

$$E_{ijkl} = E_{jikl} = E_{ijlk},$$

so we can shorten this to

$$T_{ij} = E_{ijkl}\,D_{kl}.$$

The constants α, β, κ, λ, μ, and ν must be determined from experiments (possibly by making further physical assumptions).

References

We would like to recommend a few books for further reading and we start with our favorites:

- Bernhard Schutz: Geometrical Methods of Mathematical Physics. Cambridge University Press, 1980.
- Marcelo Epstein: Differential Geometry, Basic Notions and Physical Examples. Springer, 2014.
- Theodore Frankel: The Geometry of Physics – An Introduction, 3rd Edition. Cambridge University Press, 1997, 2004, 2012.

Schutz encourages his readers to develop both a pictorial way of thinking and a feeling for geometrical tools in physics. He emphasizes the idea that tensors are geometrical objects independent of any coordinate system. The book has a large chapter on physical applications including thermodynamics, mechanics, electromagnetism, and gauge theories.

Epstein believes, just as we do, that what is true and good must also be beautiful, symmetric, and well balanced. He introduces the concept of a continuum as well as the field theories in the language of topology and modern differential geometry, and emphasizes the richness of these ideas using illustrations from continuum mechanics and materials science.

Frankel aims to provide a working knowledge of geometric techniques for a deeper understanding of both classical and modern physics and engineering at the level of advanced undergraduate and graduate students. He promotes a language with which mathematicians and scientists can communicate and includes applications to fluid dynamics, electromagnetism, thermodynamics, elasticity, and relativity.

Tensors naturally appear in the study of curves, surfaces, and manifolds, so many differential geometry books include chapters on tensor analysis.

An elementary account of the geometry of curves and surfaces, which only requires a standard background in calculus and linear algebra, and an introduction to the main ideas of differential geometry are given in:

- Barrett O'Neill: Elementary Differential Geometry, 2nd Edition. Elsevier, 1997, 2006.

For linear algebra and multilinear algebra, we recommend Strang and Greub, respectively:

- Gilbert Strang: Introduction to Linear Algebra, 5th Edition. Wellesley–Cambridge Press, 2016.
- Werner Greub: Multilinear Algebra, 2nd Edition. Springer, 1978.

https://doi.org/10.1515/9783110404265-006

The most important facts about a large number of curvilinear coordinate systems can be found in:

- Parry Moon, Domina Eberle Spencer: Field Theory Handbook. Springer, 1988.

A generalization of tensor analysis is presented in:

- Parry Moon, Domina Eberle Spencer: Theory of Holors. Cambridge University Press, 2005.

Readers who would like to learn more about the representation theory, introduced in Chapter 5, are encouraged to consult the specialized literature on the subject. We recommend:

- Jean-Paul Boehler (ed.): Application of Tensor Functions in Solid Mechanics. CISM Courses and Lectures, No. 292. Springer, 1987.
- A. J. M. Spencer: Theory of invariants. In: A. Cemal Eringen (ed.): Continuum Physics, Vol. I. Academic Press, 1971.

For problems that go beyond the scope of this book, we refer to the following books:

- A. M. Goodbody: Cartesian Tensors. Chichester: Ellis Horwood Limited, 1982.
- Jerald L. Ericksen: Tensor fields (Appendix to: Clifford Truesdell, R. A. Toupin: The Classical Field Theories). In: Handbuch der Physik, Vol. III/1. Springer, 1960.
- Mikhail Itskov: Tensor Algebra and Tensor Analysis for Engineers, 4th Edition. Springer, 2015.
- David Lovelock, Hanno Rund: Tensors, Differential Forms, and Variational Principles. Dover, 1989.

The following book is a very readable introduction to differential forms and their applications. Differential forms were first introduced by Cartan and they generalize antimetric tensors:

- Harley Flanders: Differential Forms with Applications to the Physical Sciences. Dover, 1989.

For further reading we recommend more of our favorites:

- Ralph Abraham, Jerrold E. Marsden, Tudor Ratiu: Manifolds, Tensor Analysis, and Applications, 2nd Edition. Addison Wesley, 1988.
- Philippe G. Ciarlet: An Introduction to Differential Geometry with Applications to Elasticity. Springer, 2005.
- Darryl D. Holm: Geometric Mechanics – Part I: Dynamics and Symmetry, 2nd Edition. Imperial College Press, 2011.
- Darryl D. Holm: Geometric Mechanics – Part II: Rotating, Translating and Rolling, 2nd Edition. Imperial College Press, 2011.
- Bo-Yu Hou, Bo-Yuan Hou: Differential Geometry for Physicists. World Scientific, 1997.

- Charles W. Misner, Kip S. Thorne, John Archibald Wheeler: Gravitation. Macmillan, W. H. Freemann and Company, 1973.
- Mikio Nakahara: Geometry, Topology, and Physics, 3rd Edition. CRC Press, 2017.
- M. Spivak: A Comprehensive Introduction to Differential Geometry, Vols. 1–5. Publish or perish Inc., 1979.
- Shlomo Sternberg: Curvature in Mathematics and in Physics. Dover, 2012.

Appendices

A Solutions to the Problems

Problem 1.1.

A. $a_i\,B_i = a_1\,B_1 + a_2\,B_2 + a_3\,B_3 + a_4\,B_4$.

B. $A_{ii} = A_{11} + A_{22} + A_{33} + A_{44}$.

C. $\dfrac{\partial u_i}{\partial x_i} = \dfrac{\partial u_1}{\partial x_1} + \dfrac{\partial u_2}{\partial x_2} + \dfrac{\partial u_3}{\partial x_3} + \dfrac{\partial u_4}{\partial x_4}$.

D. $\dfrac{\partial^2 \varphi}{\partial x_i^2}$ stands for $\dfrac{\partial^2 \varphi}{\partial x_i\,\partial x_i}$: $\dfrac{\partial^2 \varphi}{\partial x_i^2} = \dfrac{\partial^2 \varphi}{\partial x_1^2} + \dfrac{\partial^2 \varphi}{\partial x_2^2} + \dfrac{\partial^2 \varphi}{\partial x_3^2} + \dfrac{\partial^2 \varphi}{\partial x_4^2}$.

E. In order to gain experience, let us carry out the two summations one by one:

$$\begin{aligned} a_{ij}\,b_{ij} &= a_{1j}\,b_{1j} + a_{2j}\,b_{2j} + a_{3j}\,b_{3j} \\ &= a_{11}\,b_{11} + a_{12}\,b_{12} + a_{13}\,b_{13} + a_{21}\,b_{21} + a_{22}\,b_{22} + a_{23}\,b_{23} + a_{31}\,b_{31} \\ &\quad + a_{32}\,b_{32} + a_{33}\,b_{33}. \end{aligned}$$

F. $$\begin{aligned} a_{ii}\,b_{jj} &= (a_{11} + a_{22} + a_{33})(b_{11} + b_{22} + b_{33}) \\ &= a_{11}\,b_{11} + a_{22}\,b_{11} + a_{33}\,b_{11} + a_{11}\,b_{22} + a_{22}\,b_{22} + a_{33}\,b_{22} + a_{11}\,b_{33} \\ &\quad + a_{22}\,b_{33} + a_{33}\,b_{33}. \end{aligned}$$

G. $$\begin{aligned} \frac{\partial u_i}{\partial x_j}\,\frac{\partial u_i}{\partial x_j} &= \frac{\partial u_1}{\partial x_j}\,\frac{\partial u_1}{\partial x_j} + \frac{\partial u_2}{\partial x_j}\,\frac{\partial u_2}{\partial x_j} \\ &= \left(\frac{\partial u_1}{\partial x_1}\right)^2 + \left(\frac{\partial u_1}{\partial x_2}\right)^2 + \left(\frac{\partial u_2}{\partial x_1}\right)^2 + \left(\frac{\partial u_2}{\partial x_2}\right)^2. \end{aligned}$$

(As soon as we substitute numbers for the indices, i. e. when the summation convention does not apply anymore, we can clearly write squares.)

Problem 1.2.

A. During substitution we have to ensure two things:
   - Substitution must not lead to an index appearing more than twice in a term. We can avoid this by renaming those summation indices which appear also in one of the other terms (as free indices or as summation indices).
   - The quantity to be substituted (here u) must have the same index in the equation in which it is inserted and in the equation we use to substitute. If this is not the case, we must rename one of the two indices.

   In this problem it is easiest to rename the index k in $\varphi = u_k\,v_k$ to i: $\varphi = u_i\,v_i$; this gives the solution $\varphi = A_{ik}\,n_k\,v_i$.

B. If we want to keep i as a free index, we can write the substitution equations e. g. as $u_m = B_{mj}\,v_j$ and $C_{mi} = p_m\,q_i$. Substituting then gives $w_i = p_m\,q_i\,B_{mj}\,v_j$.

https://doi.org/10.1515/9783110404265-007

C. If we substitute τ_{ij} and d_{ij} and multiply the terms, we get

$$\phi = \tau_{ij}\, d_{ij} = \frac{\eta}{2}\left(\frac{\partial v_i}{\partial x_j} + \frac{\partial v_j}{\partial x_i}\right)\left(\frac{\partial v_i}{\partial x_j} + \frac{\partial v_j}{\partial x_i}\right)$$
$$= \frac{\eta}{2}\left(\frac{\partial v_i}{\partial x_j}\frac{\partial v_i}{\partial x_j} + 2\frac{\partial v_j}{\partial x_i}\frac{\partial v_i}{\partial x_j} + \frac{\partial v_j}{\partial x_i}\frac{\partial v_j}{\partial x_i}\right).$$

If we rename in the third term j to i and i to j, we see that the first and the third terms are equal, so we get

$$\phi = \eta\left(\frac{\partial v_i}{\partial x_j}\frac{\partial v_i}{\partial x_j} + \frac{\partial v_j}{\partial x_i}\frac{\partial v_i}{\partial x_j}\right).$$

So we get the solution to this problem as

$$\varrho\, T\,\frac{\mathrm{D}s}{\mathrm{D}t} = \eta\left(\frac{\partial v_i}{\partial x_j}\frac{\partial v_i}{\partial x_j} + \frac{\partial v_j}{\partial x_i}\frac{\partial v_i}{\partial x_j}\right) + \kappa\,\frac{\partial^2 T}{\partial x_i^2}.$$

Problem 1.3.

A. 1. Yes (i is a free index on the left and on the right).
2. No (no free index on the left, i and j are free indices on the right).
3. No (no free index on the left, i is a free index on the right).
4. Yes (i is a free index on the left and on the right).
5. Yes (m and n are free indices on the left and on the right).
6. No (i and k are free indices on the left, i and j are free indices on the right).
7. Yes (i and k are free indices on the left and on the right).
8. Yes (n is a free index on the left and on the right).
9. Yes (no free index on the left and on the right).
10. Yes (i is a free index on the left and on the right).

B. 1. $A_1\,(B_{11} + B_{22}) = C_1\,(D_{11} + D_{22}),\quad A_2\,(B_{11} + B_{22}) = C_2\,(D_{11} + D_{22})$.
4. $A_1\,B_1 = C_1,\quad A_2\,B_2 = C_2$.
5. $A_1\,B_1 = C_{11},\quad A_1\,B_2 = C_{12},\quad A_2\,B_1 = C_{21},\quad A_2\,B_2 = C_{22}$.
7. We obtain four equations:
$A_1\,B_1 = A_1\,B_1,\quad A_1\,B_2 = A_2\,B_1,\quad A_2\,B_1 = A_1\,B_2,\quad A_2\,B_2 = A_2\,B_2$.
The first equation and the last equation are trivial and the other two equations are equivalent, so that the only nontrivial statement is $A_1\,B_2 = A_2\,B_1$.
8. $\mu_{11}\,A_1 + \mu_{21}\,A_2 = c_1\,(F_{11} + F_{22}),\quad \mu_{12}\,A_1 + \mu_{22}\,A_2 = c_2\,(F_{11} + F_{22})$.
9. $A = \alpha_1\,C_{11} + \alpha_2\,C_{22}$.
10. $A_{11} = B_{11}\,C_{111} + B_{12}\,C_{221},\quad A_{22} = B_{21}\,C_{112} + B_{22}\,C_{222}$.

Problem 1.4.

A. We calculate a determinant of order three with a zero element most conveniently by an expansion with respect to the row (or column) containing the zero. Expansion with respect to the first row gives $\Delta = 1(-2+6) - 1(6-1) = -1$.

B. A determinant of order larger than three is best transformed into triangular form; see also Section 1.6.1. (In the following calculations a number next to a row means that this line is multiplied by the number and is added to the row indicated by the arrow.)

$$\Delta = \begin{array}{c} (:2) \\ \begin{vmatrix} 1 & 2 & -4 & 4 \\ 3 & 6 & 9 & 0 \\ -2 & 8 & 1 & 9 \\ 4 & 2 & -2 & 0 \end{vmatrix} \end{array} \begin{array}{l} \\ (:3) \\ \\ \\ \end{array} = 6 \begin{vmatrix} 1 & 1 & -4 & 4 \\ 1 & 1 & 3 & 0 \\ -2 & 4 & 1 & 9 \\ 4 & 1 & -2 & 0 \end{vmatrix} \begin{array}{lll} -1 & 2 & -4 \\ \leftarrow & & \\ & \leftarrow & \\ & & \leftarrow \end{array}$$

$$= 6 \begin{vmatrix} 1 & 1 & -4 & 4 \\ 0 & 0 & 7 & -4 \\ 0 & 6 & -7 & 17 \\ 0 & -3 & 14 & -16 \end{vmatrix} \begin{array}{c} \\ \uparrow \\ | \\ \downarrow \end{array} = -6 \begin{vmatrix} 1 & 1 & -4 & 4 \\ 0 & -3 & 14 & -16 \\ 0 & 6 & -7 & 17 \\ 0 & 0 & 7 & -4 \end{vmatrix} \begin{array}{c} \\ 2 \\ \leftarrow \\ \\ \end{array}$$

$$= -6 \begin{vmatrix} 1 & 1 & -4 & 4 \\ 0 & -3 & 14 & -16 \\ 0 & 0 & 21 & -15 \\ 0 & 0 & 7 & -4 \end{vmatrix} \begin{array}{c} \\ \\ (:3) \\ \\ \end{array} = -18 \begin{vmatrix} 1 & 1 & -4 & 4 \\ 0 & -3 & 14 & -16 \\ 0 & 0 & 7 & -5 \\ 0 & 0 & 7 & -4 \end{vmatrix} \begin{array}{c} \\ \\ -1 \\ \leftarrow \end{array}$$

$$= -18 \begin{vmatrix} 1 & 1 & -4 & 4 \\ 0 & -3 & 14 & -16 \\ 0 & 0 & 7 & -5 \\ 0 & 0 & 0 & 1 \end{vmatrix} = -18 \cdot 1 \cdot (-3) \cdot 7 \cdot 1 = 378.$$

Problem 1.5.

A. $\delta_{ii} = (\delta_{11}, \delta_{22}, \delta_{33}, \delta_{44}, \delta_{55}) = (1, 1, 1, 1, 1)$.

B. $\delta_{ii} = \delta_{11} + \delta_{22} + \delta_{33} + \delta_{44} + \delta_{55} = 5$.

Problem 1.6.

A. $\delta_{ij}\,\delta_{jk} = \delta_{ik}$.

B. $\delta_{i2}\,\delta_{ik}$ is equal to δ_{k2} or δ_{2k}, depending on whether we substitute the i in the first δ by the k in the second δ, or if we substitute the i in the second δ by the 2 in the first δ; since $\delta_{k2} = \delta_{2k}$, the result is the same: $\delta_{k2}\,\delta_{3k} = \delta_{32} = 0$.

C. $\delta_{1k}\,A_k = A_1$.

D. $\delta_{i2}\,\delta_{jk}\,A_{ij} = A_{2k}$.

Problem 1.7.

A. Following the arguments from Section 1.4.2, No. 2, we get:

1. for $N = 2$: $\delta^{12}_{12} = \delta^{21}_{21} = 1, \quad \delta^{12}_{21} = \delta^{21}_{12} = -1,$
2. for $N = 3$: $\delta^{12}_{12} = \delta^{21}_{21} = \delta^{13}_{13} = \delta^{31}_{31} = \delta^{23}_{23} = \delta^{32}_{32} = \ 1,$

$$\delta^{12}_{21} = \delta^{21}_{12} = \delta^{13}_{31} = \delta^{31}_{13} = \delta^{23}_{32} = \delta^{32}_{23} = -1.$$

B. δ^{ij}_{ij} is the sum of all elements with identical upper and lower indices and the solution to the sub-problem A 2 shows that this sum is six. We obtain the same result using definition (1.19):

$$\delta^{ij}_{ij} = \begin{vmatrix} \delta_{ii} & \delta_{ij} \\ \delta_{ji} & \delta_{jj} \end{vmatrix} = \delta_{ii}\,\delta_{jj} - \underbrace{\delta_{ij}\,\delta_{ji}}_{\delta_{ii}} = 3 \cdot 3 - 3 = 6.$$

Problem 1.8.

$$\begin{pmatrix} \delta_{11} & \delta_{12} \\ \delta_{21} & \delta_{22} \end{pmatrix} = \begin{pmatrix} 1 & 0 \\ 0 & 1 \end{pmatrix}, \quad \begin{pmatrix} \varepsilon_{11} & \varepsilon_{12} \\ \varepsilon_{21} & \varepsilon_{22} \end{pmatrix} = \begin{pmatrix} 0 & 1 \\ -1 & 0 \end{pmatrix}.$$

Problem 1.9.

The ε_{ijk} are different from zero only if all three indices are different. If only one index is a free index and if we have chosen a number for this index, only two elements, of the nine possible elements with this free index, are nonzero. This is used in the following.

A. $v_i = \varepsilon_{ijk}\,\omega_j\,r_k$:

$$v_1 = \varepsilon_{123}\,\omega_2\,r_3 + \varepsilon_{132}\,\omega_3\,r_2 = \omega_2\,r_3 - \omega_3\,r_2,$$
$$v_2 = \varepsilon_{231}\,\omega_3\,r_1 + \varepsilon_{213}\,\omega_1\,r_3 = \omega_3\,r_1 - \omega_1\,r_3,$$
$$v_3 = \varepsilon_{312}\,\omega_1\,r_2 + \varepsilon_{321}\,\omega_2\,r_1 = \omega_1\,r_2 - \omega_2\,r_1.$$

These three equations are usually summarized as the vector product of $\underline{\omega}$ and $\underline{r}$: $\underline{v} = \underline{\omega} \times \underline{r}$.

B. $\Omega_j = \varepsilon_{ijk}\,\dfrac{\partial v_i}{\partial x_k}$:

$$\Omega_1 = \varepsilon_{312}\,\frac{\partial v_3}{\partial x_2} + \varepsilon_{213}\,\frac{\partial v_2}{\partial x_3} = \frac{\partial v_3}{\partial x_2} - \frac{\partial v_2}{\partial x_3},$$
$$\Omega_2 = \varepsilon_{123}\,\frac{\partial v_1}{\partial x_3} + \varepsilon_{321}\,\frac{\partial v_3}{\partial x_1} = \frac{\partial v_1}{\partial x_3} - \frac{\partial v_3}{\partial x_1},$$
$$\Omega_3 = \varepsilon_{231}\,\frac{\partial v_2}{\partial x_1} + \varepsilon_{132}\,\frac{\partial v_1}{\partial x_2} = \frac{\partial v_2}{\partial x_1} - \frac{\partial v_1}{\partial x_2}.$$

These three equations are usually summarized as the curl of $\underline{v}$: $\underline{\Omega} = \text{curl}\ \underline{v}$.

Problem 1.10.

We compute the inverse of the matrix using the Gauss–Jordan algorithm (Section 1.6.2). We have

A.

1	0	2	1	0	0	−2	−1
2	1	3	0	1	0	←	
1	1	2	0	0	1		←
1	0	2	1	0	0		
0	1	−1	−2	1	0	−1	
0	1	0	−1	0	1	←	
1	0	2	1	0	0	←	
0	1	−1	−2	1	0		←
0	0	1	1	−1	1	−2	1
1	0	0	−1	2	−2		
0	1	0	−1	0	1		
0	0	1	1	−1	1		

Check:

			−1	2	−2
			−1	0	1
			1	−1	1
1	0	2	1	0	0
2	1	3	0	1	0
1	1	2	0	0	1

B.

0	1	1	−1	1	0	0	0	↑	
1	2	−1	1	0	1	0	0	↓	
1	3	−1	0	0	0	1	0		
−1	−2	1	−2	0	0	0	1		
1	2	−1	1	0	1	0	0	−1	1
0	1	1	−1	1	0	0	0		
1	3	−1	0	0	0	1	0	←	
−1	−2	1	−2	0	0	0	1		←
1	2	−1	1	0	1	0	0	←	
0	1	1	−1	1	0	0	0	−2	−1
0	1	0	−1	0	−1	1	0		←
0	0	0	−1	0	1	0	1		
1	0	−3	3	−2	1	0	0		
0	1	1	−1	1	0	0	0		
0	0	−1	0	−1	−1	1	0	−1	
0	0	0	−1	0	1	0	1		
1	0	−3	3	−2	1	0	0	←	
0	1	1	−1	1	0	0	0		←
0	0	1	0	1	1	−1	0	3	−1
0	0	0	−1	0	1	0	1	−1	
1	0	0	3	1	4	−3	0	←	
0	1	0	−1	0	−1	1	0		←
0	0	1	0	1	1	−1	0		
0	0	0	1	0	−1	0	−1	−3	1
1	0	0	0	1	7	−3	3		
0	1	0	0	0	−2	1	−1		
0	0	1	0	1	1	−1	0		
0	0	0	1	0	−1	0	−1		

Check:

				1	7	−3	3
				0	−2	1	−1
				1	1	−1	0
				0	−1	0	−1
0	1	1	−1	1	0	0	0
1	2	−1	1	0	1	0	0
1	3	−1	0	0	0	1	0
−1	−2	1	−2	0	0	0	1

Problem 1.11.

A. We have several options (as is often the case with proofs). The easiest way is to transpose both sides of (1.54), according to (1.46). We consider each i, j-element of the two matrices in the following equation:

$$\left(A_{ij}^{(-1)}\right)^{\mathrm{T}} = A_{ji}^{(-1)}, \quad \left(A_{ij}^{\mathrm{T}}\right)^{(-1)} = A_{ji}^{(-1)}.$$

The right sides are equal and we establish the claim by equating the left sides.

B. The usual calculation rules (commutativity, associativity, distributivity) clearly hold for calculations between elements, so we prove the claim for the elements (we can also say: in element notation).

We start with the definition of the inverse in the form (1.51):

$$(A_{ij}\,B_{jk})^{(-1)}\,A_{kl}\,B_{lm} = \delta_{im}.$$

To bring $A_{kl}\,B_{lm}$ to the right side, we multiply by $B_{mn}^{(-1)}$;

$$(A_{ij}\,B_{jk})^{(-1)}\,\underbrace{A_{kl}\,\underbrace{B_{lm}\,B_{mn}^{(-1)}}_{\delta_{ln}}}_{A_{kn}} = \underbrace{\delta_{im}\,B_{mn}^{(-1)}}_{B_{in}^{(-1)}},$$

$$(A_{ij}\,B_{jk})^{(-1)}\,A_{kn} = B_{in}^{(-1)}.$$

Multiplication by $A_{np}^{(-1)}$ gives

$$(A_{ij}\,B_{jk})^{(-1)}\,\underbrace{A_{kn}\,A_{np}^{(-1)}}_{\delta_{kp}} = B_{in}^{(-1)}\,A_{np}^{(-1)},$$

$$(A_{ij}\,B_{jp})^{(-1)} = B_{in}^{(-1)}\,A_{np}^{(-1)}, \quad \text{what was to be shown.}$$

Problem 1.12.

Here we use the algorithm from Section 1.6.3 to find the rank. We have A.

0	3	6	9	−3	: 3
0	0	2	−4	8	: 2
1	2	−3	4	5	
1	5	5	9	10	
0	1	2	3	−1	←
0	0	1	−2	4	←
1	2	−3	4	5	←
1	5	5	9	10	
1	2	−3	4	5	−1
0	1	2	3	−1	
0	0	1	−2	4	
1	5	5	9	10	←
1	2	−3	4	5	→ Zeros
0	1	2	3	−1	−3
0	0	1	−2	4	
0	3	8	5	5	←
1	0	0	0	0	
0	1	2	3	−1	→ Zeros
0	0	1	−2	4	−2
0	0	2	−4	8	←
1	0	0	0	0	
0	1	0	0	0	
0	0	1	−2	4	→ Zeros
0	0	0	0	0	
1	0	0	0	0	The matrix
0	1	0	0	0	has rank 3.
0	0	1	0	0	
0	0	0	0	0	

B.

1	−3	2	0	−4	2
4	−11	10	−1	←	
−2	8	−5	3		←
1	−3	2	0	→ Zeros	
0	1	2	−1	−2	
0	2	−1	3	←	
1	0	0	0		
0	1	2	−1	→ Zeros	
0	0	−5	5	: (−5)	
1	0	0	0		
0	1	0	0		
0	0	1	−1	→ Zeros	
1	0	0	0	The matrix	
0	1	0	0	has rank 3.	
0	0	1	0		

Problem 1.13.

Reflexivity: For $\underset{\sim}{A} = \underset{\sim}{B}$ we can take $\underset{\sim}{S} = \underset{\sim}{T} = \underset{\sim}{I}$: $A_{ij} = \delta_{im}\, A_{mn}\, \delta_{nj}$.
Symmetry: From $A_{ij} = S_{im}\, B_{mn}\, T_{nj}$ it follows that

$$S^{(-1)}_{pi}\, A_{ij}\, T^{(-1)}_{jq} = \underbrace{S^{(-1)}_{pi}\, S_{im}}_{\delta_{pm}}\, B_{mn}\, \underbrace{T_{nj}\, T^{(-1)}_{jq}}_{\delta_{nq}} = B_{pq}.$$

Transitivity: From $A_{ij} = S_{im}\, B_{mn}\, T_{nj}$, $B_{mn} = P_{mr}\, C_{rs}\, Q_{sn}$ it follows that

$$A_{ij} = \underbrace{S_{im}\, P_{mr}}_{=:\, U_{ir}}\, C_{rs}\, \underbrace{Q_{sn}\, T_{nj}}_{=:\, V_{sj}} = U_{ir}\, C_{rs}\, V_{sj}.$$

Here, $\underset{\sim}{U}$ and $\underset{\sim}{V}$ are the product of two regular matrices and hence they are, according to (1.13), also regular.

Problem 2.1.

Assumption:
$$\begin{aligned} A_i &= T_{ij}\, B_j, && (2.13) =: (a),\\ \widetilde{A}_i &= \widetilde{T}_{ij}\, \widetilde{B}_j, && (2.14) =: (b),\\ T_{ij} &= \alpha_{im}\, \alpha_{jn}\, \widetilde{T}_{mn}, && (2.15) =: (c),\\ B_j &= \alpha_{jk}\, \widetilde{B}_k, && (2.11) =: (d). \end{aligned}$$

Claim: $A_i = \alpha_{im}\, \widetilde{A}_m.$

Proof:
$$A_i \overset{(a)}{=} T_{ij}\, B_j \overset{(c),(d)}{=} \alpha_{im}\, \underbrace{\alpha_{jn}\, \widetilde{T}_{mn}\, \alpha_{jk}}_{\overset{(2.6)}{=}\, \delta_{nk}\, \widetilde{T}_{mn} \overset{(1.17)}{=} \widetilde{T}_{mk}}\, \widetilde{B}_k,$$

$$A_i = \alpha_{im}\, \widetilde{T}_{mk}\, \widetilde{B}_k \overset{(b)}{=} \alpha_{im}\, \widetilde{A}_m, \text{ what was to be shown.}$$

Problem 2.2.

A.

$$a^{\mathrm{T}}_{ij} = \begin{pmatrix} 1 & 4 & 2 \\ 0 & -3 & -4 \\ 2 & 6 & 5 \end{pmatrix}.$$

B.

$$a_{(ij)} = \frac{1}{2}\left[\begin{pmatrix} 1 & 0 & 2 \\ 4 & -3 & 6 \\ 2 & -4 & 5 \end{pmatrix} + \begin{pmatrix} 1 & 4 & 2 \\ 0 & -3 & -4 \\ 2 & 6 & 5 \end{pmatrix}\right]$$

$$= \frac{1}{2}\begin{pmatrix} 2 & 4 & 4 \\ 4 & -6 & 2 \\ 4 & 2 & 10 \end{pmatrix} = \begin{pmatrix} 1 & 2 & 2 \\ 2 & -3 & 1 \\ 2 & 1 & 5 \end{pmatrix},$$

$$a_{[ij]} = \frac{1}{2}\left[\begin{pmatrix} 1 & 0 & 2 \\ 4 & -3 & 6 \\ 2 & -4 & 5 \end{pmatrix} - \begin{pmatrix} 1 & 4 & 2 \\ 0 & -3 & -4 \\ 2 & 6 & 5 \end{pmatrix}\right]$$

$$= \frac{1}{2}\begin{pmatrix} 0 & -4 & 0 \\ 4 & 0 & 10 \\ 0 & -10 & 0 \end{pmatrix} = \begin{pmatrix} 0 & -2 & 0 \\ 2 & 0 & 5 \\ 0 & -5 & 0 \end{pmatrix}.$$

Check: $a_{(ij)} + a_{[ij]} = a_{ij}$,

$$\begin{pmatrix} 1 & 2 & 2 \\ 2 & -3 & 1 \\ 2 & 1 & 5 \end{pmatrix} + \begin{pmatrix} 0 & -2 & 0 \\ 2 & 0 & 5 \\ 0 & -5 & 0 \end{pmatrix} = \begin{pmatrix} 1 & 0 & 2 \\ 4 & -3 & 6 \\ 2 & -4 & 5 \end{pmatrix}.$$

Problem 2.3.

$$\underline{a}\,\underline{b} \mathrel{\hat{=}} \begin{pmatrix} -4 & 2 & 10 \\ -6 & 3 & 15 \\ -8 & 4 & 20 \end{pmatrix}, \quad \underline{b}\,\underline{a} \mathrel{\hat{=}} \begin{pmatrix} -4 & -6 & -8 \\ 2 & 3 & 4 \\ 10 & 15 & 20 \end{pmatrix}.$$

Problem 2.4.

We cannot translate the sub-problems A, B, and C into matrix notation, because the order of their tensors is greater than 2. Before translating into symbolic notation, we need to rewrite each equation so that the order of the (free) indices is the same in all terms; this is possible in all cases (B, C, and D) (otherwise translation into symbolic notation would not be possible).

A. $\underline{\underline{a}}\,\underline{b}\,\underline{c} = \underline{\underline{\underline{\underline{d}}}} \iff a_{ij}\, b_k\, c_l = d_{ijkl}.$

B. $a_i\, b_{kl}\, c_j = d_{ijkl}$,

$a_i\, c_j\, b_{kl} = d_{ijkl} \iff \underline{a}\,\underline{c}\,\underline{\underline{b}} = \underline{\underline{\underline{\underline{d}}}}.$

C. $\alpha_i\, b_{kj} = A_{ijk}$,

$\alpha_i\, b^{\mathrm{T}}_{jk} = A_{ijk} \quad\Longleftrightarrow\quad \underline{\alpha}\, \underline{\underline{b}}^{\mathrm{T}} = \underline{\underline{\underline{A}}}.$

D. $a^{\mathrm{T}}_{mn} = b_{nm}$,

$a_{nm} = b_{nm} \quad\Longleftrightarrow\quad \underline{\underline{a}} = \underline{\underline{b}} \quad\Longleftrightarrow\quad \underset{\sim}{a} = \underset{\sim}{b}.$

Problem 2.5.

A. $\underline{\underline{a}} \cdot \underline{\underline{b}} = \underline{\underline{c}} \quad\Longleftrightarrow\quad a_{ij}\, b_{jk} = c_{ik} \quad\Longleftrightarrow\quad \underset{\sim}{a}\, \underset{\sim}{b} = \underset{\sim}{c}.$

B. $\underline{\underline{b}} \cdot \underline{\underline{a}} = \underline{\underline{c}} \quad\Longleftrightarrow\quad b_{ij}\, a_{jk} = c_{ik} \quad\Longleftrightarrow\quad \underset{\sim}{b}\, \underset{\sim}{a} = \underset{\sim}{c}.$

C. $a_{ik}\, b_{ij} = c_{jk}$:

We have two ways to rewrite the left side, so that it can be translated:

$a_{ik}\, b_{ij} = b^{\mathrm{T}}_{ji}\, a_{ik}$ and $a_{ik}\, b_{ij} = a^{\mathrm{T}}_{ki}\, b_{ij}$.

They lead to different translations:

- $b^{\mathrm{T}}_{ji}\, a_{ik} = c_{jk}$ can be translated immediately, i. e.

  $b^{\mathrm{T}}_{ji}\, a_{ik} = c_{jk} \quad\Longleftrightarrow\quad \underline{\underline{b}}^{\mathrm{T}} \cdot \underline{\underline{a}} = \underline{\underline{c}} \quad\Longleftrightarrow\quad \underset{\sim}{b}^{\mathrm{T}}\, \underset{\sim}{a} = \underset{\sim}{c}.$

- In $a^{\mathrm{T}}_{ki}\, b_{ij} = c_{jk}$ we first set $c_{jk} = c^{\mathrm{T}}_{kj}$, so

  $a^{\mathrm{T}}_{ki}\, b_{ij} = c^{\mathrm{T}}_{kj} \quad\Longleftrightarrow\quad \underline{\underline{a}}^{\mathrm{T}} \cdot \underline{\underline{b}} = \underline{\underline{c}}^{\mathrm{T}} \quad\Longleftrightarrow\quad \underset{\sim}{a}^{\mathrm{T}}\, \underset{\sim}{b} = \underset{\sim}{c}^{\mathrm{T}}.$

Both translations are correct; we will show later that they can be converted into each other.

D. $a_i\, b_k\, a_i\, c_k = d$,

$a_i\, a_i\, b_k\, c_k = d \quad\Longleftrightarrow\quad \underline{a} \cdot \underline{a}\, \underline{b} \cdot \underline{c} = d \quad\Longleftrightarrow\quad (\underset{\sim}{a}^{\mathrm{T}}\, \underset{\sim}{a})\, (\underset{\sim}{b}^{\mathrm{T}}\, \underset{\sim}{c}) = d.$

E. This equation cannot be translated into symbolic notation, because symbolic notation is usually not defined for isomers of tensors of order 3.

F. $(\underline{\underline{a}} \cdot \underline{\underline{b}})^{\mathrm{T}} = \underline{\underline{c}} \quad\Longleftrightarrow\quad (a_{ij}\, b_{jk})^{\mathrm{T}} = c_{ik} \quad\Longleftrightarrow\quad (\underset{\sim}{a}\, \underset{\sim}{b})^{\mathrm{T}} = \underset{\sim}{c}.$

G. $(\underline{\underline{a}} \cdot \underline{\underline{b}} \cdot \underline{\underline{c}})^{\mathrm{T}} = \underline{\underline{d}} \quad\Longleftrightarrow\quad (a_{ij}\, b_{jk}\, c_{kl})^{\mathrm{T}} = d_{il} \quad\Longleftrightarrow\quad (\underset{\sim}{a}\, \underset{\sim}{b}\, \underset{\sim}{c})^{\mathrm{T}} = \underset{\sim}{d}.$

Problem 2.6.

In the following forms the equation can be translated:

1. $a_i\, a_i\, c_j\, b_k = d_{jk}$, 2. $c_j\, a_i\, a_i\, b_k = d_{jk}$, 3. $c_j\, b_k\, a_i\, a_i = d_{jk}$.

In case 1, we initially translate $\underline{a} \cdot \underline{a}\, \underline{c}\, \underline{b} = \underline{\underline{d}}$. Then all five associations

$$(\underline{a} \cdot \underline{a})(\underline{c}\, \underline{b}) = \underline{a} \cdot ((\underline{a}\, \underline{c})\, \underline{b}) = \underline{a} \cdot (\underline{a}\, (\underline{c}\, \underline{b})) = ((\underline{a} \cdot \underline{a})\, \underline{c})\, \underline{b} = (\underline{a} \cdot (\underline{a}\, \underline{c}))\, \underline{b}$$

are equivalent.

In case 2, we initially translate $\underline{c}\, \underline{a} \cdot \underline{a}\, \underline{b} = \underline{\underline{d}}$. Again, all five associations

$$(\underline{c}\, \underline{a}) \cdot (\underline{a}\, \underline{b}) = \underline{c}\, ((\underline{a} \cdot \underline{a})\, \underline{b}) = \underline{c}\, (\underline{a} \cdot (\underline{a}\, \underline{b})) = ((\underline{c}\, \underline{a}) \cdot \underline{a})\, \underline{b} = (\underline{c}\, (\underline{a} \cdot \underline{a}))\, \underline{b}$$

are equivalent.

In case 3, we initially translate $\underline{c}\,\underline{b}\,\underline{a}\cdot\underline{a}=\underline{\underline{d}}$. All five associations

$$(\underline{c}\,\underline{b})(\underline{a}\cdot\underline{a})=\underline{c}\,((\underline{b}\,\underline{a})\cdot\underline{a})=\underline{c}\,(\underline{b}\,(\underline{a}\cdot\underline{a}))=((\underline{c}\,\underline{b})\,\underline{a})\cdot\underline{a}=(\underline{c}\,(\underline{b}\,\underline{a}))\cdot\underline{a}$$

are again equivalent here.

So in all three cases parentheses need not to be set; all five possible ways to interpret the expression lead to the same result.

Problem 2.7.

A. $(\underline{a}\,\underline{b})^{\mathrm{T}}$ is translated to $(a_i\,b_j)^{\mathrm{T}}$, which gives transposed $(a_i\,b_j)^{\mathrm{T}}=a_j\,b_i$. To make the order of the free indices on both sides equal, we change on the right side the order of the factors $a_j\,b_i=b_i\,a_j$, which gives $(a_i\,b_j)^{\mathrm{T}}=b_i\,a_j$. This equation can be translated and we get $(\underline{a}\,\underline{b})^{\mathrm{T}}=\underline{b}\,\underline{a}$, Q. E. D.

B. Translating $\underline{a}\cdot\underline{\underline{b}}$ gives $a_i\,b_{ij}$. Now we have $a_i\,b_{ij}=b^{\mathrm{T}}_{ji}\,a_i$, which translates into $\underline{a}\cdot\underline{\underline{b}}=\underline{\underline{b}}^{\mathrm{T}}\cdot\underline{a}$, Q. E. D.

C. Translating $(\underline{\underline{a}}\cdot\underline{\underline{b}}\cdot\underline{\underline{c}})^{\mathrm{T}}$ gives $(a_{ij}\,b_{jk}\,c_{kl})^{\mathrm{T}}$. Now we have $(a_{ij}\,b_{jk}\,c_{kl})^{\mathrm{T}}=a_{lj}\,b_{jk}\,c_{ki}=c_{ki}\,b_{jk}\,a_{lj}=c^{\mathrm{T}}_{ik}\,b^{\mathrm{T}}_{kj}\,a^{\mathrm{T}}_{jl}$. Translation of the very left and the very right terms gives $(\underline{\underline{a}}\cdot\underline{\underline{b}}\cdot\underline{\underline{c}})^{\mathrm{T}}=\underline{\underline{c}}^{\mathrm{T}}\cdot\underline{\underline{b}}^{\mathrm{T}}\cdot\underline{\underline{a}}^{\mathrm{T}}$, Q. E. D.

Problem 2.8.

A. $a_{ij}\,b_{ij}=c \quad\Longleftrightarrow\quad \underline{\underline{a}}\cdot\cdot\,\underline{\underline{b}}=c.$

B. $a_{ij}\,b_{ji}=c,\quad a_{ij}\,b^{\mathrm{T}}_{ij}=c \quad\Longleftrightarrow\quad \underline{\underline{a}}\cdot\cdot\,\underline{\underline{b}}^{\mathrm{T}}=c.$

C. $a_{ij}\,b_{kjl}\,c_{ik}=d_l$:

Option 1: $a_{ij}\,c_{ik}\,b_{kjl}=d_l \quad\Longleftrightarrow\quad \underline{\underline{a}}\cdot\cdot\,(\underline{\underline{c}}\cdot\underline{\underline{\underline{b}}})=\underline{d}.$
We do not need to set parentheses in this translation, since $(\underline{\underline{a}}\cdot\cdot\,\underline{\underline{c}})\cdot\underline{\underline{\underline{b}}}$ is not defined; however, we recommend setting parentheses, so that we do not have to check this.

Option 2: $c^{\mathrm{T}}_{ki}\,a_{ij}\,b_{kjl}=d_l \quad\Longleftrightarrow\quad (\underline{\underline{c}}^{\mathrm{T}}\cdot\underline{\underline{a}})\cdot\cdot\,\underline{\underline{\underline{b}}}=\underline{d}.$
In this translation we need the parentheses, because $\underline{\underline{c}}^{\mathrm{T}}\cdot(\underline{\underline{a}}\cdot\cdot\,\underline{\underline{\underline{b}}})=\underline{d}$ is also defined, but gives something different, i. e. $c^{\mathrm{T}}_{ki}\,a_{jl}\,b_{jli}=d_k$.

D. $a_{ij}\,b_{ikl}\,c_{klj}=d \quad\Longleftrightarrow\quad \underline{\underline{a}}\cdot\cdot\,(\underline{\underline{\underline{b}}}\cdot\cdot\,\underline{\underline{\underline{c}}})=d.$
Again, we do not need parentheses here, because $(\underline{\underline{a}}\cdot\cdot\,\underline{\underline{\underline{b}}})\cdot\cdot\,\underline{\underline{\underline{c}}}$ is not defined, but we still recommend them.

Problem 2.9.

A. $\operatorname{tr}\underline{\underline{\delta}}=\delta_{ii}=\delta_{11}+\delta_{22}+\delta_{33}=3.$

B. $\delta_{ij}\,\delta_{jk}\overset{(1.17)}{=}\delta_{ik}$, so we have $\underline{\underline{\delta}}\cdot\underline{\underline{\delta}}=\underline{\underline{\delta}}$.

C. $\delta_{ij}\,\delta_{ij}=(\delta_{11})^2+(\delta_{12})^2+(\delta_{13})^2+(\delta_{21})^2+\cdots+(\delta_{33})^2=3.$

Problem 2.10.

A. $a_{ij}\, b_{ij} = b_{ij}\, a_{ij}$ can be translated into $\underline{\underline{a}} \cdot\cdot\, \underline{\underline{b}} = \underline{\underline{b}} \cdot\cdot\, \underline{\underline{a}}$; therefore the double scalar product of two tensors of order two is always commutative.

B. If we choose nine different tensors $\overset{1}{b}_{ij}, \overset{2}{b}_{ij}, \ldots, \overset{9}{b}_{ij}$ for b_{ij}, we get from $a_{ij}\, b_{ij} = 0$ a homogeneous system of linear equations for the nine unknown a_{ij}, i. e.

$$\begin{aligned} a_{11}\,\overset{1}{b}_{11} + a_{12}\,\overset{1}{b}_{12} + \cdots + a_{33}\,\overset{1}{b}_{33} &= 0, \\ a_{11}\,\overset{2}{b}_{11} + a_{12}\,\overset{2}{b}_{12} + \cdots + a_{33}\,\overset{2}{b}_{33} &= 0, \\ &\vdots \\ a_{11}\,\overset{9}{b}_{11} + a_{12}\,\overset{9}{b}_{12} + \cdots + a_{33}\,\overset{9}{b}_{33} &= 0. \end{aligned}$$

Clearly, we can cancel out the b_{ij} from $a_{ij}\, b_{ij} = 0$, if and only if the system of equations has only the trivial solution, i. e. if the nine tensors $\overset{k}{b}_{ij}$, $i, j = 1, 2, \ldots, 3$, $k = 1, \ldots, 9$ are linearly independent. Since every tensor b_{ij} can be written as a linear combination of nine linearly independent tensors $\overset{k}{b}_{ij}$, we can cancel out b_{ij} if and only if b_{ij} can take arbitrary values, and the criterion for this is that we can substitute nine linearly independent tensors.

Problem 2.11.

A. $$\begin{aligned} a_{(ij)}\, b_{[ij]} &\overset{(2.26)}{=} \frac{1}{2}\,(a_{ij} + a_{ji}) \cdot \frac{1}{2}\,(b_{ij} - b_{ji}) \\ &= \frac{1}{4}\,(a_{ij}\, b_{ij} + a_{ji}\, b_{ij} - a_{ij}\, b_{ji} - a_{ji}\, b_{ji}). \end{aligned}$$

We have $a_{ij}\, b_{ij} = a_{ji}\, b_{ji}$ and $a_{ji}\, b_{ij} = a_{ij}\, b_{ji}$, as we can see immediately, after renaming in both equations on the right side the summation index i to j and the summation index j to i, i. e. the bracket is zero, which is what we wanted to prove.

B. We want to show that, for arbitrary tensors $a_{(ij)}$, it follows from $a_{(ij)}\, b_{ij} = 0$ that b_{ij} is antimetric. It can be shown analogously that, for arbitrary values of $b_{[ij]}$, it follows from $a_{ij}\, b_{[ij]} = 0$ that a_{ij} is symmetric.

We can prove this analogously to Part A of this problem: by assumption, we have

$$\frac{1}{2}\,(a_{ij} + a_{ji})\, b_{ij} = 0, \quad a_{ij}\, b_{ij} + a_{ji}\, b_{ij} = 0,$$

where a_{ij} can take arbitrary values. Renaming the indices in the second term gives

$$a_{ij}\, b_{ij} + a_{ij}\, b_{ji} = 0, \quad a_{ij}\,(b_{ij} + b_{ji}) = 0.$$

Since a_{ij} can take any value, we can cancel out a_{ij} in the last equation and it follows that $b_{ij} + b_{ji} = 0$, i. e. b_{ij} is antimetric.

C. For $i = 1$, it follows that $\varepsilon_{123}\, a_{23} + \varepsilon_{132}\, a_{32} = 0,\;\; a_{23} - a_{32} = 0,\;\; a_{23} = a_{32}$; for $i = 2$ and $i = 3$, it correspondingly follows that $a_{31} = a_{13},\;\; a_{12} = a_{21}$. This proves the statement.

D. From six linearly independent symmetric tensors $\overset{i}{A}_{mn},\; i = 1, \ldots, 6$, we can construct any symmetric tensor B_{mn} by a linear combination $\alpha_i \overset{i}{A}_{mn} = B_{mn}$. Because of the symmetry with respect to m and n, only six of the nine equations $\alpha_i \overset{i}{A}_{mn} = B_{mn}$ are different. For given $\overset{i}{A}_{mn}$ and B_{mn}, they form a system of linear equations for the six α_i. Because the $\overset{i}{A}_{mn}$ are linearly independent, the coefficient matrix of these equations is regular and the solution to the system of equations is unique.

Problem 2.12.

A. $(\underline{a} \times \underline{b})\, \underline{c} \mathrel{\widehat{=}} \varepsilon_{ijk}\, a_j\, b_k\, c_l.$

$\underline{a} \times (\underline{b}\, \underline{c}) \mathrel{\widehat{=}} \varepsilon_{ijk}\, a_j\, b_k\, c_l.$

Both readings are defined and mean the same thing.

B. $(\underline{a} \times \underline{b}) \cdot \underline{c} \mathrel{\widehat{=}} \varepsilon_{ijk}\, a_j\, b_k\, c_i.$

The expression $\underline{a} \times (\underline{b} \cdot \underline{c})$ is not defined.

C. $(\underline{\underline{a}} \cdot \underline{b}) \times \underline{c} \mathrel{\widehat{=}} \varepsilon_{ijk}\, a_{jm}\, b_m\, c_k.$

$\underline{\underline{a}} \cdot (\underline{b} \times \underline{c}) \mathrel{\widehat{=}} a_{jm}\, \varepsilon_{mik}\, b_i\, c_k = \varepsilon_{mik}\, a_{jm}\, b_i\, c_k.$

Both readings are defined and mean something different.

D. $(\underline{a} \cdot \underline{\underline{b}}) \times \underline{c} \mathrel{\widehat{=}} \varepsilon_{ijk}\, a_m\, b_{mj}\, c_k.$

$\underline{a} \cdot (\underline{\underline{b}} \times \underline{c}) \mathrel{\widehat{=}} -a_i\, b_{ij}\, \varepsilon_{jkl}\, c_l = \varepsilon_{kjl}\, a_i\, b_{ij}\, c_l.$

Both readings are defined and mean the same thing.

Only the term C is ambiguous without parentheses.

Problem 2.13.

A. $\varepsilon_{ijk}\, a_m\, b_j\, \varepsilon_{min}\, c_k\, d_n$

$$
\begin{aligned}
&= \varepsilon_{ijk}\, b_j\, c_k\, \varepsilon_{inm}\, d_n\, a_m && \mathrel{\widehat{=}} (\underline{b} \times \underline{c}) \cdot (\underline{d} \times \underline{a}) \\
&= \varepsilon_{ijk}\, \varepsilon_{inm}\, b_j\, c_k\, d_n\, a_m \\
&= (\delta_{jn}\, \delta_{km} - \delta_{jm}\, \delta_{kn})\, b_j\, c_k\, d_n\, a_m \\
&= b_j\, c_k\, d_j\, a_k - b_j\, c_k\, d_k\, a_j \\
&= b_j\, d_j\, c_k\, a_k - b_j\, a_j\, c_k\, d_k && \mathrel{\widehat{=}} \underline{b} \cdot \underline{d}\; \underline{c} \cdot \underline{a} - \underline{b} \cdot \underline{a}\; \underline{c} \cdot \underline{d},
\end{aligned}
$$

$(\underline{b} \times \underline{c}) \cdot (\underline{d} \times \underline{a}) = \underline{b} \cdot \underline{d}\; \underline{c} \cdot \underline{a} - \underline{b} \cdot \underline{a}\; \underline{c} \cdot \underline{d}.$

Here and in the following examples, other formulations are also permissible in symbolic notation, e. g. here we can write for the left side $(\underline{c} \times \underline{b}) \cdot (\underline{a} \times \underline{d})$.

B. $\varepsilon_{ijk}\, a_j\, b_{km}\, c_m$ is a vector product with i as the only free index.

The indices i and q in ε_{qpi} are summation indices, so they belong to a vector product with the above vector product as one of its factors. We set the free index p as first index, so we have

$$
\begin{aligned}
&\varepsilon_{piq}\,\varepsilon_{ijk}\,a_j\,b_{km}\,c_m\,d_q \qquad \hat{=}\; [\underline{a}\times(\underline{\underline{b}}\cdot\underline{c})]\times\underline{d}\\
&=\varepsilon_{piq}\,\varepsilon_{kij}\,a_j\,b_{km}\,c_m\,d_q\\
&=(\delta_{pk}\,\delta_{qj}-\delta_{pj}\,\delta_{qk})\,a_j\,b_{km}\,c_m\,d_q\\
&=a_q\,b_{pm}\,c_m\,d_q-a_p\,b_{qm}\,c_m\,d_q\\
&=a_q\,d_q\,b_{pm}\,c_m-a_p\,d_q\,b_{qm}\,c_m \qquad \hat{=}\; \underline{a}\cdot\underline{d}\,\underline{\underline{b}}\cdot\underline{c}-\underline{a}\,\underline{d}\cdot\underline{\underline{b}}\cdot\underline{c},\\
&[\underline{a}\times(\underline{\underline{b}}\cdot\underline{c})]\times\underline{d} \;=\; \underline{a}\cdot\underline{d}\,\underline{\underline{b}}\cdot\underline{c}-\underline{a}\,\underline{d}\cdot\underline{\underline{b}}\cdot\underline{c}.
\end{aligned}
$$

C. $\underline{a}\times[(\underline{\underline{b}}\cdot\underline{c})\times\underline{d}] \;\hat{=}\; \varepsilon_{ijk}\,a_j\,\varepsilon_{kpq}\,b_{pm}\,c_m\,d_q$

$$
\begin{aligned}
&=\varepsilon_{ijk}\,\varepsilon_{pqk}\,a_j\,b_{pm}\,c_m\,d_q\\
&=(\delta_{ip}\,\delta_{jq}-\delta_{iq}\,\delta_{jp})\,a_j\,b_{pm}\,c_m\,d_q\\
&=a_j\,b_{im}\,c_m\,d_j-a_j\,b_{jm}\,c_m\,d_i\\
&=d_j\,a_j\,b_{im}\,c_m-d_i\,a_j\,b_{jm}\,c_m \;\hat{=}\; \underline{d}\cdot\underline{a}\,\underline{\underline{b}}\cdot\underline{c}-\underline{d}\,\underline{a}\cdot\underline{\underline{b}}\cdot\underline{c},\\
&\underline{a}\times[(\underline{\underline{b}}\cdot\underline{c})\times\underline{d}] \;=\; \underline{d}\cdot\underline{a}\,\underline{\underline{b}}\cdot\underline{c}-\underline{d}\,\underline{a}\cdot\underline{\underline{b}}\cdot\underline{c}.
\end{aligned}
$$

D. $\underline{a}\times(\underline{b}\times\underline{c}) \;\hat{=}\; \varepsilon_{ijk}\,a_j\,\varepsilon_{kmn}\,b_m\,c_n$

$$
\begin{aligned}
&=\varepsilon_{ijk}\,\varepsilon_{mnk}\,a_j\,b_m\,c_n\\
&=(\delta_{im}\,\delta_{jn}-\delta_{in}\,\delta_{jm})\,a_j\,b_m\,c_n\\
&=a_n\,b_i\,c_n-a_m\,b_m\,c_i\\
&=a_n\,c_n\,b_i-a_m\,b_m\,c_i \;\hat{=}\; \underline{a}\cdot\underline{c}\,\underline{b}-\underline{a}\cdot\underline{b}\,\underline{c},\\
&\underline{a}\times(\underline{b}\times\underline{c}) \;=\; \underline{a}\cdot\underline{c}\,\underline{b}-\underline{a}\cdot\underline{b}\,\underline{c}.
\end{aligned}
$$

E. The terms B and C are both of the form $\underline{a}\times\underline{\underline{b}}\cdot\underline{c}\times\underline{d}$ and they differ only by the order in which the two vector products have to be evaluated, in other words, by setting the parentheses. The expressions are different; therefore the associative law does not hold.

Problem 2.14.

$$a_i\,b_k-a_k\,b_i=a_m\,b_n\,(\delta_{im}\,\delta_{kn}-\delta_{km}\,\delta_{in})=a_m\,b_n\,\varepsilon_{jik}\,\varepsilon_{jmn}.$$

For translation, the order of the free indices must be the same in all terms:

$$
\begin{aligned}
&a_i\,b_k-b_i\,a_k=\varepsilon_{ikj}\,\varepsilon_{jmn}\,a_m\,b_n,\\
&\underline{a}\,\underline{b}-\underline{b}\,\underline{a}=\underline{\underline{\underline{\varepsilon}}}\cdot(\underline{a}\times\underline{b}).
\end{aligned}
$$

Problem 2.15.

A. $\operatorname{div}\operatorname{grad}\underline{\underline{a}} = \underline{\underline{b}} \quad \Longleftrightarrow \quad \frac{\partial}{\partial x_i}\frac{\partial a_{mn}}{\partial x_i} = \frac{\partial^2 a_{mn}}{\partial x_i^2} = b_{mn}.$

B. $\operatorname{grad}\operatorname{div}\underline{\underline{a}} = \underline{\underline{c}} \quad \Longleftrightarrow \quad \frac{\partial}{\partial x_i}\frac{\partial a_{mn}}{\partial x_n} = \frac{\partial^2 a_{mn}}{\partial x_i\,\partial x_n} = c_{mi}.$

C. $\frac{\partial a_i}{\partial x_j}\frac{\partial b_j}{\partial x_k} = c_{ik} \quad \Longleftrightarrow \quad (\operatorname{grad}\underline{a})\cdot(\operatorname{grad}\underline{b}) = \underline{\underline{c}}.$

D. $\frac{\partial a_i}{\partial x_j}\frac{\partial b_i}{\partial x_j} = c \quad \Longleftrightarrow \quad (\operatorname{grad}\underline{a})\cdot\cdot(\operatorname{grad}\underline{b}) = c.$

Problem 2.16.

A. $\operatorname{grad}\underline{x} \;\widehat{=}\; \frac{\partial x_i}{\partial x_j} = \delta_{ij} \quad \Longleftrightarrow \quad \operatorname{grad}\underline{x} = \underline{\underline{\delta}}.$

B. $\operatorname{div}\underline{x} \;\widehat{=}\; \frac{\partial x_i}{\partial x_i} = 3 \quad \Longleftrightarrow \quad \operatorname{div}\underline{x} = 3.$

C. $\operatorname{curl}\underline{x} \;\widehat{=}\; \frac{\partial x_i}{\partial x_k}\varepsilon_{ijk} = \delta_{ik}\,\varepsilon_{ijk} = 0 \quad \Longleftrightarrow \quad \operatorname{curl}\underline{x} = \underline{0}.$

Problem 2.17.

A. $\operatorname{div}(\lambda\,\underline{a}) \;\widehat{=}\; \frac{\partial\lambda\,a_i}{\partial x_i} = \lambda\frac{\partial a_i}{\partial x_i} + \frac{\partial\lambda}{\partial x_i}a_i,$

$\operatorname{div}(\lambda\,\underline{a}) = \lambda\operatorname{div}\underline{a} + (\operatorname{grad}\lambda)\cdot\underline{a}.$

B. $\operatorname{curl}(\lambda\,\underline{a}) \;\widehat{=}\; \frac{\partial\lambda\,a_i}{\partial x_k}\varepsilon_{ijk} = \lambda\frac{\partial a_i}{\partial x_k}\varepsilon_{ijk} + \frac{\partial\lambda}{\partial x_k}\varepsilon_{ijk}\,a_i,$

$\operatorname{curl}(\lambda\,\underline{a}) = \lambda\operatorname{curl}\underline{a} + (\operatorname{grad}\lambda)\times\underline{a}.$

C. $\operatorname{curl}\operatorname{grad}a \;\widehat{=}\; \frac{\partial}{\partial x_k}\frac{\partial a}{\partial x_i}\varepsilon_{ijk} = \frac{\partial^2 a}{\partial x_k\,\partial x_i}\varepsilon_{ijk} \overset{(2.40)}{=} 0,$

$\operatorname{curl}\operatorname{grad}a = \underline{0}.$

D. $\operatorname{div}\operatorname{curl}\underline{a} \;\widehat{=}\; \frac{\partial}{\partial x_j}\frac{\partial a_i}{\partial x_k}\varepsilon_{ijk} = \frac{\partial^2 a_i}{\partial x_j\,\partial x_k}\varepsilon_{ijk} \overset{(2.40)}{=} 0,$

$\operatorname{div}\operatorname{curl}\underline{a} = 0.$

E. $\operatorname{curl}\operatorname{curl}\underline{a} \;\widehat{=}\; \frac{\partial}{\partial x_k}\left(\frac{\partial a_m}{\partial x_n}\varepsilon_{min}\right)\varepsilon_{ijk} = \varepsilon_{inm}\,\varepsilon_{ijk}\frac{\partial^2 a_m}{\partial x_k\,\partial x_n}$

$$= (\delta_{nj}\,\delta_{mk} - \delta_{nk}\,\delta_{mj})\frac{\partial^2 a_m}{\partial x_k\,\partial x_n} = \frac{\partial^2 a_k}{\partial x_k\,\partial x_j} - \frac{\partial^2 a_j}{\partial x_k\,\partial x_k}$$

$$= \frac{\partial}{\partial x_j}\frac{\partial a_k}{\partial x_k} - \frac{\partial}{\partial x_k}\frac{\partial a_j}{\partial x_k},$$

$\operatorname{curl}\operatorname{curl}\underline{a} = \operatorname{grad}\operatorname{div}\underline{a} - \operatorname{div}\operatorname{grad}\underline{a}.$

F. We start by simplifying the given expression:

$$\varepsilon_{ijk}\,\varepsilon_{mnk}\,\frac{\partial a_{pj}}{\partial x_i}\,b_m = (\delta_{im}\,\delta_{jn}-\delta_{in}\,\delta_{jm})\,\frac{\partial a_{pj}}{\partial x_i}\,b_m = \frac{\partial a_{pn}}{\partial x_m}\,b_m - \frac{\partial a_{pm}}{\partial x_n}\,b_m$$

$$= \frac{\partial a_{pn}}{\partial x_m}\,b_m - b_m\,\frac{\partial a^{\mathrm{T}}_{mp}}{\partial x_n}.$$

Then we write the expression in a form that can be translated directly:

$$\varepsilon_{ijk}\,\varepsilon_{mnk}\,\frac{\partial a_{pj}}{\partial x_i}\,b_m = \left(\frac{\partial a_{pj}}{\partial x_i}\,\varepsilon_{jki}\right)\varepsilon_{mnk}\,b_m.$$

Setting equal the right sides gives

$$\left(\frac{\partial a_{pj}}{\partial x_i}\,\varepsilon_{jki}\right)\varepsilon_{mnk}\,b_m = \frac{\partial a_{pn}}{\partial x_m}\,b_m - b_m\,\frac{\partial a^{\mathrm{T}}_{mp}}{\partial x_n},$$

and in all the terms the order of the free indices is such that p comes before n, so the translation is

$(\mathrm{curl}\,\underline{\underline{a}})\times\underline{b} = (\mathrm{grad}\,\underline{\underline{a}})\cdot\underline{b} - \underline{b}\cdot\mathrm{grad}(\underline{\underline{a}}^{\mathrm{T}})$.

G. $$\varepsilon_{ijk}\,\varepsilon_{mnk}\,\frac{\partial a_{pj}\,b_i}{\partial x_m} = (\delta_{im}\,\delta_{jn}-\delta_{in}\,\delta_{jm})\,\frac{\partial a_{pj}\,b_i}{\partial x_m} = \frac{\partial a_{pn}\,b_m}{\partial x_m} - \frac{\partial a_{pm}\,b_n}{\partial x_m}$$

$$= a_{pn}\,\frac{\partial b_m}{\partial x_m} + \frac{\partial a_{pn}}{\partial x_m}\,b_m - a_{pm}\,\frac{\partial b_n}{\partial x_m} - \frac{\partial a_{pm}}{\partial x_m}\,b_n$$

$$= a_{pn}\,\frac{\partial b_m}{\partial x_m} + \frac{\partial a_{pn}}{\partial x_m}\,b_m - a_{pm}\left(\frac{\partial b_m}{\partial x_n}\right)^{\mathrm{T}} - \frac{\partial a_{pm}}{\partial x_m}\,b_n,$$

$$\varepsilon_{ijk}\,\varepsilon_{mnk}\,\frac{\partial a_{pj}\,b_i}{\partial x_m} = \frac{\partial a_{pj}\,(-\varepsilon_{ikj})\,b_i}{\partial x_m}\,(-\varepsilon_{knm}) = \frac{\partial a_{pj}\,\varepsilon_{ikj}\,b_i}{\partial x_m}\,\varepsilon_{knm},$$

$$\frac{\partial a_{pj}\,\varepsilon_{ikj}\,b_i}{\partial x_m}\,\varepsilon_{knm} = a_{pn}\,\frac{\partial b_m}{\partial x_m} + \frac{\partial a_{pn}}{\partial x_m}\,b_m - a_{pm}\left(\frac{\partial b_m}{\partial x_n}\right)^{\mathrm{T}} - \frac{\partial a_{pm}}{\partial x_m}\,b_n,$$

$\mathrm{curl}\,(\underline{\underline{a}}\times\underline{b}) = \underline{\underline{a}}\,\mathrm{div}\,\underline{b} + (\mathrm{grad}\,\underline{\underline{a}})\cdot\underline{b} - \underline{\underline{a}}\cdot\mathrm{grad}^{\mathrm{T}}\,\underline{b} - (\mathrm{div}\,\underline{\underline{a}})\,\underline{b}.$

Problem 2.18.

$$W = \int [\underline{x}]\underline{F}\cdot = \int (F_x\,\mathrm{d}x + F_y\,\mathrm{d}y + F_z\,\mathrm{d}z)$$

$$= \int [-\lambda x\,\mathrm{d}x - \lambda y\,\mathrm{d}y - (\lambda z + mg)\,\mathrm{d}z]$$

$$= -\int_0^{2\pi}\left[\lambda x\,\frac{\mathrm{d}x}{\mathrm{d}\varphi} + \lambda y\,\frac{\mathrm{d}y}{\mathrm{d}\varphi} + (\lambda z + mg)\,\frac{\mathrm{d}z}{\mathrm{d}\varphi}\right]\mathrm{d}\varphi.$$

We substitute the parametric representation of the helix and obtain, after elementary calculations,

$$W = -\frac{1}{2}\,\lambda\,h^2 - m\,g\,h.$$

The work of the external forces exerted on the mass point is $-\frac{1}{2}\lambda h^2 - m g h$, i. e. the mass point must do the work of $\frac{1}{2}\lambda h^2 + m g h$ to move up along the helix against the external forces.

Problem 2.19.

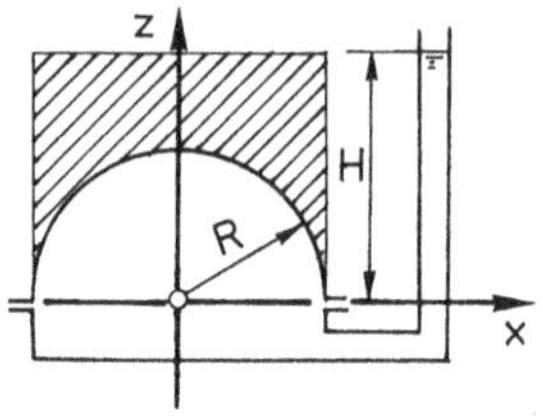

$$F_z = \int p \, \mathrm{d}A_z = \int \varrho g(H - z) \, \mathrm{d}A_z = \iint \varrho g\,(H - z\,(u, v)) \frac{\partial(x, y)}{\partial(u, v)} \, \mathrm{d}u \, \mathrm{d}v.$$

After short calculations we obtain

$$\frac{\partial(x, y)}{\partial(u, v)} = R^2 \sin u \cos u,$$

$$\begin{aligned} F_z &= \int_0^{2\pi} \mathrm{d}v \int_0^{\frac{\pi}{2}} \varrho g(H - R \cos u) R^2 \sin u \cos u \, \mathrm{d}u \\ &= \varrho g R^2 \int_0^{2\pi} \mathrm{d}v \int_0^{\frac{\pi}{2}} (H - R \cos u) \sin u \cos u \, \mathrm{d}u. \end{aligned}$$

For example, if we the substitute $\eta = \cos u$, we obtain

$$F_z = \varrho g \left(\pi R^2 H - \frac{2\pi}{3} R^3 \right).$$

In mechanics it is shown that the vertical force equals the weight of the volume above the area; in our case this is the weight of the indicated volume. This volume is the difference between a circular cylinder with radius R and height H and a hemisphere with radius R. Using this fact and the corresponding stereometric formulas we can compute the vertical force without having to solve an integral.

Problem 3.1.

From

$$a_{ij} = \begin{pmatrix} 4 & 1 & -7 \\ -3 & 5 & 8 \\ 3 & 2 & 9 \end{pmatrix}$$

we obtain its isotropic part as

$$\hat{a} = \frac{1}{3}\, a_{ii} = 6, \qquad \hat{a}\,\delta_{ij} = \begin{pmatrix} 6 & 0 & 0 \\ 0 & 6 & 0 \\ 0 & 0 & 6 \end{pmatrix},$$

its deviatoric part as

$$\mathring{a}_{ij} = a_{ij} - \hat{a}\,\delta_{ij}, \qquad \mathring{a}_{ij} = \begin{pmatrix} -2 & 1 & -7 \\ -3 & -1 & 8 \\ 3 & 2 & 3 \end{pmatrix},$$

its symmetric part as

$$a_{(ij)} = \frac{1}{2}\,(a_{ij} + a_{ji}), \qquad a_{(ij)} = \begin{pmatrix} 4 & -1 & -2 \\ -1 & 5 & 5 \\ -2 & 5 & 9 \end{pmatrix},$$

its antimetric part as

$$a_{[ij]} = \frac{1}{2}\,(a_{ij} - a_{ji}), \qquad a_{[ij]} = \begin{pmatrix} 0 & 2 & -5 \\ -2 & 0 & 3 \\ 5 & -3 & 0 \end{pmatrix},$$

and the symmetric part of its deviator as

$$\mathring{a}_{(ij)} = a_{(ij)} - \hat{a}\,\delta_{ij} = \frac{1}{2}\,(\mathring{a}_{ij} + \mathring{a}_{ji}), \qquad \mathring{a}_{(ij)} = \begin{pmatrix} -2 & -1 & -2 \\ -1 & -1 & 5 \\ -2 & 5 & 3 \end{pmatrix}.$$

Check:

$$\underset{\hat{a}\,\delta_{ij}}{\begin{pmatrix} 6 & 0 & 0 \\ 0 & 6 & 0 \\ 0 & 0 & 6 \end{pmatrix}} + \underset{\mathring{a}_{(ij)}}{\begin{pmatrix} -2 & -1 & -2 \\ -1 & -1 & 5 \\ -2 & 5 & 3 \end{pmatrix}} + \underset{a_{[ij]}}{\begin{pmatrix} 0 & 2 & -5 \\ -2 & 0 & 3 \\ 5 & -3 & 0 \end{pmatrix}} = \underset{a_{ij}.}{\begin{pmatrix} 4 & 1 & -7 \\ -3 & 5 & 8 \\ 3 & 2 & 9 \end{pmatrix}}$$

Problem 3.2.

We can prove the statement as follows:

For second-order polar tensors, we have, according to (2.17),

$$a_{ij} = \alpha_{im}\,\alpha_{jn}\,\tilde{a}_{mn}.$$

For $\tilde{a}_{mn} = -\tilde{a}_{nm}$ it then follows that

$$a_{ij} = -\,\alpha_{im}\,\alpha_{jn}\,\tilde{a}_{nm} = -\,\alpha_{jn}\,\alpha_{im}\,\tilde{a}_{nm} = -\,a_{ji}, \qquad \text{Q. E. D.}$$

Problem 3.3.

Evaluating (3.8) gives

$$A_1 = \frac{1}{2}(\varepsilon_{123}\,a_{[23]} + \varepsilon_{132}\,a_{[32]}) = a_{23}$$

and the two cyclic permutations, thus

$$(A_1, A_2, A_3) = (a_{23}, a_{31}, a_{12}).$$

Hence, the coordinates of a vector satisfy the transformation law for certain coordinates of an antimetric tensor.

Problem 3.4.

The given tensor is antimetric.

A. Consequently, its determinant is zero.

B. The cotensor satisfies, according to (3.12), $b_{ij} = A_i\, A_j$, and from the solution to Problem 3.3 it follows that $(A_1,\, A_2,\, A_3) = (a_{23},\, a_{31},\, a_{12})$, i. e. $(A_1,\, A_2,\, A_3) = (-1,\, -2,\, 3)$. Thus, we obtain the following coordinates of the cotensor:

$$b_{ij} = \begin{pmatrix} 1 & 2 & -3 \\ 2 & 4 & -6 \\ -3 & -6 & 9 \end{pmatrix}.$$

C. Since an antimetric tensor is singular, its inverse tensor does not exist.

Problem 3.5.

Let a_{ij} be the coordinate matrix of the tensor $\underline{\underline{a}}$ in a given coordinate system; then we have, according to (2.17),

$$a_{ij} = \alpha_{im}\, \alpha_{jn}\, \tilde{a}_{mn},$$

where $\tilde{a}_{mn}$ is the coordinate matrix of the tensor in another coordinate system. Furthermore, let a_{ij} be proper or improper orthogonal. Then we have, according to (1.59), in both cases

$$a_{ij}\, a_{kj} = \delta_{ik}.$$

Substituting the first equation for a_{ij} and for a_{kj} in the second equation gives

$$\alpha_{im}\, \alpha_{jn}\, \tilde{a}_{mn}\, \alpha_{kp}\, \alpha_{jq}\, \tilde{a}_{pq} = \delta_{ik},$$

$$\alpha_{im}\, \tilde{a}_{mn}\, \alpha_{kp}\, \delta_{nq}\, \tilde{a}_{pq} = \delta_{ik},$$

$$\alpha_{im}\, \alpha_{kp}\, \tilde{a}_{mn}\, \tilde{a}_{pn} = \delta_{ik}.$$

Multiplication by $\alpha_{ir}\, \alpha_{ks}$ gives

$$\underbrace{\alpha_{im}\, \alpha_{ir}}_{\delta_{mr}}\, \underbrace{\alpha_{kp}\, \alpha_{ks}}_{\delta_{ps}}\, \tilde{a}_{mn}\, \tilde{a}_{pn} = \delta_{ik}\, \alpha_{ir}\, \alpha_{ks} = \underbrace{\alpha_{kr}\, \alpha_{ks}}_{\delta_{rs}},$$

$$\tilde{a}_{rn}\, \tilde{a}_{sn} = \delta_{rs},$$

i. e. the coordinate matrix of the tensor $\underline{\underline{a}}$ is orthogonal also in the other coordinate system (and hence in every coordinate system).

Finally we need to show that $\tilde{a}_{ij}$ is proper orthogonal if a_{ij} is proper orthogonal and vice versa. To show this, we compute the determinant of the transformation equation between a_{ij} and $\tilde{a}_{ij}$ and use Theorem 1.13 for the multiplication of determinants. We have

$$\det a_{ij} = (\det \alpha_{ij})(\det \alpha_{ij})(\det \tilde{a}_{ij}).$$

Since the transformation matrix α_{ij} is orthogonal, its determinant is ±1 and it follows that

$$\det a_{ij} = \det \tilde{a}_{ij},$$

i. e. the coordinate matrices a_{ij} and $\tilde{a}_{ij}$ are either both proper orthogonal or both improper orthogonal, which proves the statement.

Problem 3.6.
Orthogonal tensors satisfy $A_{ji}^{(-1)} = A_{ji}^{\mathrm{T}} = A_{ij}$, so it follows from (1.50) that

$$B_{ij} = A_{ij} \det A_{ij}.$$

Proper orthogonal tensors have $\det A_{ij} = 1$ and hence $B_{ij} = A_{ij}$; improper orthogonal tensors have $\det A_{ij} = -1$ and thus $B_{ij} = -A_{ij}$.

Problem 3.7.

A. A tensor is single singular, if the rows of its coordinate matrix are coplanar, but not collinear. For the tensor under consideration, we see that the sum of the first two rows is the third row and that the second row is not a multiple of the first row. Hence the three rows are coplanar, but not collinear. If we do not see this, we need to convince ourselves that the determinant of the coordinate matrix is zero and that a minor is nonzero.

B. In order to find the null direction, we have to solve the homogeneous system of linear equations $a_{ij} X_j = 0$. Because the coefficient matrix of this system of equations is single singular, we can omit one equation and we can choose freely one unknown, e. g. as 1. If we omit the third equation and set $x_3 = 1$, we obtain

$$\begin{aligned} X_1 - X_2 &= 1, \\ -2X_1 + X_2 &= -3, \end{aligned} \qquad \begin{array}{rr|rrr} 1 & -1 & 1 & 2 & \\ -2 & 1 & -3 & \leftarrow & \\ \hline 1 & -1 & 1 & \leftarrow & \\ 0 & -1 & -1 & -1 & -1 \\ \hline 1 & 0 & 2 & & \\ 0 & 1 & 1 & & \end{array}, \qquad \underline{X} = (2,\ 1,\ 1),$$

i. e. the null direction (in the given coordinate system) is given by the unit vector

$$\underline{\mathring{X}} = \pm\left(\frac{1}{3}\sqrt{6},\ \frac{1}{6}\sqrt{6},\ \frac{1}{6}\sqrt{6}\right).$$

C. The columns of the coordinate matrix lie in the image plane, e. g. two linearly independent vectors in the image plane are

$$\underline{U}_1 = (1,\ -2,\ -1) \quad \text{and} \quad \underline{U}_2 = (-1,\ 1,\ 0).$$

$\underline{V}_1 = \underline{U}_1 + \alpha\, \underline{U}_2$ is perpendicular to $\underline{U}_1$ if the following holds:

$$\underline{U}_1 \cdot \underline{V}_1 = \underline{U}_1 \cdot \underline{U}_1 + \alpha\, \underline{U}_1 \cdot \underline{U}_2 = 0,$$

$$\alpha = -\frac{\underline{U}_1 \cdot \underline{U}_1}{\underline{U}_1 \cdot \underline{U}_2} = -\frac{6}{-3} = 2,$$

$$\underline{V}_1 = (-1,\ 0,\ -1).$$

The normalized vectors $\mathring{\underline{U}}_1 = (\frac{1}{6}\sqrt{6},\ -\frac{1}{3}\sqrt{6},\ -\frac{1}{6}\sqrt{6})$ and $\mathring{\underline{V}}_1 = (-\frac{1}{2}\sqrt{2},\ 0,\ -\frac{1}{2}\sqrt{2})$ form a Cartesian basis in the image plane.

Problem 3.8.

A. For the tensor to be double singular, for example, the rows of the coordinate matrix must be collinear. We see immediately that the second row is the negative of the first row and that the third row is three times the first row.

B. In order to solve the system of equations $a_{ij} X_j = 0$, we can omit two equations and we can freely choose two unknowns in the remaining equation. If we omit the second and third equations, we obtain for $X_2 = 1,\ X_3 = 0$:

$$X_1 - 1 = 0, \quad X_1 = 1$$

and for $X_2 = 0,\ X_3 = 1$:

$$X_1 + 2 = 0, \quad X_1 = -2.$$

Thus, two linearly independent null vectors are

$$\underline{X}_1 = (1,\ 1,\ 0) \quad \text{and} \quad \underline{X}_2 = (-2,\ 0,\ 1).$$

C. The unit vector normal to the null plane is in the direction of the row line, i. e.

$$\mathring{\underline{Y}} = \pm\left(\frac{1}{6}\sqrt{6},\ -\frac{1}{6}\sqrt{6},\ \frac{1}{3}\sqrt{6}\right).$$

D. The image line is the column line; its direction is given by the unit vector

$$\mathring{\underline{U}} = \pm\left(\frac{1}{11}\sqrt{11},\ -\frac{1}{11}\sqrt{11},\ \frac{3}{11}\sqrt{11}\right).$$

Problem 3.9.

A. For an orthogonal basis it immediately follows from

$$\underline{g}_1 = 3\,\underline{e}_1, \quad \underline{g}_2 = 2\,\underline{e}_2, \quad \underline{g}_3 = \underline{e}_3$$

according to Section 3.9.3 that

$$\underline{g}^1 = \frac{1}{3}\,\underline{e}_1, \quad \underline{g}^2 = \frac{1}{2}\,\underline{e}_2, \quad \underline{g}^3 = \underline{e}_3.$$

B. We first compute the three vectors of the original basis, i. e.

$$\underline{g}_1 = \underline{e}_1.$$

For $\underline{g}_2$ we have

$$\underline{g}_2 \cdot \underline{e}_3 = 0, \quad \underline{g}_2 \cdot \underline{g}_1 = \underline{g}_2 \cdot \underline{e}_1 = \cos 60^\circ = \frac{1}{2}, \quad \underline{g}_2 \cdot \underline{g}_2 = 1.$$

Using the first two equations we obtain for the third equation

$$\underline{g}_2 \cdot \underline{g}_2 = (\underline{g}_2 \cdot \underline{e}_1)^2 + (\underline{g}_2 \cdot \underline{e}_2)^2 = \frac{1}{4} + (\underline{g}_2 \cdot \underline{e}_2)^2 = 1, \quad \underline{g}_2 \cdot \underline{e}_2 = \frac{1}{2}\sqrt{3},$$

$$\underline{g}_2 = \frac{1}{2}\underline{e}_1 + \frac{1}{2}\sqrt{3}\,\underline{e}_2.$$

Finally, we have for $\underline{g}_3$

$$\underline{g}_3 \cdot \underline{g}_1 = \cos 60^\circ = \frac{1}{2}, \quad \underline{g}_3 \cdot \underline{g}_2 = \cos 60^\circ = \frac{1}{2}, \quad \underline{g}_3 \cdot \underline{g}_3 = 1.$$

This implies that

$$\underline{g}_3 \cdot \underline{e}_1 = \frac{1}{2},$$

$$\underline{g}_3 \cdot \underline{g}_2 = \frac{1}{2}\underline{g}_3 \cdot \underline{e}_1 + \frac{1}{2}\sqrt{3}\,\underline{g}_3 \cdot \underline{e}_2 = \frac{1}{4} + \frac{1}{2}\sqrt{3}\,\underline{g}_3 \cdot \underline{e}_2 = \frac{1}{2}, \quad \underline{g}_3 \cdot \underline{e}_2 = \frac{1}{6}\sqrt{3},$$

$$\underline{g}_3 \cdot \underline{g}_3 = (\underline{g}_3 \cdot \underline{e}_1)^2 + (\underline{g}_3 \cdot \underline{e}_2)^2 + (\underline{g}_3 \cdot \underline{e}_3)^2 = \frac{1}{4} + \frac{1}{12} + (\underline{g}_3 \cdot \underline{e}_3)^2 = 1,$$

$$\underline{g}_3 \cdot \underline{e}_3 = \frac{1}{3}\sqrt{6},$$

$$\underline{g}_3 = \frac{1}{2}\underline{e}_1 + \frac{1}{6}\sqrt{3}\,\underline{e}_2 + \frac{1}{3}\sqrt{6}\,\underline{e}_3.$$

The easiest method to compute the basis reciprocal to $\underline{g}_i$ is to use the Gauss–Jordan algorithm, i. e.

$$\begin{array}{c|ccc|ccc|cc}
\underline{g}_1 & 1 & 0 & 0 & 1 & 0 & 0 & -\frac{1}{2}\,| & -\frac{1}{2}\,| \\
\underline{g}_2 & \frac{1}{2} & \frac{1}{2}\sqrt{3} & 0 & 0 & 1 & 0 & \leftarrow & | \\
\underline{g}_3 & \frac{1}{2} & \frac{1}{6}\sqrt{3} & \frac{1}{3}\sqrt{6} & 0 & 0 & 1 & & \leftarrow \\
\hline
 & 1 & 0 & 0 & 1 & 0 & 0 & & \\
 & 0 & \frac{1}{2}\sqrt{3} & 0 & -\frac{1}{2} & 1 & 0 & -\frac{1}{3}\,| & \frac{2}{\sqrt{3}} = \frac{2}{3}\sqrt{3} \\
 & 0 & \frac{1}{6}\sqrt{3} & \frac{1}{3}\sqrt{6} & -\frac{1}{2} & 0 & 1 & \leftarrow & \\
\hline
 & 1 & 0 & 0 & 1 & 0 & 0 & & \\
 & 0 & 1 & 0 & -\frac{1}{3}\sqrt{3} & \frac{2}{3}\sqrt{3} & 0 & & \\
 & 0 & 0 & \frac{1}{3}\sqrt{6} & -\frac{1}{3} & -\frac{1}{3} & 1 & & \frac{3}{\sqrt{6}} = \frac{1}{2}\sqrt{6} \\
\hline
 & 1 & 0 & 0 & 1 & 0 & 0 & & \\
 & 0 & 1 & 0 & -\frac{1}{3}\sqrt{3} & \frac{2}{3}\sqrt{3} & 0 & & \\
 & 0 & 0 & 1 & -\frac{1}{6}\sqrt{6} & -\frac{1}{6}\sqrt{6} & \frac{1}{2}\sqrt{6} & & \\
 & & & & \underline{g}^1 & \underline{g}^2 & \underline{g}^3 & &
\end{array}$$

Thus, we obtain the reciprocal basis:

$$\underline{g}^1 = \underline{e}_1 - \frac{1}{3}\sqrt{3}\,\underline{e}_2 - \frac{1}{6}\sqrt{6}\,\underline{e}_3,$$

$$\underline{g}^2 = \frac{2}{3}\sqrt{3}\,\underline{e}_2 - \frac{1}{6}\sqrt{6}\,\underline{e}_3,$$

$$\underline{g}^3 = \frac{1}{2}\sqrt{6}\,\underline{e}_3.$$

Problem 3.10.
According to $(3.30)_2$, we have $\underline{g}^i\,\underline{h}_i = \underline{\underline{a}}^{\mathrm{T}}$ or $\overset{i}{g}_m\,\underset{i}{h}_n = a^{\mathrm{T}}_{mn} = a_{nm}$,

$$\begin{pmatrix} \overset{1}{g}_1 & \overset{2}{g}_1 & \overset{3}{g}_1 \\ \overset{1}{g}_2 & \overset{2}{g}_2 & \overset{3}{g}_2 \\ \overset{1}{g}_3 & \overset{2}{g}_3 & \overset{3}{g}_3 \end{pmatrix} \begin{pmatrix} \underset{1}{h_1} & \underset{1}{h_2} & \underset{1}{h_3} \\ \underset{2}{h_1} & \underset{2}{h_2} & \underset{2}{h_3} \\ \underset{3}{h_1} & \underset{3}{h_2} & \underset{3}{h_3} \end{pmatrix} = \begin{pmatrix} a_{11} & a_{21} & a_{31} \\ a_{12} & a_{22} & a_{32} \\ a_{13} & a_{23} & a_{33} \end{pmatrix}.$$

$\underline{g}^1$	$\underline{g}^2$	$\underline{g}^3$						
1	3	1	2	4	1	1		
0	1	2	3	−2	2			
−1	−3	−2	−1	3	1	←		
1	3	1	2	4	1	←		
0	1	2	3	−2	2	−3		
0	0	−1	1	7	2			
1	0	−5	−7	10	−5	←		
0	1	2	3	−2	2		←	
0	0	−1	1	7	2	−5	2	−1
1	0	0	−12	−25	−15	$\underline{h}_1$		
0	1	0	5	12	6	$\underline{h}_2$		
0	0	1	−1	−7	−2	$\underline{h}_3$		

Thus we get

$$\underline{h}_1 = (-12,\ -25,\ -15), \qquad \underline{h}_2 = (5,\ 12,\ 6), \qquad \underline{h}_3 = (-1,\ -7,\ -2).$$

The representation $\underline{\underline{a}} = \underline{h}_i\,\underline{g}^i$ is then

$$\begin{pmatrix} 2 & 3 & -1 \\ 4 & -2 & 3 \\ 1 & 2 & 1 \end{pmatrix} = \begin{pmatrix} -12 & 0 & 12 \\ -25 & 0 & 25 \\ -15 & 0 & 15 \end{pmatrix} + \begin{pmatrix} 15 & 5 & -15 \\ 36 & 12 & -36 \\ 18 & 6 & -18 \end{pmatrix} + \begin{pmatrix} -1 & -2 & 2 \\ -7 & -14 & 14 \\ -2 & -4 & 4 \end{pmatrix}.$$

Problem 3.11.

A. In order to be able to represent $\underline{\underline{a}}$ with $\underline{h}_1$ and $\underline{h}_2$ in this form, $\underline{h}_1$ and $\underline{h}_2$ have to lie in the column plane of $\underline{\underline{a}}$. The easiest way to verify this is to prove that the matrix formed by $\underline{h}_1$, $\underline{h}_2$, and two columns of $\underline{\underline{a}}$ has rank 2, so

$$
\begin{array}{rrrr|lll}
-1 & 0 & 1 & -1 & 2 & 1 & -1 \\
2 & 1 & -2 & 1 & \leftarrow & & \\
1 & 1 & -1 & 0 & & \leftarrow & \\
\hline
1 & 0 & -1 & 1 & & & \\
0 & 1 & 0 & -1 & & & \\
0 & 1 & 0 & -1 & & &
\end{array}
$$

Since the last two rows are equal, we can omit one row when we determine the rank, and since the first two rows are not collinear, the matrix has rank 2.

B. We have $\underline{h}_i \, \underline{g}^i = \underline{\underline{a}}, \quad \underset{i}{h}_m \overset{i}{g}_n = a_{mn}$,

$$
\begin{pmatrix} \underset{1}{h_1} & \underset{2}{h_1} \\ \underset{1}{h_2} & \underset{2}{h_2} \\ \underset{1}{h_3} & \underset{2}{h_3} \end{pmatrix}
\begin{pmatrix} \overset{1}{g}_1 & \overset{1}{g}_2 & \overset{1}{g}_3 \\ \overset{2}{g}_1 & \overset{2}{g}_2 & \overset{2}{g}_3 \end{pmatrix}
=
\begin{pmatrix} a_{11} & a_{12} & a_{13} \\ a_{21} & a_{22} & a_{23} \\ a_{31} & a_{32} & a_{33} \end{pmatrix}.
$$

To determine the $\overset{i}{g}_n$, the first two rows are sufficient:

$$
\begin{array}{rr|rrr|ll}
\underline{h}_1 & \underline{h}_2 & & & & & \\
-1 & 0 & 1 & -1 & -1 & 2 & -1 \\
2 & 1 & -2 & 1 & 3 & \leftarrow & \\
\hline
1 & 0 & -1 & 1 & 1 & \underline{g}^1 & \\
0 & 1 & 0 & -1 & 1 & \underline{g}^2 &
\end{array}
$$

So we obtain $\underline{g}^1 = (-1, 1, 1)$, $\underline{g}^2 = (0, -1, 1)$.

C. The representation $\underline{\underline{a}} = \underline{h}_1 \underline{g}^1 + \underline{h}_2 \underline{g}^2$ is then

$$
\begin{pmatrix} 1 & -1 & -1 \\ -2 & 1 & 3 \\ -1 & 0 & 2 \end{pmatrix}
=
\begin{pmatrix} 1 & -1 & -1 \\ -2 & 2 & 2 \\ -1 & 1 & 1 \end{pmatrix}
+
\begin{pmatrix} 0 & 0 & 0 \\ 0 & -1 & 1 \\ 0 & -1 & 1 \end{pmatrix}.
$$

Problem 3.12.

Determining the eigenvalues:

All coordinate matrices are triangular matrices. Hence, according to Section 3.11.3, No. 6, the diagonal elements are the eigenvalues: In the cases A to C, the tensor has the eigenvalue $\lambda = 1$ of multiplicity 3. In the cases D to F, the tensor has the eigenvalue $\lambda = 1$ of multiplicity 2 and the eigenvalue $\lambda = 2$ of multiplicity 1.

Determining the eigendirections:

A. The eigenvalue equation is

$$\begin{pmatrix} 0 & 1 & 0 \\ 0 & 0 & 1 \\ 0 & 0 & 0 \end{pmatrix} \begin{pmatrix} x_1 \\ x_2 \\ x_3 \end{pmatrix} = \begin{pmatrix} 0 \\ 0 \\ 0 \end{pmatrix}.$$

The coefficient matrix has rank 2, i. e. according to Section 3.11.4, Theorem 2, only one eigendirection exists and the tensor is defective. From the equation we obtain $x_2 = 0,\ x_3 = 0$. We can freely choose x_1, so the eigendirection is given by the unit vector

$$\underline{\mathring{x}} = (1,\ 0,\ 0).$$

B. The eigenvalue equation is

$$\begin{pmatrix} 0 & 0 & 0 \\ 0 & 0 & 1 \\ 0 & 0 & 0 \end{pmatrix} \begin{pmatrix} x_1 \\ x_2 \\ x_3 \end{pmatrix} = \begin{pmatrix} 0 \\ 0 \\ 0 \end{pmatrix}.$$

The coefficient matrix has rank 1, i. e. according to Section 3.11.4, Theorem 3, two distinct eigendirections exist and the tensor is again defective. Here we obtain from the equation $x_3 = 0$. We can freely choose x_1 and x_2, so two eigendirections are, for example, given by the unit vectors

$$\underline{\mathring{x}}_1 = (1,\ 0,\ 0), \quad \underline{\mathring{x}}_2 = (0,\ 1,\ 0).$$

C. The eigenvalue equation is

$$\begin{pmatrix} 0 & 0 & 0 \\ 0 & 0 & 0 \\ 0 & 0 & 0 \end{pmatrix} \begin{pmatrix} x_1 \\ x_2 \\ x_3 \end{pmatrix} = \begin{pmatrix} 0 \\ 0 \\ 0 \end{pmatrix}.$$

The coefficient matrix has rank 0, i. e. according to Section 3.11.4, Theorem 4, three linearly independent eigendirections exist, or in other words, every direction is an eigendirection. The tensor is nondefective and we can also find three mutually orthogonal eigenvectors. Accordingly, the equation does not provide any restriction for the coordinates of the eigenvectors and a set of mutually orthogonal eigenvectors is, for example,

$$\underline{\mathring{x}}_1 = (1,\ 0,\ 0), \quad \underline{\mathring{x}}_2 = (0,\ 1,\ 0), \quad \underline{\mathring{x}}_3 = (0,\ 0,\ 1).$$

D. For $\lambda = 1$ the eigenvalue equation is

$$\begin{pmatrix} 0 & 1 & 0 \\ 0 & 0 & 1 \\ 0 & 0 & 1 \end{pmatrix} \begin{pmatrix} x_1 \\ x_2 \\ x_3 \end{pmatrix} = \begin{pmatrix} 0 \\ 0 \\ 0 \end{pmatrix}.$$

The coefficient matrix has rank 2, only one eigendirection corresponds to the eigenvalue with multiplicity two, and the tensor is defective. From the equation we obtain $x_2 = 0,\ x_3 = 0;$ the eigendirection is given by the unit vector

$$\underline{\mathring{x}}_1 = (1,\ 0,\ 0).$$

For $\lambda = 2,$ the eigenvalue equation is

$$\begin{pmatrix} -1 & 1 & 0 \\ 0 & -1 & 1 \\ 0 & 0 & 0 \end{pmatrix}\begin{pmatrix} x_1 \\ x_2 \\ x_3 \end{pmatrix} = \begin{pmatrix} 0 \\ 0 \\ 0 \end{pmatrix},$$

which gives

$$-x_1 + x_2 = 0, \quad -x_2 + x_3 = 0.$$

With $x_3 = 1$ we obtain the eigenvector $\underline{x} = (1,\ 1,\ 1)$, i. e. the eigendirection is given by the unit vector

$$\underline{\mathring{x}}_2 = \left(\frac{1}{3}\sqrt{3},\ \frac{1}{3}\sqrt{3},\ \frac{1}{3}\sqrt{3}\right).$$

E. For $\lambda = 1,$ the eigenvalue equation is

$$\begin{pmatrix} 0 & 0 & 0 \\ 0 & 0 & 1 \\ 0 & 0 & 1 \end{pmatrix}\begin{pmatrix} x_1 \\ x_2 \\ x_3 \end{pmatrix} = \begin{pmatrix} 0 \\ 0 \\ 0 \end{pmatrix}.$$

The coefficient matrix has rank 1, two eigendirections correspond to the eigenvalue with multiplicity two, and the tensor is nondefective. From the equation we only obtain $x_3 = 0;$ two orthogonal eigendirections are, for example, given by the unit vectors

$$\underline{\mathring{x}}_1 = (1,\ 0,\ 0), \quad \underline{\mathring{x}}_2 = (0,\ 1,\ 0).$$

For $\lambda = 2,$ the eigenvalue equation is

$$\begin{pmatrix} -1 & 0 & 0 \\ 0 & -1 & 1 \\ 0 & 0 & 0 \end{pmatrix}\begin{pmatrix} x_1 \\ x_2 \\ x_3 \end{pmatrix} = \begin{pmatrix} 0 \\ 0 \\ 0 \end{pmatrix},$$

which gives

$$-x_1 = 0, \quad -x_2 + x_3 = 0.$$

One eigenvector is $\underline{x} = (0,\ 1,\ 1),$ i. e. the eigendirection is given by

$$\underline{\mathring{x}}_3 = \left(0,\ \frac{1}{2}\sqrt{2},\ \frac{1}{2}\sqrt{2}\right).$$

This eigendirection is not perpendicular to the eigenplane corresponding to $\lambda = 1,$ i. e. no three mutually orthogonal eigendirections exist.

F. For $\lambda = 1$, the eigenvalue equation is

$$\begin{pmatrix} 0 & 0 & 0 \\ 0 & 0 & 0 \\ 0 & 0 & 1 \end{pmatrix} \begin{pmatrix} x_1 \\ x_2 \\ x_3 \end{pmatrix} = \begin{pmatrix} 0 \\ 0 \\ 0 \end{pmatrix}.$$

The coefficient matrix has again rank 1, two eigendirections correspond to the eigenvalue with multiplicity two, and the tensor is nondefective. From the equation we obtain only $x_3 = 0$; two orthogonal eigendirections are given by the unit vectors

$$\underline{\mathring{x}}_1 = (1, 0, 0), \quad \underline{\mathring{x}}_2 = (0, 1, 0).$$

For $\lambda = 2$, the eigenvalue equation is

$$\begin{pmatrix} -1 & 0 & 0 \\ 0 & -1 & 0 \\ 0 & 0 & 0 \end{pmatrix} \begin{pmatrix} x_1 \\ x_2 \\ x_3 \end{pmatrix} = \begin{pmatrix} 0 \\ 0 \\ 0 \end{pmatrix},$$

which gives

$$-x_1 = 0, \quad -x_2 = 0.$$

The eigendirection is given by the unit vector

$$\underline{\mathring{x}}_3 = (0, 0, 1).$$

This eigendirection is perpendicular to the eigenplane corresponding to $\lambda = 1$, i. e. three mutually orthogonal eigendirections exist.

Problem 3.13.

A principal axes transformation requires finding three mutually perpendicular eigendirections. Such a triple of eigendirections exists only for symmetric tensors. See also the results of the last problem.

We start by finding the eigenvalues of the tensor. The characteristic equation is

$$\begin{vmatrix} 5-\lambda & 0 & 4 \\ 0 & 9-\lambda & 0 \\ 4 & 0 & 5-\lambda \end{vmatrix} = 0.$$

Expanding with respect to the second row gives

$$(9-\lambda)[(5-\lambda)(5-\lambda) - 16] = 0,$$

and after elementary computations

$$\lambda_1 = \lambda_2 = 9, \quad \lambda_3 = 1.$$

From the eigenvalue equation for the eigenvalue $\lambda = 9$ with multiplicity two, we obtain the orthonormal eigenvectors

$$\underline{q}_1 = \left(\frac{1}{2}\sqrt{2},\, 0,\, \frac{1}{2}\sqrt{2}\right) \quad \text{and} \quad \underline{q}_2 = (0,\, 1,\, 0),$$

and from the eigenvalue equation for the eigenvalue $\lambda = 1$, we obtain the normalized eigenvector

$$\underline{q}_3 = \left(\frac{1}{2}\sqrt{2},\, 0,\, -\frac{1}{2}\sqrt{2}\right),$$

which is orthogonal to the other two eigenvectors. We still have to check if the three eigendirections form, in this order, a right-handed system. For this to be true, the value of the determinant

$$\begin{vmatrix} \frac{1}{2}\sqrt{2} & 0 & \frac{1}{2}\sqrt{2} \\ 0 & 1 & 0 \\ \frac{1}{2}\sqrt{2} & 0 & -\frac{1}{2}\sqrt{2} \end{vmatrix}$$

must be positive. This is clearly not the case, i. e. we have to reverse the direction of the third eigendirection, and thus we obtain as a principal axes system

$$\underline{q}_1 = \left(\frac{1}{2}\sqrt{2},\, 0,\, \frac{1}{2}\sqrt{2}\right), \quad \underline{q}_2 = (0,\, 1,\, 0), \quad \underline{q}_3 = \left(-\frac{1}{2}\sqrt{2},\, 0,\, \frac{1}{2}\sqrt{2}\right).$$

The coordinate matrix of the tensor in this principal axes system is

$$\hat{a}_{ij} = \begin{pmatrix} 9 & 0 & 0 \\ 0 & 9 & 0 \\ 0 & 0 & 1 \end{pmatrix}.$$

Problem 3.14.

We assume that the generalized eigenvalue problem has a complex eigenvalue and a complex eigendirection. Then, similarly to Section 3.11.3, No. 2, it follows that

$$a_{ij}\,(X_j + \mathrm{i}Y_j) = (\lambda + \mathrm{i}\mu)\, b_{ij}\,(X_j + \mathrm{i}Y_j)\,.$$

Splitting into real and imaginary part gives

$$a_{ij}\, X_j = b_{ij}\,(\lambda\, X_j - \mu\, Y_j)\,, \quad a_{ij}\, Y_j = b_{ij}\,(\mu\, X_j + \lambda\, Y_j)\,.$$

Contraction of the first equation with Y_i and of the second equation with X_i and computing the difference gives

$$\underbrace{a_{ij}\, Y_i\, X_j - a_{ij}\, X_i\, Y_j}_{0} = \lambda\,(\underbrace{b_{ij}\, Y_i\, X_j - b_{ij}\, X_i\, Y_j}_{0}) - \mu\,(b_{ij}\, Y_i\, Y_j + b_{ij}\, X_i\, X_j)\,.$$

The first two expressions are zero, because a_{ij} and b_{ij} are symmetric, and what remains is

$$\mu\,(b_{ij}\,Y_i\,Y_j + b_{ij}\,X_i\,X_j) = 0.$$

To obtain $\mu = 0$ from this equation, the expression in parentheses must be nonzero, which is the case if the tensor $\underline{\underline{b}}$ is positive or negative definite.

Problem 3.15.
Solving the characteristic equation gives

$$\lambda_1 = \pm 1, \quad \lambda_2 = \cos\vartheta + \mathrm{i}\,\sin\vartheta, \quad \lambda_3 = \cos\vartheta - \mathrm{i}\,\sin\vartheta.$$

Problem 3.16.

A. The two orthogonality relations yield six equations, which can be solved for the six unknowns α, β, γ, δ, ε, ζ. Here $\alpha_{ki}\,\alpha_{kj} = \delta_{ij}$ (a statement about the sums of the squares and the sums of the products of the columns) is more useful than $\alpha_{ik}\,\alpha_{jk} = \delta_{ij}$ (an analogous statement about the row sums) because in the first case the six equations are partially decoupled.

   From $\alpha_{k1}\,\alpha_{k1} = 1$ it follows that $\alpha = \pm\frac{1}{2}$.

   From $\alpha_{k1}\,\alpha_{k2} = 0$ and $\alpha_{k2}\,\alpha_{k2} = 1$ it follows, independently of the sign of α, that

$$\beta = \pm\frac{1}{2}\sqrt{2}, \quad \gamma = 0.$$

   From $\alpha_{k1}\,\alpha_{k3} = 0$, $\alpha_{k2}\,\alpha_{k3} = 0$, and $\alpha_{k3}\,\alpha_{k3} = 1$ it follows, independently of the sign of α and β, that

$$\delta = \zeta = \pm\frac{1}{2}, \quad \varepsilon = \pm\frac{1}{2}\sqrt{2}.$$

   Thus the general form of the transformation matrix is

$$\alpha_{ij} = \begin{pmatrix} \pm\frac{1}{2} & \pm\frac{1}{2}\sqrt{2} & \pm\frac{1}{2} \\ \mp\frac{1}{2}\sqrt{2} & 0 & \pm\frac{1}{2}\sqrt{2} \\ \pm\frac{1}{2} & \mp\frac{1}{2}\sqrt{2} & \pm\frac{1}{2} \end{pmatrix},$$

   where we can choose in each column either the upper or the lower sign, independently of the other columns.

B. Thus the problem has $2^3 = 8$ solutions. We can characterize one of these solutions e. g. by specifying the signs of α, β, and δ, which corresponds to choosing the upper or the lower sign in the three columns. According to (2.4), the columns of α_{ij} represent the coordinates of the rotated basis vectors $\underline{\tilde{e}}_j$ with respect to the original basis $\underline{e}_i$; every sign change of a column can be interpreted as reversing the orientation of the corresponding basis vector.

Problem 3.17.

To decide whether to use (3.78) or (3.82) to compute the angle of rotation, we first have to check if the transformation matrix is proper orthogonal or improper orthogonal. The determinant is +1, so the transformation matrix describes a rotation, and (3.78) yields $\cos\vartheta = 0$, $\vartheta = \pm\frac{\pi}{2}$. We choose $\vartheta = +\frac{\pi}{2}$. Then $\sin\vartheta = 1$, and from (3.79) it follows after a short computation that

$$\underline{n} = \left(-\frac{1}{2}\sqrt{2},\ 0,\ -\frac{1}{2}\sqrt{2}\right).$$

The transformation matrix describes a rotation of 90° about an axis in the x, z-plane which forms with both the negative x-axis and the negative z-axis an angle of 45°. This is equivalent to a rotation of −90° about an axis in the x, z-plane which forms with both the positive x-axis and the positive z-axis an angle of 45°, respectively. This is the case for $\vartheta = -\frac{\pi}{2}$, $\sin\vartheta = -1$.

Problem 3.18.

We can prove this as follows. Let λ be an eigenvalue and let X_i be an eigenvector of a_{ij} corresponding to λ. Then we have

$$a_{ij}\, X_j = \lambda\, X_i.$$

Multiplication by $a_{ki}^{(-1)}$ gives

$$\underbrace{a_{ki}^{(-1)}\, a_{ij}}_{\delta_{kj}}\, X_j = \lambda\, a_{ki}^{(-1)}\, X_i, \quad X_k = \lambda\, a_{ki}^{(-1)}\, X_i, \quad a_{ki}^{(-1)}\, X_i = \lambda^{-1}\, X_k, \text{ Q. E. D.}$$

Problem 3.19.

We can prove this by mathematical induction:

Basic step: $n = 0$.

For $n = 0$, according to (3.87), this statement is satisfied.

Inductive step:

Assumption: For $n = k$ we have $(\underline{\underline{A}}^k)^{\mathrm{T}} = (\underline{\underline{A}}^{\mathrm{T}})^k$. (a)

1. Statement: For $n = k + 1$ we have $(\underline{\underline{A}}^{k+1})^{\mathrm{T}} = (\underline{\underline{A}}^{\mathrm{T}})^{k+1}$.

2. Statement: For $n = k - 1$ we have $(\underline{\underline{A}}^{k-1})^{\mathrm{T}} = (\underline{\underline{A}}^{\mathrm{T}})^{k-1}$.

Proof of statement 1:

$$(\underline{\underline{A}}^{k+1})^{\mathrm{T}} \overset{(3.86)}{=} (\underline{\underline{A}} \cdot \underline{\underline{A}}^k)^{\mathrm{T}} \overset{(2.35)}{=} (\underline{\underline{A}}^k)^{\mathrm{T}} \cdot \underline{\underline{A}}^{\mathrm{T}} \overset{(a)}{=} (\underline{\underline{A}}^{\mathrm{T}})^k \cdot \underline{\underline{A}}^{\mathrm{T}}$$

$$\overset{(3.86)}{=} (\underline{\underline{A}}^{\mathrm{T}})^{k+1}, \text{ Q. E. D.}$$

Proof of the statement 2:

$$(\underline{\underline{A}}^{k-1})^{\mathrm{T}} \overset{(3.86)}{=} (\underline{\underline{A}}^{-1} \cdot \underline{\underline{A}}^{k})^{\mathrm{T}} \overset{(2.35)}{=} (\underline{\underline{A}}^{k})^{\mathrm{T}} \cdot (\underline{\underline{A}}^{-1})^{\mathrm{T}} \overset{(a),\,(1.54)}{=} (\underline{\underline{A}}^{\mathrm{T}})^{k} \cdot (\underline{\underline{A}}^{\mathrm{T}})^{-1}$$

$$\overset{(3.86)}{=} (\underline{\underline{A}}^{\mathrm{T}})^{k-1}, \quad \text{Q. E. D.}$$

Problem 3.20.
We have $\underline{U} = \underline{\underline{a}} \cdot \underline{X}$ and we set $\underline{V} := \underline{\underline{a}} \cdot \underline{U}$. Then we have

$$\underline{W} = \underline{\underline{a}}^{3} \cdot \underline{X} = \underline{\underline{a}}^{2} \cdot \underbrace{\underline{\underline{a}} \cdot \underline{X}}_{\underline{U}} = \underline{\underline{a}}^{2} \cdot \underline{U} = \underline{\underline{a}} \cdot \underbrace{\underline{\underline{a}} \cdot \underline{U}}_{\underline{V}} = \underline{\underline{a}} \cdot \underline{V}.$$

The tensor $\underline{\underline{a}}$ rotates (in each case through the angle ϑ) first the vector $\underline{X}$ into the vector $\underline{U}$, then the vector $\underline{U}$ into the vector $\underline{V}$, and finally the vector $\underline{V}$ into the vector $\underline{W}$, i. e. $\underline{\underline{a}}^{3}$ rotates the vector $\underline{X}$ through the angle $3\,\vartheta$ into the vector $\underline{W}$.

Problem 3.21.
Problem 3.13 shows that the tensor has the following eigenvalues and the following corresponding normalized eigenvectors, which form a right-handed coordinate system:

$$\begin{aligned} \lambda_1 = \lambda_2 = 9, \quad & \underline{q}_1 = \left(\frac{1}{2}\sqrt{2},\ 0,\ \frac{1}{2}\sqrt{2}\right), \\ & \underline{q}_2 = (0,\ 1,\ 0), \\ \lambda_3 = 1, \quad & \underline{q}_3 = \left(-\frac{1}{2}\sqrt{2},\ 0,\ \frac{1}{2}\sqrt{2}\right). \end{aligned}$$

A. Since the original tensor is symmetric and all its eigenvalues are positive, the tensor is positive definite.
B. We can compute the coordinates of the positive definite square root $\underline{\underline{b}} = \sqrt{\underline{\underline{a}}}$, using (3.90), from

$$b_{ij} = \sqrt{\lambda_{\underline{k}}}\, \overset{k}{q}_i\, \overset{k}{q}_j .$$

We can also write this equation in the form

$$\begin{pmatrix} \sqrt{\lambda_1}\,\overset{1}{q}_1 & \sqrt{\lambda_2}\,\overset{2}{q}_1 & \sqrt{\lambda_3}\,\overset{3}{q}_1 \\ \sqrt{\lambda_1}\,\overset{1}{q}_2 & \sqrt{\lambda_2}\,\overset{2}{q}_2 & \sqrt{\lambda_3}\,\overset{3}{q}_2 \\ \sqrt{\lambda_1}\,\overset{1}{q}_3 & \sqrt{\lambda_2}\,\overset{2}{q}_3 & \sqrt{\lambda_3}\,\overset{3}{q}_3 \end{pmatrix} \begin{pmatrix} \overset{1}{q}_1 & \overset{1}{q}_2 & \overset{1}{q}_3 \\ \overset{2}{q}_1 & \overset{2}{q}_2 & \overset{2}{q}_3 \\ \overset{3}{q}_1 & \overset{3}{q}_2 & \overset{3}{q}_3 \end{pmatrix}$$

or in the form

$$b_{ij} = \sqrt{\lambda_1}\begin{pmatrix} \overset{1}{q}_1 \\ \overset{1}{q}_2 \\ \overset{1}{q}_3 \end{pmatrix}\left(\overset{1}{q}_1, \overset{1}{q}_2, \overset{1}{q}_3\right) + \sqrt{\lambda_2}\begin{pmatrix} \overset{2}{q}_1 \\ \overset{2}{q}_2 \\ \overset{2}{q}_3 \end{pmatrix}\left(\overset{2}{q}_1, \overset{2}{q}_2, \overset{2}{q}_3\right)$$

$$+ \sqrt{\lambda_3}\begin{pmatrix} \overset{3}{q}_1 \\ \overset{3}{q}_2 \\ \overset{3}{q}_3 \end{pmatrix}\left(\overset{3}{q}_1, \overset{3}{q}_2, \overset{3}{q}_3\right).$$

Both ways lead to

$$b_{ij} = \begin{pmatrix} 2 & 0 & 1 \\ 0 & 3 & 0 \\ 1 & 0 & 2 \end{pmatrix}.$$

We can verify this result by showing that the square of b_{ij} is a_{ij}.

Problem 3.22.
Straightforward computations give

$$a_{ik}\, a^{\mathrm{T}}_{kj} = \begin{pmatrix} 9 & 0 & 0 \\ 0 & 9 & 0 \\ 0 & 0 & 9 \end{pmatrix}, \quad V_{ij} = \sqrt{a_{ik}\, a^{\mathrm{T}}_{kj}} = \begin{pmatrix} 3 & 0 & 0 \\ 0 & 3 & 0 \\ 0 & 0 & 3 \end{pmatrix}.$$

Using the Gauss–Jordan algorithm we obtain from $V_{ik}\, R_{kj} = a_{ij}$

$$R_{ij} = \begin{pmatrix} \frac{2}{3} & \frac{2}{3} & \frac{1}{3} \\ \frac{1}{3} & -\frac{2}{3} & \frac{2}{3} \\ \frac{2}{3} & -\frac{1}{3} & -\frac{2}{3} \end{pmatrix},$$

and we can easily verify that $V_{ik}\, R_{kj} = a_{ij}$.

Problem 4.1.
A. We see from the figure that

$$\begin{aligned} OB &= x_1, & OA &= OC\ \cos\alpha = u^1\ \cos\alpha, \\ BP &= x_2, & AB &= CP\ \sin\beta = u^2\ \sin\beta, \\ OC &= u^1, & BD &= OC\ \sin\alpha = u^1\ \sin\alpha, \\ CP &= u^2, & DP &= CP\ \cos\beta = u^2\ \cos\beta, \end{aligned}$$

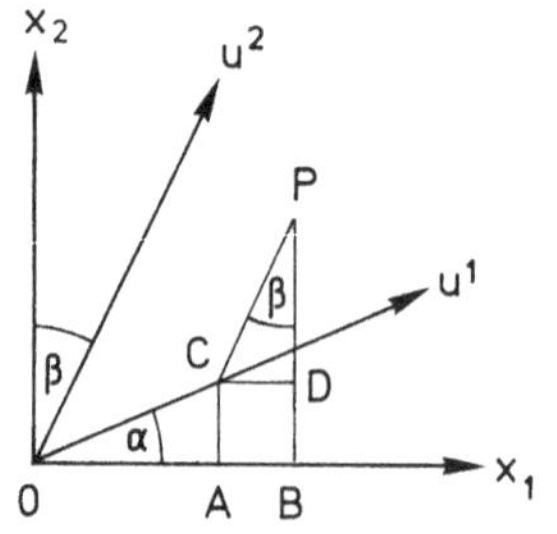

$$x_1 = u^1 \cos\alpha + u^2 \sin\beta,$$
$$x_2 = u^1 \sin\alpha + u^2 \cos\beta,$$

$$\begin{array}{l|c|c} x_1 = u^1 \cos\alpha + u^2 \sin\beta & \cos\beta & -\sin\alpha \\ x_2 = u^1 \sin\alpha + u^2 \cos\beta & -\sin\beta & \cos\alpha \end{array},$$

$$\begin{aligned} &x_1 \cos\beta - x_2 \sin\beta \\ &\quad = u^1(\cos\alpha \cos\beta - \sin\alpha \sin\beta) = u^1 \cos(\alpha+\beta), \\ &-x_1 \sin\alpha + x_2 \cos\alpha \\ &\quad = u^2(-\sin\alpha \sin\beta + \cos\alpha \cos\beta) = u^2 \cos(\alpha+\beta), \end{aligned}$$

$$u^1 = \frac{\cos\beta}{\cos(\alpha+\beta)} x_1 - \frac{\sin\beta}{\cos(\alpha+\beta)} x_2,$$

$$u^2 = -\frac{\sin\alpha}{\cos(\alpha+\beta)} x_1 + \frac{\cos\alpha}{\cos(\alpha+\beta)} x_2.$$

B. $$\underset{i}{g}_j = \frac{\partial x_j}{\partial u^i} \quad \text{gives} \quad \underline{g}_1 = \left(\frac{\partial x_1}{\partial u^1}, \frac{\partial x_2}{\partial u^1}\right) = (\cos\alpha, \sin\alpha),$$

$$\underline{g}_2 = \left(\frac{\partial x_1}{\partial u^2}, \frac{\partial x_2}{\partial u^2}\right) = (\sin\beta, \cos\beta);$$

$$\overset{i}{g}_j = \frac{\partial u^i}{\partial x_j} \quad \text{gives} \quad \underline{g}^1 = \left(\frac{\partial u^1}{\partial x_1}, \frac{\partial u^1}{\partial x_2}\right)$$

$$= \left(\frac{\cos\beta}{\cos(\alpha+\beta)}, -\frac{\sin\beta}{\cos(\alpha+\beta)}\right),$$

$$\underline{g}^2 = \left(\frac{\partial u^2}{\partial x_1}, \frac{\partial u^2}{\partial x_2}\right)$$

$$= \left(-\frac{\sin\alpha}{\cos(\alpha+\beta)}, \frac{\cos\alpha}{\cos(\alpha+\beta)}\right).$$

In this case the covariant basis vectors are unit vectors, but the contravariant basis vectors are not.

Problem 4.2.

$$x = R\cos\varphi:\quad \frac{\partial x}{\partial R} = \cos\varphi,\quad \frac{\partial x}{\partial \varphi} = -R\sin\varphi,\quad \frac{\partial x}{\partial z} = 0;$$

$$y = R\sin\varphi:\quad \frac{\partial y}{\partial R} = \sin\varphi,\quad \frac{\partial y}{\partial \varphi} = R\cos\varphi,\quad \frac{\partial y}{\partial z} = 0;$$

$$z = z:\quad \frac{\partial z}{\partial R} = 0,\quad \frac{\partial z}{\partial \varphi} = 0,\quad \frac{\partial z}{\partial z} = 1;$$

$$R = \sqrt{x^2+y^2}:\quad \begin{aligned} \frac{\partial R}{\partial x} &= \frac{2x}{2\sqrt{x^2+y^2}} = \frac{x}{R} = \cos\varphi,\\ \frac{\partial R}{\partial y} &= \frac{2y}{2\sqrt{x^2+y^2}} = \frac{y}{R} = \sin\varphi,\\ \frac{\partial R}{\partial z} &= 0; \end{aligned}$$

$$\varphi = \arctan\frac{y}{x}:\quad \begin{aligned} \frac{\partial \varphi}{\partial x} &= \frac{1}{1+(\frac{y}{x})^2}\left(-\frac{y}{x^2}\right) = -\frac{x^2}{x^2+y^2}\frac{y}{x^2}\\ &= -\frac{y}{R^2} = -\frac{1}{R}\sin\varphi,\\ \frac{\partial \varphi}{\partial y} &= \frac{1}{1+(\frac{y}{x})^2}\frac{1}{x} = \frac{x^2}{x^2+y^2}\frac{1}{x}\\ &= \frac{x}{R^2} = \frac{1}{R}\cos\varphi,\\ \frac{\partial \varphi}{\partial z} &= 0; \end{aligned}$$

$$\underset{i}{g}_j = \frac{\partial x_j}{\partial u^i}\ \text{ gives }\ \begin{aligned} \underline{g}_1 &= \left(\frac{\partial x}{\partial R}, \frac{\partial y}{\partial R}, \frac{\partial z}{\partial R}\right) = (\cos\varphi, \sin\varphi, 0),\\ \underline{g}_2 &= \left(\frac{\partial x}{\partial \varphi}, \frac{\partial y}{\partial \varphi}, \frac{\partial z}{\partial \varphi}\right) = (-R\sin\varphi, R\cos\varphi, 0),\\ \underline{g}_3 &= \left(\frac{\partial x}{\partial z}, \frac{\partial y}{\partial z}, \frac{\partial z}{\partial z}\right) = (0, 0, 1); \end{aligned}$$

$$\overset{i}{g}_j = \frac{\partial u^i}{\partial x_j}\ \text{ gives }\ \begin{aligned} \underline{g}^1 &= \left(\frac{\partial R}{\partial x}, \frac{\partial R}{\partial y}, \frac{\partial R}{\partial z}\right) = (\cos\varphi, \sin\varphi, 0),\\ \underline{g}^2 &= \left(\frac{\partial \varphi}{\partial x}, \frac{\partial \varphi}{\partial y}, \frac{\partial \varphi}{\partial z}\right) = (-\frac{1}{R}\sin\varphi, \frac{1}{R}\cos\varphi, 0),\\ \underline{g}^3 &= \left(\frac{\partial z}{\partial x}, \frac{\partial z}{\partial y}, \frac{\partial z}{\partial z}\right) = (0, 0, 1). \end{aligned}$$

Problem 4.3.

Equation (4.14) relates contravariant coordinates and Cartesian coordinates. We have

$$T^{ij} = \underline{g}^i \cdot \underline{\underline{T}} \cdot \underline{g}^j = \overset{i}{g}_m \, \widehat{T}_{mn} \, \overset{j}{g}_n .$$

In order to apply this equation, we first have to compute the Cartesian coordinates of the basis reciprocal to $\underline{g}_i$. We get, for example using the Gauss–Jordan algorithm,

$$\underline{g}^1 = (1,\,-1,\,0), \quad \underline{g}^2 = (0,\,1,\,-1), \quad \underline{g}^3 = (0,\,0,\,1),$$

and further, using Falk's scheme, in two steps

$$\begin{array}{ccc|ccc|ccc}
 & & & \widehat{T}_{11} & \widehat{T}_{12} & \widehat{T}_{13} & \overset{1}{g}_1 & \overset{2}{g}_1 & \overset{3}{g}_1 \\
 & & & \widehat{T}_{21} & \widehat{T}_{22} & \widehat{T}_{23} & \overset{1}{g}_2 & \overset{2}{g}_2 & \overset{3}{g}_2 \\
 & & & \widehat{T}_{31} & \widehat{T}_{32} & \widehat{T}_{33} & \overset{1}{g}_3 & \overset{2}{g}_3 & \overset{3}{g}_3 \\
\hline
\overset{1}{g}_1 & \overset{1}{g}_2 & \overset{1}{g}_3 & & & & T^{11} & T^{12} & T^{13} \\
\overset{2}{g}_1 & \overset{2}{g}_2 & \overset{2}{g}_3 & & & & T^{21} & T^{22} & T^{23} \\
\overset{3}{g}_1 & \overset{3}{g}_2 & \overset{3}{g}_3 & & & & T^{31} & T^{32} & T^{33}
\end{array}, \qquad T^{ij} = \begin{pmatrix} 2 & -6 & 3 \\ -1 & 6 & -2 \\ 1 & -3 & 1 \end{pmatrix}.$$

Problem 4.4.

If we identify the tilde coordinates in (4.19), i. e. in

$$\tilde{a}^i = \frac{\partial \tilde{u}^i}{\partial u^m} \, a^m, \qquad \tilde{a}_i = \frac{\partial u^m}{\partial \tilde{u}^i} \, a_m,$$

with cylindrical coordinates and the no-tilde coordinates with Cartesian coordinates, and then use the solutions to Problem 4.2, we get

$$\begin{aligned}
a^1 &= \frac{\partial R}{\partial x} \, a_x + \frac{\partial R}{\partial y} \, a_y + \frac{\partial R}{\partial z} \, a_z = a_x \cos\varphi + a_y \sin\varphi, \\
a^2 &= \frac{\partial \varphi}{\partial x} \, a_x + \frac{\partial \varphi}{\partial y} \, a_y + \frac{\partial \varphi}{\partial z} \, a_z = -a_x \frac{\sin\varphi}{R} + a_y \frac{\cos\varphi}{R}, \\
a^3 &= \frac{\partial z}{\partial x} \, a_x + \frac{\partial z}{\partial y} \, a_y + \frac{\partial z}{\partial z} \, a_z = a_z; \\
a_1 &= \frac{\partial x}{\partial R} \, a_x + \frac{\partial y}{\partial R} \, a_y + \frac{\partial z}{\partial R} \, a_z = a_x \cos\varphi + a_y \sin\varphi, \\
a_2 &= \frac{\partial x}{\partial \varphi} \, a_x + \frac{\partial y}{\partial \varphi} \, a_y + \frac{\partial z}{\partial \varphi} \, a_z = -a_x R \sin\varphi + a_y R \cos\varphi, \\
a_3 &= \frac{\partial x}{\partial z} \, a_x + \frac{\partial y}{\partial z} \, a_y + \frac{\partial z}{\partial z} \, a_z = a_z.
\end{aligned}$$

Conversely, if we identify the tilde coordinates with Cartesian coordinates and the no-tilde coordinates with cylindrical coordinates, we get

$$
\begin{aligned}
a_x &= \frac{\partial x}{\partial R} a^1 + \frac{\partial x}{\partial \varphi} a^2 + \frac{\partial x}{\partial z} a^3 = a^1 \cos\varphi - a^2 R \sin\varphi \\
&= \frac{\partial R}{\partial x} a_1 + \frac{\partial \varphi}{\partial x} a_2 + \frac{\partial z}{\partial x} a_3 = a_1 \cos\varphi - a_2 \frac{\sin\varphi}{R}, \\
a_y &= \frac{\partial y}{\partial R} a^1 + \frac{\partial y}{\partial \varphi} a^2 + \frac{\partial y}{\partial z} a^3 = a^1 \sin\varphi + a^2 R \cos\varphi \\
&= \frac{\partial R}{\partial y} a_1 + \frac{\partial \varphi}{\partial y} a_2 + \frac{\partial z}{\partial y} a_3 = a_1 \sin\varphi + a_2 \frac{\cos\varphi}{R}, \\
a_z &= \frac{\partial z}{\partial R} a^1 + \frac{\partial z}{\partial \varphi} a^2 + \frac{\partial z}{\partial z} a^3 = a^3 \\
&= \frac{\partial R}{\partial z} a_1 + \frac{\partial \varphi}{\partial z} a_2 + \frac{\partial z}{\partial z} a_3 = a_3.
\end{aligned}
$$

Problem 4.5.

A. With the solutions to Problem 4.1, we obtain

$$
A^i = \frac{\partial u^i}{\partial x_j} \hat{A}_j \quad \text{gives} \quad
\begin{aligned}
A^1 &= \frac{\cos\beta}{\cos(\alpha+\beta)} A_x - \frac{\sin\beta}{\cos(\alpha+\beta)} A_y, \\
A^2 &= -\frac{\sin\alpha}{\cos(\alpha+\beta)} A_x + \frac{\cos\alpha}{\cos(\alpha+\beta)} A_y;
\end{aligned}
$$

$$
\hat{A}_i = \frac{\partial x^i}{\partial u^j} A^j \quad \text{gives} \quad
\begin{aligned}
A_x &= A^1 \cos\alpha + A^2 \sin\beta, \\
A_y &= A^1 \sin\alpha + A^2 \cos\beta;
\end{aligned}
$$

$$
A_i = \frac{\partial x_j}{\partial u^i} \hat{A}_j \quad \text{gives} \quad
\begin{aligned}
A_1 &= A_x \cos\alpha + A_y \sin\alpha, \\
A_2 &= A_x \sin\beta + A_y \cos\beta.
\end{aligned}
$$

B.

$$\underline{A} = \overrightarrow{PQ}$$

$$
\begin{aligned}
A^1 &= JI = PR, \quad A^1 \underline{g}_1 = \overrightarrow{PR}, \\
A^2 &= HC = RQ, \quad A^2 \underline{g}_2 = \overrightarrow{RQ},
\end{aligned}
$$

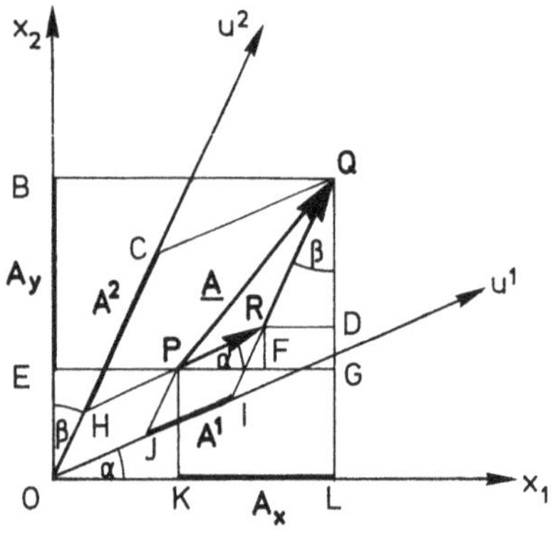

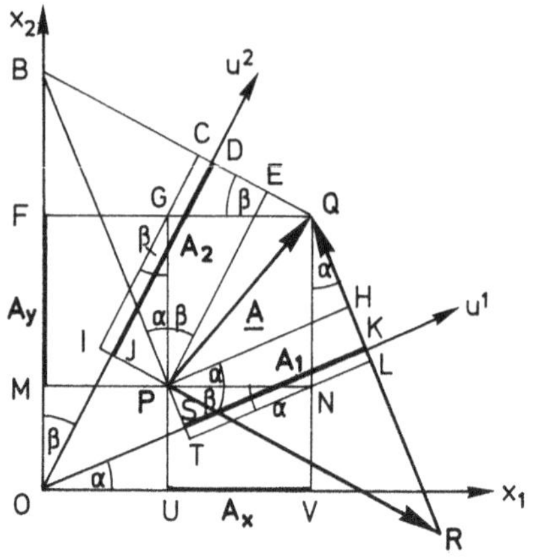

$$A_x = KL = PG = PF + FG,$$

$$PF = PR\,\cos\alpha = A^1\,\cos\alpha,$$

$$FG = RD = RQ\,\sin\beta = A^2\,\sin\beta,$$

$$A_y = EB = GQ = GD + DQ,$$

$$GD = FR = PR\,\sin\alpha = A^1\,\sin\alpha,$$

$$DQ = RQ\,\sin\beta = A^2\,\sin\beta,$$

$$A^1\,\underline{g}_1 = \overrightarrow{PR}, \quad |A^1\,\underline{g}_1| = PR = A^1,$$

$$A^2\,\underline{g}_2 = \overrightarrow{RQ}, \quad |A^2\,\underline{g}_2| = RQ = A^2.$$

$$\underline{A} = \overrightarrow{PQ}$$

$$A_x = UV = PN = GQ,$$

$$A_y = MF = PG = NQ,$$

$$A_1 = SK = PH = TL = TN + NL,$$

$$TN = PN\,\cos\alpha = A_x\,\cos\alpha, \quad NL = NQ\,\sin\alpha = A_y\,\sin\alpha,$$

$$A_2 = JD = PE = IC = IG + GC,$$

$$IG = PG\,\cos\beta = A_y\,\cos\beta, \quad GC = GQ\,\sin\beta = A_x\,\sin\beta,$$

$$A_1\,\underline{g}^1 = \overrightarrow{PR}, \qquad |A_1\,\underline{g}^1| = PR = \frac{PH}{\cos(\alpha+\beta)} = \frac{A_1}{\cos(\alpha+\beta)},$$

$$A_2\,\underline{g}^2 = \overrightarrow{RQ} = \overrightarrow{PB}, \quad |A_2\,\underline{g}^2| = PB = \frac{PE}{\cos(\alpha+\beta)} = \frac{A_2}{\cos(\alpha+\beta)}.$$

Problem 4.6.
From the solution to Problem 4.2 we know

$$g_{11} = 1, \quad g_{22} = R^2, \quad g_{33} = 1, \quad g^{11} = 1, \quad g^{22} = \frac{1}{R^2}, \quad g^{33} = 1.$$

The off-diagonal elements are all zero, because cylindrical coordinates are orthogonal.

Problem 4.7.
According to (4.19), the transformation equations are

$$\tilde{g}_{ij} = \frac{\partial u^m}{\partial \tilde{u}^i} \frac{\partial u^n}{\partial \tilde{u}^j} g_{mn}.$$

Computing the determinant and using (1.13) yields

$$\det(\tilde{g}_{ij}) = \det\left(\frac{\partial u^m}{\partial \tilde{u}^i}\right) \det\left(\frac{\partial u^n}{\partial \tilde{u}^j}\right) \det(g_{mn}),$$

$$\tilde{g} = \left[\frac{\partial(u^1, u^2, u^3)}{\partial(\tilde{u}^1, \tilde{u}^2, \tilde{u}^3)}\right]^2 g,$$

A quantity satisfying this transformation law is called a pseudoscalar of weight 2.

The determinant of the Cartesian coordinates of a second-order tensor is, according to (3.7), a scalar (and is called the determinant of the tensor); the determinant of its covariant coordinates clearly is a pseudoscalar of weight 2. Analogously, we can show that the determinant of the mixed coordinates of a second-order tensor is a scalar and that the determinant of its contravariant coordinates is a pseudoscalar of weight −2.

Problem 4.8.
A tensor of order three has $3^3 = 27$ Cartesian coordinates, each of which can be translated into $2^3 = 8$ different holonomic coordinates, so the ε-tensor has $8 \cdot 27 = 216$ holonomic coordinates. Because cylindrical coordinates are orthogonal, all the coordinates of the ε-tensor which match in two indices are zero, so only $8 \cdot 6 = 48$ coordinates are nonzero.

The easiest way to compute these coordinates is from (4.33):

$$e^{ijk} = [\underline{g}^i, \underline{g}^j, \underline{g}^k], \quad e^{ij}{}_k = [\underline{g}^i, \underline{g}^j, \underline{g}_k], \quad \text{etc.}$$

Since the basis vectors are mutually orthogonal to each other, the magnitudes of the triple products are equal to the product of the lengths of the vectors spanning the triple products, and a coordinate is positive or negative, depending on whether the three vectors of the triple product form a right-handed system or a left-handed system.

Problem 4.9.
$e^{ijk}\, e^m{}_{nk} = g^{im}\, \delta^j_n - \delta^i_n\, g^{jm}$.

Problem 4.10.

Possible translations are:

A.
$$\underline{\underline{a}} \cdot \underline{\underline{b}} = \underline{\underline{c}} \iff a_{ij}\, b^j{}_k = c_{ik},$$
$$\underline{\underline{b}} \cdot \underline{\underline{a}} = \underline{\underline{c}} \iff b^{ij}\, a_{jk} = c^i{}_k,$$
$$\widehat{a}_{ik}\, \widehat{b}_{ij} = \widehat{c}_{jk} \iff a_{ik}\, b^i{}_j = c_{jk},$$
$$\widehat{a}_i\, \widehat{b}_k\, \widehat{a}_i\, \widehat{c}_k = d \iff a_i\, b_k\, a^i\, c^k = d,$$
$$\widehat{a}_{ijk}\, \widehat{b}_i\, \widehat{c}_k = \widehat{d}_j \iff a_{ijk}\, b^i\, c^k = d_j,$$
$$\widehat{a}_{ijk}\, \widehat{b}_i\, \widehat{c}_j = \widehat{d}_k \iff a_{ijk}\, b^i\, c^j = d_k,$$
$$(\underline{\underline{a}} \cdot \underline{\underline{b}})^{\mathrm{T}} = \underline{\underline{c}} \iff (a_{ij}\, b^{jk})^{\mathrm{T}} = c_i{}^k,$$
$$(\underline{\underline{a}} \cdot \underline{\underline{b}} \cdot \underline{\underline{c}})^{\mathrm{T}} = \underline{\underline{\underline{d}}} \iff (a_i{}^j\, b_j{}^k\, c_k{}^l)^{\mathrm{T}} = d_i{}^l.$$

B. $e^{ijk}\, a_j\, e_{kmn}\, b^m\, c^n = a_j\, c^j\, b^i - a_j\, b^j\, c^i.$

Problem 4.11.

Since cylindrical coordinates are orthogonal, our considerations from Section 4.3, No. 3 apply here as well.

A. From the solution to Problem 4.2 it follows immediately, after normalizing the basis vectors, that

$$\underline{g}_{\langle 1\rangle} = (\cos\varphi,\ \sin\varphi,\ 0), \quad \underline{g}_{\langle 2\rangle} = (-\sin\varphi,\ \cos\varphi,\ 0), \quad \underline{g}_{\langle 3\rangle} = (0,\ 0,\ 1).$$

B. According to (4.48), we have $a_{\langle i\rangle} = a^i\,\sqrt{g_{\underline{ii}}}$, and with $g_{11} = 1,\ g_{22} = R^2,\ g_{33} = 1$, it follows that

$$\begin{aligned} a_r &= a_x \cos\varphi + a_y \sin\varphi, & a_x &= a_r \cos\varphi - a_\varphi \sin\varphi,\\ a_\varphi &= -a_x \sin\varphi + a_y \cos\varphi, & a_y &= a_r \sin\varphi + a_\varphi \cos\varphi,\\ a_z &= a_z, & a_z &= a_z. \end{aligned}$$

C. Since the $\underline{g}_{\langle i\rangle}$ are locally Cartesian and since the coordinates of $\underline{\underline{\delta}}$ and $\underline{\underline{\underline{\varepsilon}}}$ are equal in all Cartesian coordinate systems, it follows that $g_{\langle ij\rangle} = \delta_{ij},\ \ e_{\langle ijk\rangle} = \varepsilon_{ijk}$.

D.
$$(\underline{\underline{a}} \cdot \underline{\underline{b}})^{\mathrm{T}} = \underline{\underline{c}} \iff (a_{\langle ij\rangle}\, b_{\langle jk\rangle})^{\mathrm{T}} = c_{\langle ik\rangle},$$
$$a_{ik}\, b^i{}_j = c_{jk} \iff a_{\langle ik\rangle}\, b_{\langle ij\rangle} = c_{\langle jk\rangle},$$
$$\underline{a} \times (\underline{b} \times \underline{c}) = \underline{a} \cdot \underline{c}\, \underline{b} - \underline{a} \cdot \underline{b}\, \underline{c} \iff \varepsilon_{ijk}\, a_{\langle j\rangle}\, \varepsilon_{kmn}\, b_{\langle m\rangle}\, c_{\langle n\rangle}$$
$$= a_{\langle j\rangle}\, c_{\langle j\rangle}\, b_{\langle i\rangle} - a_{\langle j\rangle}\, b_{\langle j\rangle}\, c_{\langle i\rangle}.$$

Problem 4.12.

The easiest way to compute the Christoffel symbols is by (4.60).

For cylindrical coordinates we get

$$g_{11} = 1, \quad g_{11,i} = 0,$$

$$g_{22} = R^2, \quad g_{22,1} = 2R, \quad g_{22,2} = g_{22,3} = 0,$$

$$g_{33} = 1, \quad g_{33,i} = 0,$$

$$g^{11} = 1, \quad g^{22} = \frac{1}{R^2}, \quad g^{33} = 1.$$

This gives

$$\Gamma^1_{ip} = \frac{1}{2} g^{11}(g_{1i,p} + g_{1p,i} - g_{ip,1}),$$

which is nonzero only for $i = p = 2$, and then we have $\Gamma^1_{22} = -R$;

$$\Gamma^2_{ip} = \frac{1}{2} g^{22}(g_{2i,p} + g_{2p,i} - g_{ip,2}),$$

which is nonzero only for $i = 2, \ p = 1$ and $i = 1, \ p = 2$, so we have $\Gamma^2_{12} = \Gamma^2_{21} = \frac{1}{R}$;

$$\Gamma^3_{ip} = \frac{1}{2} g^{33}(g_{3i,p} + g_{3p,i} - g_{ip,3}),$$

which is zero for all i and p.

So the only nonzero Christoffel symbols are $\Gamma^1_{22} = -R, \ \Gamma^2_{12} = \Gamma^2_{21} = \frac{1}{R}$.

Problem 4.13.

We can derive the transformation law as follows. Starting with equation $(4.58)_1$ in the tilde system, we substitute the transformation equations $(4.18)_1$ for the covariant basis vectors, so we have

$$\tilde{\Gamma}^i_{jk}\,\tilde{\underline{g}}_i = \frac{\partial \tilde{\underline{g}}_j}{\partial \tilde{u}^k} = \frac{\partial}{\partial \tilde{u}^k}\left(\frac{\partial u^m}{\partial \tilde{u}^j}\,\underline{g}_m\right) = \frac{\partial u^m}{\partial \tilde{u}^j}\frac{\partial \underline{g}_m}{\partial \tilde{u}^k} + \frac{\partial^2 u^m}{\partial \tilde{u}^j\,\partial \tilde{u}^k}\,\underline{g}_m;$$

with $\frac{\partial \underline{g}_m}{\partial \tilde{u}^k} = \frac{\partial \underline{g}_m}{\partial u^n}\frac{\partial u^n}{\partial \tilde{u}^k} \overset{(4.58)}{=} \frac{\partial u^n}{\partial \tilde{u}^k}\,\Gamma^l_{mn}\,\underline{g}_l$ and $\underline{g}_l \overset{(4.18)}{=} \frac{\partial \tilde{u}^i}{\partial u^l}\,\tilde{\underline{g}}_i$

we obtain

$$\tilde{\Gamma}^i_{jk}\,\tilde{\underline{g}}_i = \frac{\partial u^m}{\partial \tilde{u}^j}\frac{\partial u^n}{\partial \tilde{u}^k}\frac{\partial \tilde{u}^i}{\partial u^l}\,\Gamma^l_{mn}\,\tilde{\underline{g}}_i + \frac{\partial^2 u^m}{\partial \tilde{u}^j\,\partial \tilde{u}^k}\frac{\partial \tilde{u}^i}{\partial u^m}\,\tilde{\underline{g}}_i,$$

$$\tilde{\Gamma}^i_{jk} = \frac{\partial \tilde{u}^i}{\partial u^l}\frac{\partial u^m}{\partial \tilde{u}^j}\frac{\partial u^n}{\partial \tilde{u}^k}\,\Gamma^l_{mn} + \frac{\partial^2 u^m}{\partial \tilde{u}^j\,\partial \tilde{u}^k}\frac{\partial \tilde{u}^i}{\partial u^m}.$$

The first term is the transformation law for single contravariant and twice covariant tensor coordinates. The second term shows that Christoffel symbols are not tensor coordinates. If the no-tilde coordinates are Cartesian, the Christoffel symbols vanish in the no-tilde coordinate system and the transformation law reduces to Definition 4.56 of the Christoffel symbols. Starting from equation $(4.58)_2$, we analogously obtain the equivalent transformation law

$$\tilde{\Gamma}^i_{jk} = \frac{\partial \tilde{u}^i}{\partial u^l} \frac{\partial u^m}{\partial \tilde{u}^j} \frac{\partial u^n}{\partial \tilde{u}^k} \Gamma^l_{mn} - \frac{\partial^2 \tilde{u}^i}{\partial u^m \partial u^n} \frac{\partial u^m}{\partial \tilde{u}^j} \frac{\partial u^n}{\partial \tilde{u}^k}.$$

Problem 4.14.

A. $$\underline{a} \times \operatorname{curl} \underline{a} = \underline{b} \iff \varepsilon_{ijk}\, \hat{a}_j \frac{\partial \hat{a}_m}{\partial x_n} \varepsilon_{mkn} = \hat{b}_i$$
$$\iff e^{ijk}\, a_j\, a_m|_n\, e^{m}{}_{k}{}^{n} = b^i.$$

B. $$\frac{\partial \hat{a}_i}{\partial x_j} + \frac{\partial \hat{a}_j}{\partial x_i} = \hat{b}_{ij} \iff a_i|_j + a_j|_i = b_{ij} \iff \operatorname{grad} \underline{a} + \operatorname{grad}^{\mathrm{T}} \underline{a} = \underline{\underline{b}}.$$

C. $$a^i\, b_j|_i = c_j \iff \hat{a}_i \frac{\partial \hat{b}_j}{\partial x_i} = \hat{c}_j \iff (\operatorname{grad} \underline{b}) \cdot \underline{a} = \underline{c}.$$

D. $$e^i{}_{jk}\, e_{imn}\, a^j\, b^k\, c^m\, d^n = e^i{}_{jk}\, a^j\, b^k\, e_{imn}\, c^m\, d^n$$
$$= (g_{jm}\, g_{kn} - g_{jn}\, g_{km})\, a^j\, b^k\, c^m\, d^n = a^j\, c_j\, b^k\, d_k - a^j\, d_j\, b^k\, c_k,$$
$$\iff \varepsilon_{ijk}\, \varepsilon_{imn}\, \hat{a}_j\, \hat{b}_k\, \hat{c}_m\, \hat{d}_n = \hat{a}_j\, \hat{c}_j\, \hat{b}_k\, \hat{d}_k - \hat{a}^j\, \hat{d}_j\, \hat{b}_k\, \hat{c}_k,$$
$$\iff (\underline{a} \times \underline{b}) \cdot (\underline{c} \times \underline{d}) = \underline{a} \cdot \underline{c}\; \underline{b} \cdot \underline{d} - \underline{a} \cdot \underline{d}\; \underline{b} \cdot \underline{c}.$$

Problem 4.15.

A. $$(\operatorname{grad} a)_i \overset{(4.73)}{=} a|_i \overset{(4.61)}{=} a_{,i},$$
$$(\operatorname{grad} a)_R \overset{(B.6)}{=} (\operatorname{grad} a)_1 = a_{,1} = \frac{\partial a}{\partial R},$$
$$(\operatorname{grad} a)_\varphi = \frac{1}{R} (\operatorname{grad} a)_2 = \frac{1}{R} a_{,2} = \frac{1}{R} \frac{\partial a}{\partial \varphi},$$
$$(\operatorname{grad} a)_z = (\operatorname{grad} a)_3 = a_{,3} = \frac{\partial a}{\partial z}.$$

B. $$\operatorname{div} \underline{a} \overset{(4.76)}{=} a^i|_i \overset{(4.61)}{=} a^i{}_{,i} + \Gamma^i_{mi}\, a^m \overset{(B.13)}{=} a^1{}_{,1} + a^2{}_{,2} + a^3{}_{,3} + \Gamma^2_{12}\, a^1$$
$$= \frac{\partial a^1}{\partial R} + \frac{\partial a^2}{\partial \varphi} + \frac{\partial a^3}{\partial z} + \frac{1}{R} a^1 \overset{(B.6)}{=} \frac{\partial a_R}{\partial R} + \frac{\partial}{\partial \varphi} \frac{a_\varphi}{R} + \frac{\partial a_z}{\partial z} + \frac{a_R}{R}$$
$$= \underbrace{\frac{\partial a_R}{\partial R} + \frac{a_R}{R}}_{\frac{1}{R} \frac{\partial}{\partial R} (R\, a_R)} + \frac{1}{R} \frac{\partial a_\varphi}{\partial \varphi} + \frac{\partial a_z}{\partial z}.$$

C. $(\text{curl}\,\underline{a})^j \overset{(4.79)}{=} e^{ijk}\, a_{i,k}$,

$$(\text{curl}\,\underline{a})_R \overset{(B.6)}{=} (\text{curl}\,\underline{a})^1 = e^{312}\, a_{3,2} + e^{213}\, a_{2,3} \overset{(B.11)}{=} \frac{1}{R}\left(\frac{\partial a_3}{\partial \varphi} - \frac{\partial a_2}{\partial z}\right)$$

$$\overset{(B.6)}{=} \frac{1}{R}\left[\frac{\partial a_z}{\partial \varphi} - \frac{\partial}{\partial z}(R\, a_\varphi)\right] = \frac{1}{R}\frac{\partial a_z}{\partial \varphi} - \frac{\partial a_\varphi}{\partial z};$$

analogously we find

$$(\text{curl}\,\underline{a})_\varphi = \frac{\partial a_R}{\partial z} - \frac{\partial a_z}{\partial R} \quad \text{and} \quad (\text{curl}\,\underline{a})_z = \underbrace{\frac{\partial a_\varphi}{\partial R} + \frac{a_\varphi}{R}}_{\frac{1}{R}\frac{\partial}{\partial R}(R\, a_\varphi)} - \frac{1}{R}\frac{\partial a_R}{\partial \varphi}.$$

D. $(\text{grad}\,\underline{a})^i{}_j \overset{(4.73)}{=} a^i|_j \overset{(4.61)}{=} a^i{}_{,j} + \Gamma^i_{mj}\, a^m$,

$$(\text{grad}\,\underline{a})_{RR} \overset{(B.9)}{=} (\text{grad}\,\underline{a})^1{}_1 = a^1{}_{,1} + \Gamma^1_{m1}\, a^m \overset{(B.13)}{=} a^1{}_{,1} = \frac{\partial a^1}{\partial R} \overset{(B.6)}{=} \frac{\partial a_R}{\partial R};$$

analogously we find the remaining coordinates; compare with (B.17).

E. Here we have two options: analogously to the earlier problems we compute

$$[(\text{grad}\,\underline{a}) \cdot \underline{b}]_i = a_i|_m\, b^m, \quad \text{etc.},$$

or we use the fact that cylindrical coordinates are orthogonal, and thus the scalar product is an algebraic operation where the factors are composed of the physical cylindrical coordinates, similarly as in Cartesian coordinates. Using the solutions to part D and from (B.17), respectively, we obtain

$$\begin{aligned}[(\text{grad}\,\underline{a}) \cdot \underline{b}]_R &= (\text{grad}\,\underline{a})_{RR}\, b_R + (\text{grad}\,\underline{a})_{R\varphi}\, b_\varphi + (\text{grad}\,\underline{a})_{Rz}\, b_z \\ &= \frac{\partial a_R}{\partial R}\, b_R + \left(\frac{1}{R}\frac{\partial a_R}{\partial \varphi} - \frac{a_\varphi}{R}\right) b_\varphi + \frac{\partial a_R}{\partial z}\, b_z \\ &= \frac{\partial a_R}{\partial R}\, b_R + \frac{\partial a_R}{\partial \varphi}\frac{b_\varphi}{R} + \frac{\partial a_R}{\partial z}\, b_z - \frac{a_\varphi\, b_\varphi}{R}\end{aligned}$$

and analogously the remaining coordinates; compare with (B.18).

Problem 4.16.

A. $\dfrac{\partial^2 \widehat{a}_m}{\partial x_k^2} = \widehat{b}_m \quad \Longleftrightarrow \quad a_m|_{kp}\, g^{kp} = b_m \quad \Longleftrightarrow \quad \operatorname{div}\operatorname{grad} \underline{a} = \Delta\, \underline{a} = \underline{b}.$

B. $\dfrac{\partial^2 \widehat{a}_k}{\partial x_m\, \partial x_k} = \widehat{b}_m \quad \Longleftrightarrow \quad a^k|_{km} = b_m \quad \Longleftrightarrow \quad \operatorname{grad}\operatorname{div} \underline{a} = \underline{b}.$

C. $\mathrm{d}\,\underline{\underline{a}} = (\operatorname{grad} \underline{\underline{a}}) \cdot \mathrm{d}\,\underline{x} \quad \Longleftrightarrow \quad \delta a_{ij} = a_{ij}|_k\, \mathrm{d}\,u^k \quad \Longleftrightarrow \quad \mathrm{d}\,\widehat{a}_{ij} = \dfrac{\partial \widehat{a}_{ij}}{\partial x_k}\, \mathrm{d}\,x_k.$

Problem 4.17.

Similarly as in the last problem, we have two options:
- We first perform the computations in holonomic coordinates and translate the result into physical coordinates.
- We compute the Laplace operator as the product of the physical coordinates of the gradient and the divergence, which we have computed earlier.

A. Option 1:

$$\begin{aligned}\Delta\, a &\overset{(4.84)}{=} g^{mn}\, a|_{mn} \overset{(4.81)}{=} g^{mn}\, a|_m|_n \\ &\overset{(4.61)}{=} g^{mn}[(a|_m)_{,n} - \Gamma^j_{mn}\, a|_j] \overset{(4.61)}{=} g^{mn}(a_{,mn} - \Gamma^j_{mn}\, a_{,j}).\end{aligned}$$

In cylindrical coordinates we have with (B.10) and (B.13)

$$\begin{aligned}\Delta\, a &= g^{11}\, a_{,11} + g^{22}(a_{,22} - \Gamma^1_{22}\, a_{,1}) + g^{33}\, a_{,33} \\ &= \frac{\partial^2 a}{\partial R^2} + \frac{1}{R^2}\left(\frac{\partial^2 a}{\partial \varphi^2} + R\,\frac{\partial a}{\partial R}\right) + \frac{\partial^2 a}{\partial z^2},\end{aligned}$$

$$\Delta\, a = \underbrace{\frac{\partial^2 a}{\partial R^2} + \frac{1}{R}\frac{\partial a}{\partial R}}_{\frac{1}{R}\frac{\partial}{\partial R}\left(R\frac{\partial a}{\partial R}\right)} + \frac{1}{R^2}\frac{\partial^2 a}{\partial \varphi^2} + \frac{\partial^2 a}{\partial z^2}.$$

Option 2:

$$\underline{b} = \operatorname{grad} a \quad \Longleftrightarrow \quad b_R = \frac{\partial a}{\partial R}, \quad b_\varphi = \frac{1}{R}\frac{\partial a}{\partial \varphi}, \quad b_z = \frac{\partial a}{\partial z},$$

$$\begin{aligned}\Delta\, a = \operatorname{div} \underline{b} &= \frac{\partial b_R}{\partial R} + \frac{b_R}{R} + \frac{1}{R}\frac{\partial b_\varphi}{\partial \varphi} + \frac{\partial b_z}{\partial z} \\ &= \frac{\partial^2 a}{\partial R^2} + \frac{1}{R}\frac{\partial a}{\partial R} + \frac{1}{R^2}\frac{\partial^2 a}{\partial \varphi^2} + \frac{\partial^2 a}{\partial z^2}.\end{aligned}$$

B. We compute only the R-coordinate and refer to (B.21) for the other two coordinates.
Option 1:

$$(\Delta \underline{a})_i \overset{(4.84)}{=} g^{mn}\, a_i|_{mn} \overset{(4.81)}{=} g^{mn}\, a_i|_m|_n, \qquad \text{with } b_{im} := a_i|_m,$$

$$(\Delta \underline{a})_i = g^{mn}\, b_{im}|_n,$$

$$\begin{aligned} a_i|_{mn} &= b_{im}|_n \overset{(4.61)}{=} b_{im,n} - \Gamma^j_{in}\, b_{jm} - \Gamma^j_{mn}\, b_{ij} \\ &\overset{(4.61)}{=} (a_{i,m} - \Gamma^j_{im}\, a_j)_{,n} - \Gamma^j_{in}(a_{j,m} - \Gamma^k_{jm}\, a_k) \\ &\quad - \Gamma^j_{mn}(a_{i,j} - \Gamma^k_{ij}\, a_k) \\ &= a_{i,mn} - \Gamma^j_{im,n}\, a_j - \Gamma^j_{im}\, a_{j,n} - \Gamma^j_{in}\, a_{j,m} \\ &\quad + \Gamma^j_{in}\, \Gamma^k_{jm}\, a_k - \Gamma^j_{mn}\, a_{i,j} + \Gamma^j_{mn}\, \Gamma^k_{ij}\, a_k. \end{aligned}$$

Since we have to contract this expression with g^{mn} and since g^{mn} is nonzero in cylindrical coordinates only if $m = n$, only $a_i|_{\underline{mm}}$ remains. We have

$$a_i|_{\underline{mm}} = a_{i,\underline{mm}} - \Gamma^j_{im,\underline{m}}\, a_j - 2\Gamma^j_{im}\, a_{j,\underline{m}} - \Gamma^j_{\underline{mm}}\, a_{i,j} + \Gamma^j_{im}\, \Gamma^k_{j\underline{m}}\, a_k + \Gamma^j_{\underline{mm}}\, \Gamma^k_{ij}\, a_k.$$

Taking (B.13) into account we obtain

$$\begin{aligned} a_1|_{11} &= a_{1,11} = \frac{\partial^2 a_1}{\partial R^2}, \\ a_1|_{22} &= a_{1,22} - \Gamma^2_{12,2}\, a_2 - 2\Gamma^2_{12}\, a_{2,2} - \Gamma^1_{22}\, a_{1,1} + \Gamma^2_{12}\, \Gamma^1_{22}\, a_1 \\ &= \frac{\partial^2 a_1}{\partial \varphi^2} - \frac{2}{R}\frac{\partial a_2}{\partial \varphi} + R\frac{\partial a_1}{\partial R} - a_1, \\ a_1|_{33} &= a_{1,33} = \frac{\partial^2 a_1}{\partial z^2}, \end{aligned}$$

$$\begin{aligned} (\Delta \underline{a})_1 &= g^{11}\, a_1|_{11} + g^{22}\, a_1|_{22} + g^{33}\, a_1|_{33} \\ &= \frac{\partial^2 a_1}{\partial R^2} + \frac{1}{R^2}\left(\frac{\partial^2 a_1}{\partial \varphi^2} - \frac{2}{R}\frac{\partial a_2}{\partial \varphi} + R\frac{\partial a_1}{\partial R} - a_1\right) + \frac{\partial^2 a_1}{\partial z^2}, \\ (\Delta \underline{a})_R &= \frac{\partial^2 a_R}{\partial R^2} + \frac{1}{R}\frac{\partial a_R}{\partial R} - \frac{a_R}{R^2} + \frac{1}{R^2}\frac{\partial^2 a_R}{\partial \varphi^2} + \frac{\partial^2 a_R}{\partial z^2} - \frac{2}{R^2}\frac{\partial a_\varphi}{\partial \varphi}. \end{aligned}$$

Option 2: From $\underline{\underline{b}} = \operatorname{grad} \underline{a}$, $\Delta \underline{a} = \operatorname{div} \underline{\underline{b}}$ it follows that

$$
\begin{aligned}
(\Delta \underline{a})_R = (\operatorname{div} \underline{\underline{b}})_R &= \frac{\partial b_{RR}}{\partial R} + \frac{1}{R}\frac{\partial b_{R\varphi}}{\partial \varphi} + \frac{\partial b_{Rz}}{\partial z} + \frac{b_{RR} - b_{\varphi\varphi}}{R} \\
&= \frac{\partial^2 a_R}{\partial R^2} + \frac{1}{R}\frac{\partial}{\partial \varphi}\left(\frac{1}{R}\frac{\partial a_R}{\partial \varphi} - \frac{a_\varphi}{R}\right) + \frac{\partial^2 a_R}{\partial z^2} \\
&\quad + \frac{1}{R}\left(\frac{\partial a_R}{\partial R} - \frac{1}{R}\frac{\partial a_\varphi}{\partial \varphi} - \frac{a_R}{R}\right) \\
&= \frac{\partial^2 a_R}{\partial R^2} + \frac{1}{R}\frac{\partial a_R}{\partial R} - \frac{a_R}{R^2} + \frac{1}{R^2}\frac{\partial^2 a_R}{\partial \varphi^2} + \frac{\partial^2 a_R}{\partial z^2} - \frac{2}{R^2}\frac{\partial a_\varphi}{\partial \varphi}.
\end{aligned}
$$

Problem 4.18.

A. According to (4.61), we have

$$
\begin{aligned}
a_i|_k &= a_{i,k} - \Gamma^m_{ik}\, a_m =: b_{ik}, \\
a_i|_{kl} &= b_{ik}|_l = b_{ik,l} - \Gamma^n_{il}\, b_{nk} - \Gamma^n_{kl}\, b_{in} \\
&= (a_{i,k} - \Gamma^m_{ik}\, a_m)_{,l} - \Gamma^n_{il}(a_{n,k} - \Gamma^m_{nk}\, a_m) - \Gamma^n_{kl}(a_{i,n} - \Gamma^m_{in}\, a_m) \\
&= \underbrace{a_{i,kl}}_{1} - \Gamma^m_{ik,l}\, a_m \underbrace{- \Gamma^m_{ik}\, a_{m,l}}_{2} \underbrace{- \Gamma^n_{il}\, a_{n,k}}_{3} + \Gamma^n_{il}\,\Gamma^m_{nk}\, a_m \\
&\quad \underbrace{- \Gamma^n_{kl}\, a_{i,n}}_{4} \underbrace{+ \Gamma^n_{kl}\,\Gamma^m_{in}\, a_m}_{5}, \\
a_i|_{lk} &= \underbrace{a_{i,lk}}_{1} - \Gamma^m_{il,k}\, a_m \underbrace{- \Gamma^m_{il}\, a_{m,k}}_{3} \underbrace{- \Gamma^n_{ik}\, a_{n,l}}_{2} + \Gamma^n_{ik}\,\Gamma^m_{nl}\, a_m \\
&\quad \underbrace{- \Gamma^n_{lk}\, a_{i,n}}_{4} \underbrace{+ \Gamma^n_{lk}\,\Gamma^m_{in}\, a_m}_{5}.
\end{aligned}
$$

Here equal terms are labeled by numbers, and we get

$$
a_i|_{kl} - a_i|_{lk} = (\Gamma^m_{il,k} - \Gamma^m_{ik,l} + \Gamma^n_{il}\,\Gamma^m_{nk} - \Gamma^n_{ik}\,\Gamma^m_{nl})\, a_m.
$$

B. On the left side are the entirely covariant coordinates of a third-order tensor, and with the quotient rule the terms in parentheses represent the single contravariant and triple covariant coordinates $R^m{}_{ikl}$ of a fourth-order tensor. According to (4.83), the tensor on the left side is the zero tensor for arbitrary a_m, i. e. we have $R^m{}_{ikl} a_m = 0$ for arbitrary a_m, and this implies that $R^m{}_{ikl} = 0$.

Problem 4.19.

To derive the transformation law for the contravariant basis vectors, we cannot refer to the Cartesian basis, as we did in Section 4.2.2, because only the relation between the surface coordinates u^α and $\tilde{u}^\alpha$ is one-to-one, but the relation (4.102) between u^α or

$\tilde{u}^\alpha$ and the Cartesian coordinates x_i of the three-dimensional space is not one-to-one. Instead, we have to use the transformation law (4.105) for the covariant basis vectors and the orthogonality relations (4.106) and (4.107).

Since the relation between the surface coordinates u^α and $\tilde{u}^\alpha$ is one-to-one, we can substitute the two equations (4.104) into each other, i. e.

$$u^\beta = u^\beta\left[\,\tilde{u}^\alpha(u^\gamma)\,\right],$$

and then we obtain, using the chain rule, analogous to (4.17), the relation

$$\frac{\partial u^\beta}{\partial u^\gamma} = \frac{\partial u^\beta}{\partial \tilde{u}^\alpha}\frac{\partial \tilde{u}^\alpha}{\partial u^\gamma} = \delta^\beta_\gamma.$$

Therefore, contraction of (4.105) with $\partial\tilde{u}^\alpha/\partial u^\gamma$ gives

$$\frac{\partial \tilde{u}^\alpha}{\partial u^\gamma}\,\underline{\tilde{g}}_\alpha = \underbrace{\frac{\partial \tilde{u}^\alpha}{\partial u^\gamma}\frac{\partial u^\beta}{\partial \tilde{u}^\alpha}}_{\delta^\beta_\gamma}\,\underline{\tilde{g}}_\beta = \underline{g}_\gamma.$$

Substitution of $\underline{g}_\gamma$ into the orthogonality relation (4.107) gives

$$\underline{\underline{\delta}} = \underline{g}_\gamma\,\underline{g}^\gamma + \underline{n}\,\underline{n} = \frac{\partial \tilde{u}^\alpha}{\partial u^\gamma}\,\underline{\tilde{g}}_\alpha\,\underline{g}^\gamma + \underline{n}\,\underline{n},$$

so scalar multiplication with $\underline{\tilde{g}}^\beta$ from the left gives

$$\underbrace{\underline{\tilde{g}}^\beta\cdot\underline{\underline{\delta}}}_{\underline{\tilde{g}}^\beta} = \frac{\partial \tilde{u}^\alpha}{\partial u^\gamma}\underbrace{\underline{\tilde{g}}^\beta\cdot\underline{\tilde{g}}_\alpha}_{\delta^\beta_\alpha}\,\underline{g}^\gamma + \underbrace{\underline{\tilde{g}}^\beta\cdot\underline{n}}_{0}\,\underline{n},$$

$$\underline{\tilde{g}}^\beta = \frac{\partial \tilde{u}^\beta}{\partial u^\gamma}\,\underline{g}^\gamma.$$

This is the desired transformation law for the contravariant basis vectors, analogous to (4.18).

To derive the transformation law for the covariant coordinates a_α of a surface vector $\underline{a}$, we compute the scalar product of the relation

$$\underline{a} = \tilde{a}_\beta\,\underline{\tilde{g}}^\beta = a_\beta\,\underline{g}^\beta$$

with $\underline{\tilde{g}}_\alpha$ and substitute on the right side the transformation law (4.105), so we have

$$\tilde{a}_\beta\,\underbrace{\underline{\tilde{g}}^\beta\cdot\underline{\tilde{g}}_\alpha}_{\delta^\beta_\alpha} = a_\beta\,\underline{g}^\beta\cdot\underline{\tilde{g}}_\alpha = a_\beta\,\underbrace{\underline{g}^\beta\cdot\underline{g}_\gamma}_{\delta^\beta_\gamma}\,\frac{\partial u^\gamma}{\partial \tilde{u}^\alpha}.$$

This gives the transformation law for the covariant coordinates of a surface vector, analogous to (4.19),

$$\tilde{a}_\alpha = \frac{\partial u^\beta}{\partial \tilde{u}^\alpha}\,a_\beta.$$

Problem 4.20.
Writing the covariant metric coefficients as scalar products of the covariant basis vectors, we have in two different surface coordinate systems

$$\tilde{g}_{\alpha\beta} = \underline{\tilde{g}}_{\alpha} \cdot \underline{\tilde{g}}_{\beta} \quad \text{and} \quad g_{\alpha\beta} = \underline{g}_{\alpha} \cdot \underline{g}_{\beta}.$$

Substituting the transformations law (4.105) into the left equation gives

$$\tilde{g}_{\alpha\beta} = \left(\frac{\partial u^{\gamma}}{\partial \tilde{u}^{\alpha}}\, \underline{g}_{\gamma}\right) \cdot \left(\frac{\partial u^{\delta}}{\partial \tilde{u}^{\beta}}\, \underline{g}_{\delta}\right) = \frac{\partial u^{\gamma}}{\partial \tilde{u}^{\alpha}} \frac{\partial u^{\delta}}{\partial \tilde{u}^{\beta}} \underbrace{\underline{g}_{\gamma} \cdot \underline{g}_{\delta}}_{g_{\gamma\delta}} = \frac{\partial u^{\gamma}}{\partial \tilde{u}^{\alpha}} \frac{\partial u^{\delta}}{\partial \tilde{u}^{\beta}}\, g_{\gamma\delta}.$$

Comparing this equation with (4.19) shows that it is the transformation law for the entirely covariant coordinates of a second-order tensor, here in particular of a surface tensor, because the summation runs only from 1 to 2.

Problem 4.21.
The scalar product of $\underline{a} = a^{\alpha}\, \underline{g}_{\alpha} = a_{\alpha}\, \underline{g}^{\alpha}$, once with $\underline{g}^{\beta}$ and once with $\underline{g}_{\beta}$, together with (4.106), (4.110) and (4.114), yields

$$a^{\alpha}\, \underline{g}_{\alpha} \cdot \underline{g}^{\beta} = a_{\alpha}\, \underline{g}^{\alpha} \cdot \underline{g}^{\beta}, \quad a^{\alpha}\, \delta_{\alpha}^{\beta} = a_{\alpha}\, g^{\alpha\beta}, \quad a^{\beta} = g^{\beta\alpha}\, a_{\alpha},$$

$$a^{\alpha}\, \underline{g}_{\alpha} \cdot \underline{g}_{\beta} = a_{\alpha}\, \underline{g}^{\alpha} \cdot \underline{g}_{\beta}, \quad a^{\alpha}\, g_{\alpha\beta} = a_{\alpha}\, \delta_{\beta}^{\alpha}, \quad a_{\beta} = g_{\beta\alpha}\, a^{\alpha}.$$

Problem 4.22.
If we take the Cartesian coordinates as surface parameters $u^1 = x$, $u^2 = y$, the parametric representation of the surface $z = x\,y$ is given as

$$\underline{x} = x\, \underline{e}_1 + y\, \underline{e}_2 + x\,y\, \underline{e}_2.$$

The covariant basis vectors are, using (4.100),

$$\underline{g}_1 = \frac{\partial x_i}{\partial x}\, \underline{e}_i = \underline{e}_1 + y\, \underline{e}_3, \quad \underline{g}_2 = \frac{\partial x_i}{\partial y}\, \underline{e}_i = \underline{e}_2 + x\, \underline{e}_3.$$

For the next computations we need the metric coefficients. We compute the covariant metric coefficients $g_{\alpha\beta}$, using (4.110), as follows:

$$g_{11} = \underline{g}_1 \cdot \underline{g}_1 = 1 + y^2, \quad g_{22} = \underline{g}_2 \cdot \underline{g}_2 = 1 + x^2, \quad g_{12} = g_{21} = \underline{g}_1 \cdot \underline{g}_2 = x\,y.$$

The determinant is

$$g = \det g_{\alpha\beta} = g_{11}\, g_{22} - g_{12}\, g_{21} = 1 + x^2 + y^2.$$

Using (4.116), we obtain the contravariant metric coefficients $g^{\alpha\beta}$ as

$$g^{11} = \frac{g_{22}}{g} = \frac{1 + x^2}{g}, \quad g^{22} = \frac{g_{11}}{g} = \frac{1 + y^2}{g}, \quad g^{12} = g^{21} = -\frac{g_{12}}{g} = \frac{-x\,y}{g}.$$

To compute the covariant coordinates of the curvature tensor $b_{\alpha\beta} = \underline{n} \cdot \underline{g}_{\alpha,\beta}$, we need, according to (4.127), the normal vector $\underline{n}$ and the partial derivatives $\underline{g}_{\alpha,\beta}$ of the covariant basis vectors.

We compute the vector product $\underline{g}_1 \times \underline{g}_2$,

$$\underline{g}_1 \times \underline{g}_2 = -y\,\underline{e}_1 - x\,\underline{e}_2 + \underline{e}_3,$$

and obtain, from (4.101), with $|\underline{g}_1 \times \underline{g}_2| = \sqrt{y^2 + x^2 + 1} = \sqrt{g}$, the normal vector, i. e.

$$\underline{n} = \frac{\underline{g}_1 \times \underline{g}_2}{|\underline{g}_1 \times \underline{g}_2|} = \frac{1}{\sqrt{g}}\left(-y\,\underline{e}_1 - x\,\underline{e}_2 + \underline{e}_3\right).$$

Taking the partial derivatives of the covariant basis vectors gives

$$\underline{g}_{1,1} = \frac{\partial \underline{g}_1}{\partial x} = \underline{0}, \quad \underline{g}_{2,2} = \frac{\partial \underline{g}_2}{\partial y} = \underline{0}, \quad \underline{g}_{1,2} = \underline{g}_{2,1} = \frac{\partial \underline{g}_1}{\partial y} = \underline{e}_3.$$

Hence we obtain for the covariant coordinates $b_{\alpha\beta}$ of the curvature tensor

$$b_{11} = \underline{n} \cdot \underline{g}_{1,1} = 0, \quad b_{22} = \underline{n} \cdot \underline{g}_{2,2} = 0, \quad b_{12} = b_{21} = \underline{n} \cdot \underline{g}_{1,2} = \frac{1}{\sqrt{g}}.$$

For the mixed coordinates $b^{\alpha}{}_{\beta} = g^{\alpha\gamma}\, b_{\gamma\beta}$ we obtain

$$b^1{}_1 = g^{1\gamma}\, b_{\gamma 1} = g^{12}\, b_{21} = -\frac{x\,y}{g\sqrt{g}}, \quad b^1{}_2 = g^{1\gamma}\, b_{\gamma 2} = g^{11}\, b_{12} = \frac{1 + x^2}{g\sqrt{g}},$$

$$b^2{}_1 = g^{2\gamma}\, b_{\gamma 1} = g^{22}\, b_{21} = \frac{1 + y^2}{g\sqrt{g}}, \quad b^2{}_2 = g^{2\gamma}\, b_{\gamma 2} = g^{21}\, b_{12} = -\frac{x\,y}{g\sqrt{g}}.$$

We note that the matrix of the covariant coordinates $b_{\alpha\beta}$ of the curvature tensor is symmetric, but the matrix of the mixed coordinates $b^{\alpha}{}_{\beta}$ is not symmetric. We further note that the metric coefficients $g_{12} = g_{21}$ are zero only if $x = 0$ or $y = 0$. Thus, in general, $\underline{g}_1 \cdot \underline{g}_2 \neq 0$, i. e. the x- and y-coordinate curves do not form an orthogonal grid on the surface $z = x\,y$, but only in the x, y-plane.

Problem 4.23.

Using the covariant coordinates of the curvature tensor, we obtain the mean curvature as $H = \frac{1}{2}\, b^{\alpha}{}_{\alpha} = \frac{1}{2}\, g^{\alpha\beta}\, b_{\beta\alpha}$ and the Gaussian curvature as $K = \det b^{\alpha}{}_{\beta} = \det(g^{\alpha\gamma}\, b_{\gamma\beta}) = (\det g^{\alpha\gamma})(\det b_{\gamma\beta}) = \frac{1}{g} \det b_{\gamma\delta}$ (compare (1.13) and (4.27)). Substituting (4.113), (4.116), and (4.131) gives

$$H = \frac{1}{2}\left(g^{11}\, b_{11} + 2 g^{12}\, b_{12} + g^{22}\, b_{22}\right) = \frac{G\,L - 2 F\,M + E\,N}{2(E\,G - F^2)},$$

$$K = \frac{b_{11}\, b_{22} - b_{12}\, b_{12}}{g_{11}\, g_{22} - g_{12}\, g_{12}} = \frac{L\,N - M^2}{E\,G - F^2}.$$

Problem 4.24.
Using the results of Problem 4.22, we obtain for the mean curvature H and the Gaussian curvature K

$$H = \frac{1}{2}\left(b^1{}_1 + b^2{}_2\right) = \frac{-x\,y}{g\sqrt{g}}, \quad K = b^1{}_1\,b^2{}_2 - b^1{}_2\,b^2{}_1 = \frac{-\left(1+x^2+y^2\right)}{g^3} = -\frac{1}{g^2}.$$

Solving the characteristic equation

$$k^2 - 2H\,k + K = k^2 + \frac{2\,x\,y}{g\sqrt{g}}\,k - \frac{1}{g^2} = 0,$$

we obtain the principal curvatures as

$$k_{1,2} = \frac{-x\,y \pm \sqrt{x^2y^2 + g}}{g\sqrt{g}} = \frac{-x\,y \pm \sqrt{(1+x^2)\,(1+y^2)}}{g\sqrt{g}}.$$

The principal curvature directions are determined by $(b^\alpha{}_\beta - k\,\delta^\alpha_\beta)t^\beta = 0$, and with $g_{11} = 1+y^2, g_{22} = 1+x^2, g_{12} = x\,y$ we have

$$\left[\frac{1}{g\sqrt{g}}\begin{pmatrix} -g_{12} & g_{22} \\ g_{11} & -g_{12}\end{pmatrix} - \frac{-g_{12} \pm \sqrt{g_{11}\,g_{22}}}{g\sqrt{g}}\begin{pmatrix} 1 & 0 \\ 0 & 1\end{pmatrix}\right]\begin{pmatrix} t^1 \\ t^2\end{pmatrix} = \begin{pmatrix} 0 \\ 0\end{pmatrix},$$

$$\begin{pmatrix} \mp\sqrt{g_{11}\,g_{22}} & g_{22} \\ g_{11} & \mp\sqrt{g_{11}\,g_{22}}\end{pmatrix}\begin{pmatrix} t^1 \\ t^2\end{pmatrix} = \begin{pmatrix} 0 \\ 0\end{pmatrix}.$$

Solving these equations we obtain the principal curvature directions (in nonnormalized form) as

$$\underset{1}{t^\beta} = \begin{pmatrix} \sqrt{g_{11}\,g_{22}} \\ g_{11}\end{pmatrix}, \quad \underset{2}{t^\beta} = \begin{pmatrix} -\sqrt{g_{11}\,g_{22}} \\ g_{11}\end{pmatrix}.$$

It is easy to verify by substitution that $g_{\alpha\beta}\,\underset{1}{t^\alpha}\,\underset{2}{t^\beta} = 0$, i.e. the principal curvature directions are perpendicular to each other.

Problem 4.25.
If we set $m = \mu, i = \alpha, j = \beta, k = \gamma$ for the indices of the three-dimensional Riemann curvature tensor in Problem 4.18 and write out the third terms of the summation over n, we have

$$0 = \underbrace{\Gamma^\mu_{\alpha\gamma,\beta} - \Gamma^\mu_{\alpha\beta,\gamma} + \Gamma^\nu_{\alpha\gamma}\,\Gamma^\mu_{\nu\beta} - \Gamma^\nu_{\alpha\beta}\,\Gamma^\mu_{\nu\gamma}}_{R^\mu{}_{\alpha\beta\gamma}} + \Gamma^3_{\alpha\gamma}\,\Gamma^\mu_{3\beta} - \Gamma^3_{\alpha\beta}\,\Gamma^\mu_{3\gamma}.$$

From this we obtain, using $(4.138)_1$ and (4.139), the integrability condition (4.143):

$$R^\mu{}_{\alpha\beta\gamma} = \underbrace{\Gamma^3_{\alpha\beta}}_{b_{\alpha\beta}}\,\underbrace{\Gamma^\mu_{3\gamma}}_{-b^\mu{}_\gamma} - \underbrace{\Gamma^3_{\alpha\gamma}}_{b_{\alpha\gamma}}\,\underbrace{\Gamma^\mu_{3\beta}}_{-b^\mu{}_\beta} = b^\mu{}_\beta\,b_{\alpha\gamma} - b^\mu{}_\gamma\,b_{\alpha\beta}.$$

Setting $m = 3, i = \alpha, j = \beta, k = \gamma$ and using (4.138) and (4.139), we further obtain

$$0 = \underbrace{\Gamma^3_{\alpha\gamma,\beta}}_{b_{\alpha\gamma,\beta}} - \underbrace{\Gamma^3_{\alpha\beta,\gamma}}_{b_{\alpha\beta,\gamma}} + \Gamma^\nu_{\alpha\gamma}\underbrace{\Gamma^3_{\nu\beta}}_{b_{\nu\beta}} - \Gamma^\nu_{\alpha\beta}\underbrace{\Gamma^3_{\nu\gamma}}_{b_{\nu\gamma}} + \underbrace{\Gamma^3_{\alpha\gamma}}_{b_{\alpha\gamma}}\underbrace{\Gamma^3_{3\beta}}_{0} - \underbrace{\Gamma^3_{\alpha\beta}}_{b_{\alpha\beta}}\underbrace{\Gamma^3_{3\gamma}}_{0},$$

$$b_{\alpha\beta,\gamma} - \Gamma^\nu_{\alpha\gamma}\, b_{\nu\beta} = b_{\alpha\gamma,\beta} - \Gamma^\nu_{\alpha\beta}\, b_{\nu\gamma};$$

this is the integrability condition (4.144).

Problem 4.26.
From the definition (4.146) it follows, for two different surface coordinate systems, with (4.33) and $\underline{\tilde{g}}_3 = \underline{g}_3 = \underline{n}$, that

$$\tilde{e}_{\mu\nu} = [\underline{\tilde{g}}_\mu, \underline{\tilde{g}}_\nu, \underline{n}], \quad e_{\mu\nu} = [\underline{g}_\mu, \underline{g}_\nu, \underline{n}].$$

Substituting the transformation law (4.105) into the left equation gives

$$\tilde{e}_{\mu\nu} = \left[\frac{\partial u^\alpha}{\partial \tilde{u}^\mu}\underline{g}_\alpha, \frac{\partial u^\beta}{\partial \tilde{u}^\nu}\underline{g}_\beta, \underline{n}\right] = \frac{\partial u^\alpha}{\partial \tilde{u}^\mu}\frac{\partial u^\beta}{\partial \tilde{u}^\nu}[\underline{g}_\mu, \underline{g}_\nu, \underline{n}] = \frac{\partial u^\alpha}{\partial \tilde{u}^\mu}\frac{\partial u^\beta}{\partial \tilde{u}^\nu} e_{\alpha\beta}.$$

This is, similar to Problem 4.20, the transformation law for the entirely covariant coordinates of a surface tensor of second order.

Problem 4.27.
We can compute the Gaussian curvature from the Riemann curvature tensor. According to (4.147), we have $K = R_{1212}/g$. Using (4.142), we further obtain

$$\begin{aligned} R_{1212} &= g_{1\mu}(\Gamma^\mu_{22,1} - \Gamma^\mu_{21,2} + \Gamma^\nu_{22}\Gamma^\mu_{\nu 1} - \Gamma^\nu_{21}\Gamma^\mu_{\nu 2}) \\ &= g_{1\mu}(\Gamma^\mu_{22,1} - \Gamma^\mu_{21,2} + \Gamma^1_{22}\Gamma^\mu_{11} + \Gamma^2_{22}\Gamma^\mu_{21} - \Gamma^1_{21}\Gamma^\mu_{12} - \Gamma^2_{21}\Gamma^\mu_{22}). \end{aligned}$$

A spherical surface is in spherical coordinates given by $r = \text{const},$ and correspondingly, a cylindrical surface is in cylindrical coordinates given by $R = \text{const}.$ Thus, for further calculations we conveniently use the formulas from Appendix B. Both coordinate systems are orthogonal, i. e. when summing over μ only g_{11} is nonzero. We note that in the appendix the radial direction is in both cases the 1-direction; however, in surface theory the radial direction is the 3-direction, i. e. we need to rename the indices $1 \to 2$ and $2 \to 3$. This gives

$$\begin{aligned} K\,g &= g_{11}\,(\Gamma^1_{22,1} - \Gamma^1_{21,2} + \Gamma^1_{22}\Gamma^1_{11} + \Gamma^2_{22}\Gamma^1_{21} - \Gamma^1_{21}\Gamma^1_{12} - \Gamma^2_{21}\Gamma^1_{22}) \\ &\mathrel{\hat{=}} g_{22}\,(\Gamma^2_{33,2} - \Gamma^2_{32,3} + \Gamma^2_{33}\Gamma^2_{22} + \Gamma^3_{33}\Gamma^2_{32} - \Gamma^2_{32}\Gamma^2_{23} - \Gamma^3_{32}\Gamma^2_{33}). \end{aligned}$$

In spherical coordinates, using (B.33), (B.35), and (B.36), we have $g_{22} = r^2, g = r^4\sin^2\vartheta, \Gamma^2_{33} = -\sin\vartheta\cos\vartheta, \Gamma^3_{32} = \cot\vartheta;$ the other Christoffel symbols are zero. So we obtain the Gaussian curvature of a spherical surface as

$$K = \frac{g_{22}}{g}\left(\Gamma^2_{33,2} - \Gamma^3_{32}\,\Gamma^2_{33}\right) = \frac{r^2}{r^4\,\sin^2\vartheta}\Big(\underbrace{\frac{\partial}{\partial\vartheta}(-\sin\vartheta\,\cos\vartheta)}_{-\cos^2\vartheta+\sin^2\vartheta} + \underbrace{\sin\vartheta\,\cos\vartheta\,\cot\vartheta}_{\cos^2\vartheta}\Big) = \frac{1}{r^2}.$$

In cylindrical coordinates, according to (B.13), all Christoffel symbols that appear here are zero, so the Gaussian curvature of a cylinder surface is $K = 0$.

Alternatively, we can find the Gaussian curvature from its definition (4.133). We start with $K = \det b^\alpha{}_\beta$. Using (4.139), we have

$$b^\alpha{}_\beta = g^{\alpha\mu}\,b_{\mu\beta} = g^{\alpha\mu}\,\Gamma^3_{\mu\beta},$$

and thus, using (1.13) and (4.27), we obtain

$$K = \det b^\alpha{}_\beta = (\det g^{\alpha\mu})\,(\det b_{\mu\beta}) = \frac{1}{g}\,\det\Gamma^3_{\mu\beta}.$$

In order to further obtain expressions in spherical and cylindrical coordinates, we rename the indices $1 \to 2$, $2 \to 3$, $3 \to 1$ and obtain

$$K = \frac{1}{g_{22}\,g_{33}}\begin{vmatrix}\Gamma^1_{22} & \Gamma^1_{23}\\ \Gamma^1_{32} & \Gamma^1_{33}\end{vmatrix}.$$

In spherical coordinates, we have using (B.33) and (B.36)

$$K = \frac{1}{r^4\,\sin^2\vartheta}\begin{vmatrix}-r & 0\\ 0 & -r\,\sin^2\vartheta\end{vmatrix} = \frac{1}{r^2};$$

in cylindrical coordinates, we have, using (B.10) and (B.13),

$$K = \frac{1}{R^2}\begin{vmatrix}-R & 0\\ 0 & 0\end{vmatrix} = 0.$$

On a cylinder surface we have $K = 0$. This means we can unfold the surface into a plane, whereas such an unfolding is not possible with a spherical surface, because $K \neq 0$.

Problem 5.1.
If $\underline{\underline{T}}$ is a polar tensor, then its coordinates transform according to (2.17), so

$$T_{ij} = \alpha_{im}\,\alpha_{jn}\,\tilde{T}_{mn}.$$

Setting the indices equal, $i = j$, and using the orthogonality relation (2.6), we obtain for the trace of $\underline{\underline{T}}$

$$\operatorname{tr}\underline{\underline{T}} = T_{ii} = \alpha_{im}\,\alpha_{in}\,\tilde{T}_{mn} = \delta_{mn}\,\tilde{T}_{mn} = \tilde{T}_{mm},$$

i. e. the trace is independent of the underlying coordinate system and thus it is an invariant of the tensor $\underline{\underline{T}}$.

Similarly, we obtain for the trace of $\underline{\underline{T}}^2$

$$\begin{aligned}\operatorname{tr}\underline{\underline{T}}^2 &= T_{ik}\,T_{ki} = \alpha_{im}\,\alpha_{kn}\,\tilde{T}_{mn}\,\alpha_{kp}\,\alpha_{iq}\,\tilde{T}_{pq} = \alpha_{im}\,\alpha_{iq}\,\alpha_{kn}\,\alpha_{kp}\,\tilde{T}_{mn}\,\tilde{T}_{pq}\\ &= \delta_{mq}\,\delta_{np}\,\tilde{T}_{mn}\,\tilde{T}_{pq} = \tilde{T}_{qp}\,\tilde{T}_{pq}.\end{aligned}$$

If $\underline{\underline{T}}$ is polar, then $\operatorname{tr}\underline{\underline{T}}$ and $\operatorname{tr}\underline{\underline{T}}^2$ are also polar.

If $\underline{\underline{T}}$ is axial, then similar computations, using the transformation law (2.18), give

$$T_{ii} = \left(\det \underset{\sim}{\alpha}\right)\,\tilde{T}_{mm}$$

and

$$T_{ij}\,T_{ji} = \tilde{T}_{qp}\,\tilde{T}_{pq}.$$

Here $\operatorname{tr}\underline{\underline{T}}^2$ is a polar scalar, but $\operatorname{tr}\underline{\underline{T}}$ is an axial scalar, and hence, according to our convention from Chapter 5, $\operatorname{tr}\underline{\underline{T}}$ does not count as an invariant.

Problem 5.2.
All odd powers of antimetric tensors are antimetric, i. e. the trace of these powers is zero; but the even powers are symmetric. Furthermore, according to (5.8), the trace of a scalar product of a symmetric and an antimetric tensor is zero, so we see, by comparing with (5.15), that we have the following nonzero invariants: $\operatorname{tr}\underline{\underline{A}}^2$, $\operatorname{tr}\underline{\underline{B}}^2$, $\operatorname{tr}(\underline{\underline{A}}\cdot\underline{\underline{B}})$, $\operatorname{tr}(\underline{\underline{A}}^2\cdot\underline{\underline{B}}^2)$, $\operatorname{tr}(\underline{\underline{A}}^2\cdot\underline{\underline{B}}\cdot\underline{\underline{A}}\cdot\underline{\underline{B}}^2)$.

The invariant of degree six can be shown to be reducible, in the same way as we showed this for two symmetric tensors in Section 5.3.3, No. 5.

Only the investigation of the invariant of degree four requires more effort. If we write the antimetric tensor $\underline{\underline{A}}$, using $(3.8)_2$, in terms of its corresponding vector $\underline{a}$, we obtain, using the Grassmann identity, for the coordinates of the square $\underline{\underline{A}}^2$

$$A_{ij}\,A_{jk} = \varepsilon_{ijl}\,\varepsilon_{jkm}\,a_l\,a_m = (\delta_{lk}\,\delta_{im} - \delta_{lm}\,\delta_{ik})\;a_l\,a_m = a_i\,a_k - a_m\,a_m\,\delta_{ik}.$$

The same holds for the coordinates of the square $\underline{\underline{B}}^2$, so we compute the coordinates of $\underline{\underline{A}}^2\cdot\underline{\underline{B}}^2$. We have

$$\begin{aligned}A_{ij}\,A_{jk}\,B_{kp}\,B_{pq} &= (a_i\,a_k - a_m\,a_m\,\delta_{ik})\,(b_k\,b_q - b_l\,b_l\,\delta_{kq})\\ &= a_k\,b_k\,a_i\,b_q - a_m\,a_m\,b_i\,b_q - b_l\,b_l\,a_i\,a_q + a_m\,a_m\,b_l\,b_l\,\delta_{iq}.\end{aligned}$$

Then the trace is

$$\operatorname{tr}(\underline{\underline{A}}^2\cdot\underline{\underline{B}}^2) = A_{ij}\,A_{jk}\,B_{kp}\,B_{pi} = a_k\,b_k\,a_i\,b_i + a_m\,a_m\,b_i\,b_i. \tag{a}$$

Using $(3.8)_1$, we can reintroduce the coordinates of the antrimetric tensors on the right side. Using the Grassmann identity and taking antimetry into account, we get initially

$$a_k\, b_k = \frac{1}{4}\, \varepsilon_{klm}\, \varepsilon_{kpq}\, A_{lm}\, B_{pq} = \frac{1}{4}\, (\delta_{lp}\, \delta_{mq} - \delta_{lq}\, \delta_{mp})\, A_{lm}\, B_{pq}$$
$$= \frac{1}{4}\, (A_{pq}\, B_{pq} - A_{qp}\, B_{pq}) = -\frac{1}{2}\, A_{qp}\, B_{pq} = -\frac{1}{2}\, \mathrm{tr}(\underline{\underline{A}} \cdot \underline{\underline{B}}).$$

Similarly, we obtain $a_m\, a_m = -\frac{1}{2}\, \mathrm{tr}\, \underline{\underline{A}}^2$ and $b_i\, b_i = -\frac{1}{2}\, \mathrm{tr}\, \underline{\underline{B}}^2$; then, from (a) it follows that $\mathrm{tr}(\underline{\underline{A}}^2 \cdot \underline{\underline{B}}^2)$ can be written in terms of the lower-order invariants and thus is reducible, i. e.

$$\mathrm{tr}(\underline{\underline{A}}^2 \cdot \underline{\underline{B}}^2) = \frac{1}{4}\, \mathrm{tr}^2\, (\underline{\underline{A}} \cdot \underline{\underline{B}}) + \frac{1}{4}\, \mathrm{tr}\, \underline{\underline{A}}^2\, \mathrm{tr}\, \underline{\underline{B}}^2.$$

Hence the integrity basis for two antimetric tensors consists of only three irreducible invariants:

- the only irreducible invariants, according to Section 5.3.3, No. 4, of the tensors $\underline{\underline{A}}$ and $\underline{\underline{B}}$ themselves: $I_1 = \mathrm{tr}\, \underline{\underline{A}}^2, \quad I_2 = \mathrm{tr}\, \underline{\underline{B}}^2$;
- the only irreducible simultaneous invariant of $\underline{\underline{A}}$ and $\underline{\underline{B}}$: $I_3 = \mathrm{tr}(\underline{\underline{A}} \cdot \underline{\underline{B}})$.

According to Section 3.3, we can uniquely associate a vector to an antimetric tensor, and two vectors have, according to Section 5.3.1, three irreducible invariants, i. e. the same number of invariants as two antimetric tensors.

Problem 5.3.
We can compute the following invariants from the coordinates of a symmetric tensor $\underline{\underline{S}}$ and a vector $\underline{u}$:

- the basic invariants of the symmetric tensor $\underline{\underline{S}}$:

  $$I_1 = \mathrm{tr}\, \underline{\underline{S}}, \quad I_2 = \mathrm{tr}\, \underline{\underline{S}}^2, \quad I_3 = \mathrm{tr}\, \underline{\underline{S}}^3;$$

- the square of the vector $\underline{u}$:

  $$I_4 = \underline{u} \cdot \underline{u};$$

- the simultaneous invariants of $\underline{\underline{S}}$ und $\underline{u}$:

  $$I_5 = \underline{u} \cdot \underline{\underline{S}} \cdot \underline{u}, \quad I_6 = \underline{u} \cdot \underline{\underline{S}}^2 \cdot \underline{u}, \quad I_7 = [\underline{u}, \underline{\underline{S}} \cdot \underline{u}, \underline{\underline{S}}^2 \cdot \underline{u}].$$

Simultaneous invariants which contain $\underline{\underline{S}}^3$ can be reduced to I_1 to I_7, using the Cayley–Hamilton theorem (3.94). Scalar quantities such as $\varepsilon_{ijk}\, u_i\, S_{jk}$ and $\varepsilon_{ijk}\, u_i\, S^{(2)}_{jk}$ are zero, because of the symmetry of $\underline{\underline{S}}$ and due to (2.40), so they do not belong to the simultaneous invariants.

If $\underline{\underline{S}}$ is polar and $\underline{u}$ is axial, then I_1 to I_7 are polar, i. e. the integrity basis then consists of seven irreducible invariants. If, on the other hand, both $\underline{\underline{S}}$ and $\underline{u}$ are polar, only I_1 to

I_6 are polar, but $I_7 = \varepsilon_{ijk} u_i S_{jl} u_l S^{(2)}_{km} u_m$ is axial because of the ε-tensor, i. e. in this case the integrity basis has only six irreducible invariants.

We can, according to Section 3.3, associate an axial vector to a polar antimetric tensor and, according to Section 5.3.3, No. 6, a polar antimetric tensor and a polar symmetric tensor have seven irreducible invariants, i. e. the same number as a polar symmetric tensor and an axial vector.

Problem 5.4.

We obtain from $\underline{u} = f(\underline{\underline{T}})$, by introducing an auxiliary vector, $\underline{u} \cdot \underline{h} = f(\underline{\underline{T}}, \underline{h})$. Simultaneous invariants of $\underline{\underline{T}}$ and $\underline{h}$ which are linear in $\underline{h}$ can only be computed using the ε-tensor.

If $\underline{\underline{T}}$ is symmetric, then, according to (2.40), we have $\varepsilon_{ijk} T_{jk} h_i = 0$; the same result is obtained for the tensor powers $\underline{\underline{T}}^2, \underline{\underline{T}}^3$, which are also symmetric. So a representation $\underline{u} = f(\underline{\underline{T}})$, in which a vector $\underline{u}$ depends on a symmetric tensor $\underline{\underline{T}}$, does not exist.

If $\underline{\underline{T}}$ is antimetric, then $\underline{\underline{T}}^2$ is symmetric and $\underline{\underline{T}}^3$ is again antimetric; but using the Cayley–Hamilton theorem (3.94), $\underline{\underline{T}}^3$ can be written in terms of $\underline{\underline{T}}$. So with $\varepsilon_{ijk} T_{jk} h_i$ only one irreducible invariant exists which is linear in $\underline{h}$. If $\underline{\underline{T}}$ is polar, then $\underline{\underline{\underline{\varepsilon}}} \cdot\cdot \underline{\underline{T}}$ is axial because of the ε-tensor. Thus, a representation $\underline{u} = f(\underline{\underline{T}})$ of a vector $\underline{u}$, which depends on an antimetric tensor $\underline{\underline{T}}$, only exists if the vector $\underline{u}$ is also axial, and it is

$$\underline{u} = k(\operatorname{tr} \underline{\underline{T}}^2)\, \underline{\underline{\underline{\varepsilon}}} \cdot\cdot \underline{\underline{T}},$$

or, in terms of the axial vector $\underline{t}$ associated with $\underline{\underline{T}}$,

$$\underline{u} = k^*(\underline{t} \cdot \underline{t})\, \underline{t}.$$

The coefficients k and k^* are polar scalar-valued functions of the polar invariants $\operatorname{tr} \underline{\underline{T}}^2$ and $\underline{t} \cdot \underline{t}$, respectively.

Problem 5.5.

We get from $\underline{\underline{T}} = f(\underline{v})$, after introducing an auxiliary tensor, $\underline{\underline{T}} \cdot\cdot \underline{\underline{H}} = f(\underline{v}, \underline{\underline{H}})$. So, three invariants exist which are linear in $\underline{\underline{H}}$:

$$H_{ii} = \underline{\underline{\delta}} \cdot\cdot \underline{\underline{H}}, \quad v_i v_j H_{ij} = \underline{v}\,\underline{v} \cdot\cdot \underline{\underline{H}}, \quad \varepsilon_{ijk} v_k H_{ij} = (\underline{\underline{\underline{\varepsilon}}} \cdot \underline{v}) \cdot\cdot \underline{\underline{H}}.$$

If $\underline{v}$ is polar, then $\underline{v}\,\underline{v}$ is also polar, but $\underline{\underline{\underline{\varepsilon}}} \cdot \underline{v}$ is axial because of the ε-tensor. Since we are looking for a representation for a polar tensor $\underline{\underline{T}}$, we have only the two generators $\underline{\underline{\delta}}$ and $\underline{v}\,\underline{v}$, and the representation is

$$\underline{\underline{T}} = k_1(\underline{v} \cdot \underline{v})\, \underline{\underline{\delta}} + k_2(\underline{v} \cdot \underline{v})\, \underline{v}\,\underline{v}.$$

If $\underline{v}$ is axial, then both $\underline{v}\,\underline{v}$ and $\underline{\underline{\underline{\varepsilon}}} \cdot \underline{v}$ are polar, and the representation is

$$\underline{\underline{T}} = k_1(\underline{v} \cdot \underline{v})\, \underline{\underline{\delta}} + k_2(\underline{v} \cdot \underline{v})\, \underline{v}\,\underline{v} + k_3(\underline{v} \cdot \underline{v})\, \underline{\underline{\underline{\varepsilon}}} \cdot \underline{v}.$$

B Cylindrical Coordinates and Spherical Coordinates

The most commonly used curvilinear coordinate systems in practice are the cylindrical and spherical coordinates. We summarize here the most important formulas for these two coordinate systems. The radial cylindrical coordinate is denoted by R and the radial spherical coordinate is denoted by r.

B.1 Cylindrical Coordinates

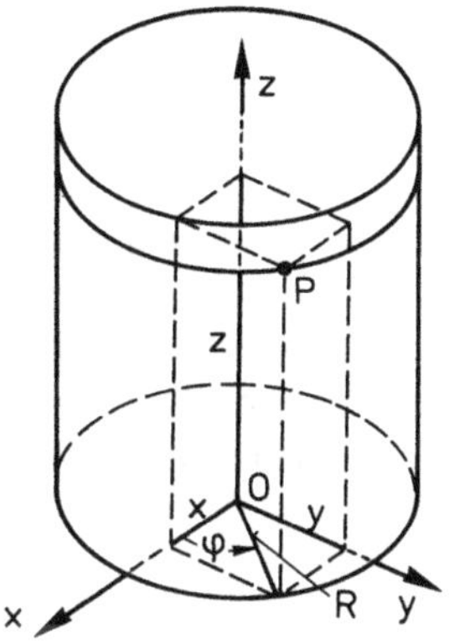

B.1.1 Transformation Equations for Point Coordinates

The transformation equations are

$$x = R\cos\varphi, \qquad y = R\sin\varphi, \tag{B.1}$$

$$\begin{aligned} R &= \sqrt{x^2 + y^2}, \quad 0 \leqq R < \infty, \\ \varphi &= \arctan\frac{y}{x}, \quad 0 \leqq \varphi < 2\pi. \end{aligned} \tag{B.2}$$

https://doi.org/10.1515/9783110404265-008

B.1.2 Bases

The Cartesian coordinates of the covariant basis, the contravariant basis, and the physical basis are

$$\begin{aligned}
&\underline{g}_1 = \underline{g}^1 = \underline{g}_{<1>} = \{\cos\varphi,\ \sin\varphi,\ 0\};\\
&\underline{g}_2 = \{-R\sin\varphi,\ R\cos\varphi,\ 0\},\\
&\underline{g}^2 = \left\{-\frac{1}{R}\sin\varphi,\ \frac{1}{R}\cos\varphi,\ 0\right\},\\
&\underline{g}_{<2>} = \{-\sin\varphi,\ \cos\varphi,\ 0\};\\
&\underline{g}_3 = \underline{g}^3 = \underline{g}_{<3>} = \{0,\ 0,\ 1\}.
\end{aligned} \tag{B.3}$$

B.1.3 Transformation Equations for Tensor Coordinates

In the following formulas, we denote the Cartesian coordinates, e. g. of a vector, by a_x, a_y, a_z, its covariant cylindrical coordinates by a_1, a_2, a_3, its contravariant cylindrical coordinates by a^1, a^2, a^3, and its physical cylindrical coordinates by a_R, a_φ, a_z:

$$\begin{aligned}
a_R &= a_x\cos\varphi + a_y\sin\varphi,\\
a_\varphi &= a_y\cos\varphi - a_x\sin\varphi,\\
a_z &= a_z;
\end{aligned} \tag{B.4}$$

$$\begin{aligned}
a_x &= a_R\cos\varphi - a_\varphi\sin\varphi,\\
a_y &= a_\varphi\cos\varphi + a_R\sin\varphi,\\
a_z &= a_z;
\end{aligned} \tag{B.5}$$

$$\begin{aligned}
a_1 &= a^1 = a_R,\\
a_2 &= R\,a_\varphi, \qquad a^2 = \tfrac{1}{R}\,a_\varphi,\\
a_3 &= a^3 = a_z;
\end{aligned} \tag{B.6}$$

$$\begin{aligned}
a_{RR} &= a_{xx}\cos^2\varphi + (a_{xy} + a_{yx})\cos\varphi\sin\varphi + a_{yy}\sin^2\varphi,\\
a_{R\varphi} &= a_{xy}\cos^2\varphi - (a_{xx} - a_{yy})\cos\varphi\sin\varphi - a_{yx}\sin^2\varphi,\\
a_{Rz} &= a_{xz}\cos\varphi + a_{yz}\sin\varphi,\\
a_{\varphi R} &= a_{yx}\cos^2\varphi - (a_{xx} - a_{yy})\cos\varphi\sin\varphi - a_{xy}\sin^2\varphi,\\
a_{\varphi\varphi} &= a_{yy}\cos^2\varphi - (a_{xy} + a_{yx})\cos\varphi\sin\varphi + a_{xx}\sin^2\varphi,\\
a_{\varphi z} &= a_{yz}\cos\varphi - a_{xz}\sin\varphi,\\
a_{zR} &= a_{zx}\cos\varphi + a_{zy}\sin\varphi,\\
a_{z\varphi} &= a_{zy}\cos\varphi - a_{zx}\sin\varphi,\\
a_{zz} &= a_{zz};
\end{aligned} \tag{B.7}$$

$$\begin{aligned}
a_{xx} &= a_{RR}\cos^2\varphi - (a_{R\varphi} + a_{\varphi R})\cos\varphi\,\sin\varphi + a_{\varphi\varphi}\sin^2\varphi,\\
a_{xy} &= a_{R\varphi}\cos^2\varphi + (a_{RR} - a_{\varphi\varphi})\cos\varphi\,\sin\varphi - a_{\varphi R}\sin^2\varphi,\\
a_{xz} &= a_{Rz}\cos\varphi - a_{\varphi z}\sin\varphi,\\
a_{yx} &= a_{\varphi R}\cos^2\varphi + (a_{RR} - a_{\varphi\varphi})\cos\varphi\,\sin\varphi - a_{R\varphi}\sin^2\varphi,\\
a_{yy} &= a_{\varphi\varphi}\cos^2\varphi + (a_{R\varphi} + a_{\varphi R})\cos\varphi\,\sin\varphi + a_{RR}\sin^2\varphi,\\
a_{yz} &= a_{\varphi z}\cos\varphi + a_{Rz}\sin\varphi,\\
a_{zx} &= a_{zR}\cos\varphi - a_{z\varphi}\sin\varphi,\\
a_{zy} &= a_{z\varphi}\cos\varphi + a_{zR}\sin\varphi,\\
a_{zz} &= a_{zz};
\end{aligned} \tag{B.8}$$

$$\begin{aligned}
&a_{11} = a^{11} = a_1{}^1 = a^1{}_1 = a_{RR},\\
&a_{12} = a^1{}_2 = R\,a_{R\varphi}, \qquad a^{12} = a_1{}^2 = \frac{1}{R}\,a_{R\varphi},\\
&a_{13} = a^{13} = a_1{}^3 = a^1{}_3 = a_{Rz},\\
&a_{21} = a_2{}^1 = R\,a_{\varphi R}, \qquad a^{21} = a^2{}_1 = \frac{1}{R}\,a_{\varphi R},\\
&a_{22} = R^2\,a_{\varphi\varphi}, \qquad a^{22} = \frac{1}{R^2}\,a_{\varphi\varphi}, \qquad a_2{}^2 = a^2{}_2 = a_{\varphi\varphi},\\
&a_{23} = a_2{}^3 = R\,a_{\varphi z}, \qquad a^{23} = a^2{}_3 = \frac{1}{R}\,a_{\varphi z},\\
&a_{31} = a^{31} = a_3{}^1 = a^3{}_1 = a_{zR},\\
&a_{32} = a^3{}_2 = R\,a_{z\varphi}, \qquad a^{32} = a_3{}^2 = \frac{1}{R}\,a_{z\varphi},\\
&a_{33} = a^{33} = a_3{}^3 = a^3{}_3 = a_{zz}.
\end{aligned} \tag{B.9}$$

B.1.4 δ-Tensor and ε-Tensor

The only nonzero coordinates are

$$g_{11} = g^{11} = 1, \qquad g_{22} = R^2, \qquad g^{22} = \frac{1}{R^2}, \qquad g_{33} = g^{33} = 1, \tag{B.10}$$

$$e_{123} = e_{12}{}^3 = e^1{}_{23} = e^1{}_2{}^3 = R, \qquad e^{123} = e_1{}^{23} = e^{12}{}_3 = e_1{}^2{}_3 = \frac{1}{R} \tag{B.11}$$

and in addition the ε-tensor also contains the corresponding permutations. Also

$$g := \det g_{ij} = R^2. \tag{B.12}$$

B.1.5 Christoffel Symbols

The only nonzero Christoffel symbols are

$$\Gamma^2_{12} = \Gamma^2_{21} = \frac{1}{R}, \qquad \Gamma^1_{22} = -R. \tag{B.13}$$

B.1.6 Differential Operators

For each operator the physical coordinates are given:

$$\operatorname{grad} a = \left\{ \frac{\partial a}{\partial R}, \frac{1}{R}\frac{\partial a}{\partial \varphi}, \frac{\partial a}{\partial z} \right\}; \tag{B.14}$$

$$\operatorname{div} \underline{a} = \frac{\partial a_R}{\partial R} + \frac{a_R}{R} + \frac{1}{R}\frac{\partial a_\varphi}{\partial \varphi} + \frac{\partial a_z}{\partial z}; \tag{B.15}$$

$$\operatorname{curl} \underline{a} = \left\{ \frac{1}{R}\frac{\partial a_z}{\partial \varphi} - \frac{\partial a_\varphi}{\partial z}, \frac{\partial a_R}{\partial z} - \frac{\partial a_z}{\partial R}, \frac{\partial a_\varphi}{\partial R} - \frac{1}{R}\frac{\partial a_R}{\partial \varphi} + \frac{a_\varphi}{R} \right\}; \tag{B.16}$$

$$\begin{aligned}
(\operatorname{grad} \underline{a})_{RR} &= \frac{\partial a_R}{\partial R},\\
(\operatorname{grad} \underline{a})_{R\varphi} &= \frac{1}{R}\frac{\partial a_R}{\partial \varphi} - \frac{a_\varphi}{R},\\
(\operatorname{grad} \underline{a})_{Rz} &= \frac{\partial a_R}{\partial z},\\
(\operatorname{grad} \underline{a})_{\varphi R} &= \frac{\partial a_\varphi}{\partial R},\\
(\operatorname{grad} \underline{a})_{\varphi\varphi} &= \frac{1}{R}\frac{\partial a_\varphi}{\partial \varphi} + \frac{a_R}{R},\\
(\operatorname{grad} \underline{a})_{\varphi z} &= \frac{\partial a_\varphi}{\partial z},\\
(\operatorname{grad} \underline{a})_{zR} &= \frac{\partial a_z}{\partial R},\\
(\operatorname{grad} \underline{a})_{z\varphi} &= \frac{1}{R}\frac{\partial a_z}{\partial \varphi},\\
(\operatorname{grad} \underline{a})_{zz} &= \frac{\partial a_z}{\partial z};
\end{aligned} \tag{B.17}$$

$$\begin{aligned}
[(\operatorname{grad} \underline{a}) \cdot \underline{b}]_R &= \frac{\partial a_R}{\partial R} b_R + \frac{\partial a_R}{\partial \varphi}\frac{b_\varphi}{R} + \frac{\partial a_R}{\partial z} b_z - \frac{a_\varphi b_\varphi}{R},\\
[(\operatorname{grad} \underline{a}) \cdot \underline{b}]_\varphi &= \frac{\partial a_\varphi}{\partial R} b_R + \frac{\partial a_\varphi}{\partial \varphi}\frac{b_\varphi}{R} + \frac{\partial a_\varphi}{\partial z} b_z + \frac{a_R b_\varphi}{R},\\
[(\operatorname{grad} \underline{a}) \cdot \underline{b}]_z &= \frac{\partial a_z}{\partial R} b_R + \frac{\partial a_z}{\partial \varphi}\frac{b_\varphi}{R} + \frac{\partial a_z}{\partial z} b_z;
\end{aligned} \tag{B.18}$$

$$
\begin{aligned}
(\operatorname{div} \underline{\underline{a}})_R &= \frac{\partial a_{RR}}{\partial R} + \frac{1}{R}\frac{\partial a_{R\varphi}}{\partial \varphi} + \frac{\partial a_{Rz}}{\partial z} + \frac{a_{RR} - a_{\varphi\varphi}}{R},\\
(\operatorname{div} \underline{\underline{a}})_\varphi &= \frac{\partial a_{\varphi R}}{\partial R} + \frac{1}{R}\frac{\partial a_{\varphi\varphi}}{\partial \varphi} + \frac{\partial a_{\varphi z}}{\partial z} + \frac{a_{R\varphi} + a_{\varphi R}}{R},\\
(\operatorname{div} \underline{\underline{a}})_z &= \frac{\partial a_{zR}}{\partial R} + \frac{1}{R}\frac{\partial a_{z\varphi}}{\partial \varphi} + \frac{\partial a_{zz}}{\partial z} + \frac{a_{zR}}{R};
\end{aligned}
\tag{B.19}
$$

$$
\Delta a = \frac{\partial^2 a}{\partial R^2} + \frac{1}{R}\frac{\partial a}{\partial R} + \frac{1}{R^2}\frac{\partial^2 a}{\partial \varphi^2} + \frac{\partial^2 a}{\partial z^2};
\tag{B.20}
$$

$$
\begin{aligned}
(\Delta \underline{a})_R &= \frac{\partial^2 a_R}{\partial R^2} + \frac{1}{R}\frac{\partial a_R}{\partial R} - \frac{a_R}{R^2} + \frac{1}{R^2}\frac{\partial^2 a_R}{\partial \varphi^2} + \frac{\partial^2 a_R}{\partial z^2} - \frac{2}{R^2}\frac{\partial a_\varphi}{\partial \varphi},\\
(\Delta \underline{a})_\varphi &= \frac{\partial^2 a_\varphi}{\partial R^2} + \frac{1}{R}\frac{\partial a_\varphi}{\partial R} - \frac{a_\varphi}{R^2} + \frac{1}{R^2}\frac{\partial^2 a_\varphi}{\partial \varphi^2} + \frac{\partial^2 a_\varphi}{\partial z^2} + \frac{2}{R^2}\frac{\partial a_R}{\partial \varphi},\\
(\Delta \underline{a})_z &= \frac{\partial^2 a_z}{\partial R^2} + \frac{1}{R}\frac{\partial a_z}{\partial R} + \frac{1}{R^2}\frac{\partial^2 a_z}{\partial \varphi^2} + \frac{\partial^2 a_z}{\partial z^2}.
\end{aligned}
\tag{B.21}
$$

B.1.7 Curve-, Area-, and Volume Elements

The elements are

$$
\begin{aligned}
\mathrm{d}u^i &= (\mathrm{d}R,\ \mathrm{d}\varphi,\ \mathrm{d}z),\\
\mathrm{d}A_i &= (R\,\mathrm{d}\varphi\,\mathrm{d}z,\ R\,\mathrm{d}z\,\mathrm{d}R,\ R\,\mathrm{d}R\,\mathrm{d}\varphi),\\
\mathrm{d}V &= R\,\mathrm{d}R\,\mathrm{d}\varphi\,\mathrm{d}z.
\end{aligned}
\tag{B.22}
$$

In (B.23) the physical coordinates are given:

$$
\begin{aligned}
\mathrm{d}\underline{x} &= (\mathrm{d}R,\ R\,\mathrm{d}\varphi,\ \mathrm{d}z),\\
\mathrm{d}\underline{A} &= (R\,\mathrm{d}\varphi\,\mathrm{d}z,\ \mathrm{d}z\,\mathrm{d}R,\ R\,\mathrm{d}R\,\mathrm{d}\varphi).
\end{aligned}
\tag{B.23}
$$

B.2 Spherical Coordinates

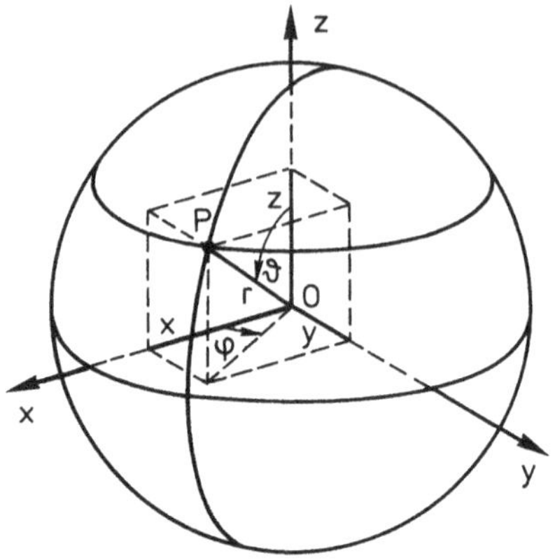

B.2.1 Transformation Equations for Point Coordinates

The transformation equations are

$$\begin{aligned} x &= r \sin\vartheta \cos\varphi, \\ y &= r \sin\vartheta \sin\varphi, \\ z &= r \cos\vartheta, \end{aligned} \tag{B.24}$$

$$\begin{aligned} r &= \sqrt{x^2 + y^2 + z^2}, \quad 0 \leqq r < \infty, \\ \vartheta &= \arccos \frac{z}{\sqrt{x^2 + y^2 + z^2}}, \quad 0 \leqq \vartheta \leqq \pi, \\ \varphi &= \arctan \frac{y}{x}, \quad 0 \leqq \varphi < 2\pi. \end{aligned} \tag{B.25}$$

B.2.2 Bases

The Cartesian coordinates of the covariant basis, the contravariant basis, and the physical basis are

$$\underline{g}_1 = \underline{g}^1 = \underline{g}_{<1>} = \{\sin\vartheta\,\cos\varphi,\ \sin\vartheta\,\sin\varphi,\ \cos\vartheta\};$$

$$\underline{g}_2 = \{r\cos\vartheta\,\cos\varphi,\ r\cos\vartheta\,\sin\varphi,\ -r\sin\vartheta\},$$

$$\underline{g}^2 = \left\{\frac{1}{r}\cos\vartheta\,\cos\varphi,\ \frac{1}{r}\cos\vartheta\,\sin\varphi,\ -\frac{1}{r}\sin\vartheta\right\},$$

$$\underline{g}_{<2>} = \{\cos\vartheta\,\cos\varphi,\ \cos\vartheta\,\sin\varphi,\ -\sin\vartheta\}; \tag{B.26}$$

$$\underline{g}_3 = \{-r\sin\vartheta\,\sin\varphi,\ r\sin\vartheta\,\cos\varphi,\ 0\},$$

$$\underline{g}^3 = \left\{-\frac{\sin\varphi}{r\sin\vartheta},\ \frac{\cos\varphi}{r\sin\vartheta},\ 0\right\},$$

$$\underline{g}_{<3>} = \{-\sin\varphi,\ \cos\varphi,\ 0\}.$$

B.2.3 Transformation Equations for Tensor Coordinates

We denote the Cartesian coordinates, e. g. of a vector, by a_x, a_y, a_z, its covariant spherical coordinates by a_1, a_2, a_3, its contravariant spherical coordinates by a^1, a^2, a^3, and its physical spherical coordinates by a_r, a_ϑ, a_φ:

$$\begin{aligned} a_r &= a_x \sin\vartheta\,\cos\varphi + a_y \sin\vartheta\,\sin\varphi + a_z \cos\vartheta, \\ a_\vartheta &= a_x \cos\vartheta\,\cos\varphi + a_y \cos\vartheta\,\sin\varphi - a_z \sin\vartheta, \\ a_\varphi &= -a_x \sin\varphi + a_y \cos\varphi; \end{aligned} \tag{B.27}$$

$$\begin{aligned} a_x &= a_r \sin\vartheta\,\cos\varphi + a_\vartheta \cos\vartheta\,\cos\varphi - a_\varphi \sin\varphi, \\ a_y &= a_r \sin\vartheta\,\sin\varphi + a_\vartheta \cos\vartheta\,\sin\varphi + a_\varphi \cos\varphi, \\ a_z &= a_r \cos\vartheta - a_\vartheta \sin\vartheta; \end{aligned} \tag{B.28}$$

$$\begin{aligned} a_1 &= a^1 = a_r, \\ a_2 &= r\,a_\vartheta, \qquad a^2 = \frac{1}{r}\,a_\vartheta, \\ a_3 &= r\sin\vartheta\,a_\varphi, \qquad a^3 = \frac{1}{r\sin\vartheta}\,a_\varphi; \end{aligned} \tag{B.29}$$

$$
\begin{aligned}
a_{rr} &= a_{xx}\ \sin^2\vartheta\ \cos^2\varphi + a_{xy}\ \sin^2\vartheta\ \cos\varphi\ \sin\varphi \\
&\quad + a_{xz}\ \cos\vartheta\ \sin\vartheta\ \cos\varphi + a_{yx}\ \sin^2\vartheta\ \cos\varphi\ \sin\varphi \\
&\quad + a_{yy}\ \sin^2\vartheta\ \sin^2\varphi + a_{yz}\ \cos\vartheta\ \sin\vartheta\ \cos\varphi \\
&\quad + a_{zx}\ \cos\vartheta\ \sin\vartheta\ \cos\varphi + a_{zy}\ \cos\vartheta\ \sin\vartheta\ \sin\varphi + a_{zz}\ \cos^2\vartheta, \\
a_{r\vartheta} &= a_{xx}\ \cos\vartheta\ \sin\vartheta\ \cos^2\varphi + a_{xy}\ \cos\vartheta\ \sin\vartheta\ \cos\varphi\ \sin\varphi \\
&\quad - a_{xz}\ \sin^2\vartheta\ \cos\varphi + a_{yx}\ \cos\vartheta\ \sin\vartheta\ \cos\varphi\ \sin\varphi \\
&\quad + a_{yy}\ \cos\vartheta\ \sin\vartheta\ \sin^2\varphi - a_{yz}\ \sin^2\vartheta\ \sin\varphi \\
&\quad + a_{zx}\ \cos^2\vartheta\ \cos\varphi + a_{zy}\ \cos^2\vartheta\ \sin\varphi - a_{zz}\ \cos\vartheta\ \sin\vartheta, \\
a_{r\varphi} &= -a_{xx}\ \sin\vartheta\ \cos\varphi\ \sin\varphi + a_{xy}\ \sin\vartheta\ \cos^2\varphi \\
&\quad - a_{yx}\ \sin\vartheta\ \sin^2\varphi + a_{yy}\ \sin\vartheta\ \cos\varphi\ \sin\varphi \\
&\quad - a_{zx}\ \cos\vartheta\ \sin\varphi + a_{zy}\ \cos\vartheta\ \cos\varphi, \\
a_{\vartheta r} &= a_{xx}\ \cos\vartheta\ \sin\vartheta\ \cos^2\varphi + a_{xy}\ \cos\vartheta\ \sin\vartheta\ \cos\varphi\ \sin\varphi \\
&\quad + a_{xz}\ \cos^2\vartheta\ \cos\varphi + a_{yx}\ \cos\vartheta\ \sin\vartheta\ \cos\varphi\ \sin\varphi \\
&\quad + a_{yy}\ \cos\vartheta\ \sin\vartheta\ \sin^2\varphi + a_{yz}\ \cos^2\vartheta\ \sin\varphi \\
&\quad - a_{zx}\ \sin^2\vartheta\ \cos\varphi - a_{zy}\ \sin^2\vartheta\ \sin\varphi - a_{zz}\ \cos\vartheta\ \sin\vartheta, \\
a_{\vartheta\vartheta} &= a_{xx}\ \cos^2\vartheta\ \cos^2\varphi + a_{xy}\ \cos^2\vartheta\ \cos\varphi\ \sin\varphi \\
&\quad - a_{xz}\ \cos\vartheta\ \sin\vartheta\ \cos\varphi + a_{yx}\ \cos^2\vartheta\ \cos\varphi\ \sin\varphi \\
&\quad + a_{yy}\ \cos^2\vartheta\ \sin^2\varphi - a_{yz}\ \cos\vartheta\ \sin\vartheta\ \sin\varphi \\
&\quad - a_{zx}\ \cos\vartheta\ \sin\vartheta\ \cos\varphi - a_{zy}\ \cos\vartheta\ \sin\vartheta\ \sin\varphi + a_{zz}\ \sin^2\vartheta, \\
a_{\vartheta\varphi} &= -a_{xx}\ \cos\vartheta\ \cos\varphi\ \sin\varphi + a_{xy}\ \cos\vartheta\ \cos^2\varphi \\
&\quad - a_{yx}\ \cos\vartheta\ \sin^2\varphi + a_{yy}\ \cos\vartheta\ \cos\varphi\ \sin\varphi \\
&\quad + a_{zx}\ \sin\vartheta\ \sin\varphi - a_{zy}\ \sin\vartheta\ \cos\varphi, \\
a_{\varphi r} &= -a_{xx}\ \sin\vartheta\ \cos\varphi\ \sin\varphi - a_{xy}\ \sin\vartheta\ \sin^2\varphi \\
&\quad - a_{xz}\ \cos\vartheta\ \sin\varphi + a_{yx}\ \sin\vartheta\ \cos^2\varphi \\
&\quad + a_{yy}\ \sin\vartheta\ \cos\varphi\ \sin\varphi + a_{yz}\ \sin\vartheta\ \cos\varphi, \\
a_{\varphi\vartheta} &= -a_{xx}\ \cos\vartheta\ \cos\varphi\ \sin\varphi - a_{xy}\ \cos\vartheta\ \sin^2\varphi \\
&\quad + a_{xz}\ \sin\vartheta\ \sin\varphi + a_{yx}\ \cos\vartheta\ \cos^2\varphi \\
&\quad + a_{yy}\ \cos\vartheta\ \cos\varphi\ \sin\varphi - a_{yz}\ \sin\vartheta\ \cos\varphi, \\
a_{\varphi\varphi} &= a_{xx}\ \sin^2\varphi - a_{xy}\ \cos\varphi\ \sin\varphi \\
&\quad - a_{yx}\ \cos\varphi\ \sin\varphi + a_{yy}\ \cos^2\varphi;
\end{aligned}
\tag{B.30}
$$

$$
\begin{aligned}
a_{xx} &= a_{rr}\ \sin^2\vartheta\ \cos^2\varphi + a_{r\vartheta}\ \cos\vartheta\ \sin\vartheta\ \cos^2\varphi \\
&\quad - a_{r\varphi}\ \sin\vartheta\ \cos\varphi\ \sin\varphi + a_{\vartheta r}\ \cos\vartheta\ \sin\vartheta\ \cos^2\varphi \\
&\quad + a_{\vartheta\vartheta}\ \cos^2\vartheta\ \cos^2\varphi - a_{\vartheta\varphi}\ \cos\vartheta\ \cos\varphi\ \sin\varphi \\
&\quad - a_{\varphi r}\ \sin\vartheta\ \cos\varphi\ \sin\varphi - a_{\varphi\vartheta}\ \cos\vartheta\ \cos\varphi\ \sin\varphi + a_{\varphi\varphi}\ \sin^2\varphi, \\
a_{xy} &= a_{rr}\ \sin^2\vartheta\ \cos\varphi\sin\varphi + a_{r\vartheta}\ \cos\vartheta\ \sin\vartheta\ \cos\varphi\ \sin\varphi \\
&\quad + a_{r\varphi}\ \sin\vartheta\ \cos^2\varphi - a_{\vartheta r}\ \cos\vartheta\ \sin\vartheta\ \cos\varphi\ \sin\varphi \\
&\quad + a_{\vartheta\vartheta}\ \cos^2\vartheta\ \cos\varphi\ \sin\varphi + a_{\vartheta\varphi}\ \cos\vartheta\ \cos^2\varphi \\
&\quad - a_{\varphi r}\ \sin\vartheta\ \sin^2\varphi - a_{\varphi\vartheta}\ \cos\vartheta\ \sin^2\varphi - a_{\varphi\varphi}\ \cos\varphi\ \sin\varphi, \\
a_{xz} &= a_{rr}\ \cos\vartheta\ \sin\vartheta\ \cos\varphi - a_{r\vartheta}\ \sin^2\vartheta\ \cos\varphi \\
&\quad + a_{\vartheta r}\ \cos^2\vartheta\ \cos\varphi - a_{\vartheta\vartheta}\ \cos\vartheta\ \sin\vartheta\ \cos\varphi \\
&\quad - a_{\varphi r}\ \cos\vartheta\ \sin\varphi + a_{\varphi\vartheta}\ \sin\vartheta\ \sin\varphi, \\
a_{yx} &= a_{rr}\ \sin^2\vartheta\ \cos\varphi\ \sin\varphi + a_{r\vartheta}\ \cos\vartheta\ \sin\vartheta\ \cos\varphi\ \sin\varphi \\
&\quad - a_{r\varphi}\ \sin\vartheta\ \sin^2\varphi + a_{\vartheta r}\ \cos\vartheta\ \sin\vartheta\ \cos\varphi\ \sin\varphi \\
&\quad + a_{\vartheta\vartheta}\ \cos^2\vartheta\ \cos\varphi\ \sin\varphi - a_{\vartheta\varphi}\ \cos\vartheta\ \sin^2\varphi \\
&\quad + a_{\varphi r}\ \sin\vartheta\ \cos^2\varphi + a_{\varphi\vartheta}\ \cos\vartheta\ \cos^2\varphi - a_{\varphi\varphi}\ \cos\varphi\ \sin\varphi, \\
a_{yy} &= a_{rr}\ \sin^2\vartheta\ \sin^2\varphi + a_{r\vartheta}\ \cos\vartheta\ \sin\vartheta\ \sin^2\varphi \\
&\quad + a_{r\varphi}\ \sin\vartheta\ \cos\varphi\ \sin\varphi + a_{\vartheta r}\ \cos\vartheta\ \sin\vartheta\ \sin^2\varphi \\
&\quad + a_{\vartheta\vartheta}\ \cos^2\vartheta\ \sin^2\varphi + a_{\vartheta\varphi}\ \cos\vartheta\ \cos\varphi\ \sin\varphi \\
&\quad + a_{\varphi r}\ \sin\vartheta\ \cos\varphi\ \sin\varphi + a_{\varphi\vartheta}\ \sin\vartheta\ \cos\varphi\ \sin\varphi + a_{\varphi\varphi}\ \cos^2\varphi, \\
a_{yz} &= a_{rr}\ \cos\vartheta\ \sin\vartheta\ \sin\varphi - a_{r\vartheta}\ \sin^2\vartheta\ \sin\varphi \\
&\quad + a_{\vartheta r}\ \cos^2\vartheta\ \sin\varphi - a_{\vartheta\vartheta}\ \cos\vartheta\ \sin\vartheta\ \sin\varphi \\
&\quad + a_{\varphi r}\ \cos\vartheta\ \cos\varphi - a_{\varphi\vartheta}\ \sin\vartheta\ \cos\varphi, \\
a_{zx} &= a_{rr}\ \cos\vartheta\ \sin\vartheta\ \cos\varphi + a_{r\vartheta}\ \cos^2\vartheta\ \cos\varphi \\
&\quad - a_{r\varphi}\ \cos\vartheta\ \sin\varphi - a_{\vartheta r}\ \sin^2\vartheta\ \cos\varphi \\
&\quad - a_{\vartheta\vartheta}\ \cos\vartheta\ \sin\vartheta\ \cos\varphi + a_{\vartheta\varphi}\ \sin\vartheta\ \sin\varphi, \\
a_{zy} &= a_{rr}\ \cos\vartheta\ \sin\vartheta\ \sin\varphi + a_{r\vartheta}\ \cos^2\vartheta\ \sin\varphi \\
&\quad + a_{r\varphi}\ \cos\vartheta\ \cos\varphi - a_{\vartheta r}\ \sin^2\vartheta\ \sin\varphi \\
&\quad - a_{\vartheta\vartheta}\ \cos\vartheta\ \sin\vartheta\ \sin\varphi - a_{\vartheta\varphi}\ \sin\vartheta\ \cos\varphi, \\
a_{zz} &= a_{rr}\ \cos^2\vartheta - a_{r\vartheta}\ \cos\vartheta\ \sin\vartheta \\
&\quad - a_{\vartheta r}\ \cos\vartheta\ \sin\vartheta + a_{\vartheta\vartheta}\ \sin^2\vartheta;
\end{aligned}
\tag{B.31}
$$

$$
\begin{aligned}
&a_{11} = a^{11} = a_1{}^1 = a^1{}_1 = a_{rr},\\
&a_{12} = a^1{}_2 = r\, a_{r\vartheta}, \qquad a^{12} = a_1{}^2 = \frac{1}{r}\, a_{r\vartheta},\\
&a_{13} = a^1{}_3 = r \sin\vartheta\, a_{r\varphi}, \qquad a^{13} = a_1{}^3 = \frac{1}{r\sin\vartheta}\, a_{r\varphi},\\
&a_{21} = a_2{}^1 = r\, a_{\vartheta r}, \qquad a^{21} = a^2{}_1 = \frac{1}{r}\, a_{\vartheta r},\\
&a_{22} = r^2\, a_{\vartheta\vartheta}, \qquad a^{22} = \frac{1}{r^2}\, a_{\vartheta\vartheta}, \qquad a_2{}^2 = a^2{}_2 = a_{\vartheta\vartheta},\\
&a_{23} = r^2 \sin\vartheta\, a_{\vartheta\varphi}, \qquad a^{23} = \frac{1}{r^2\sin\vartheta}\, a_{\vartheta\varphi},\\
&a_2{}^3 = \frac{1}{\sin\vartheta}\, a_{\vartheta\varphi}, \qquad a^2{}_3 = \sin\vartheta\, a_{\vartheta\varphi},\\
&a_{31} = a_3{}^1 = r\sin\vartheta\, a_{\varphi r}, \qquad a^{31} = a^3{}_1 = \frac{1}{r\sin\vartheta}\, a_{\varphi r},\\
&a_{32} = r^2 \sin\vartheta\, a_{\varphi\vartheta}, \qquad a^{32} = \frac{1}{r^2\sin\vartheta}\, a_{\varphi\vartheta},\\
&a_3{}^2 = \sin\vartheta\, a_{\varphi\vartheta}, \qquad a^3{}_2 = \frac{1}{\sin\vartheta}\, a_{\varphi\vartheta},\\
&a_{33} = r^2 \sin^2\vartheta\, a_{\varphi\varphi}, \qquad a^{33} = \frac{1}{r^2\sin^2\vartheta}\, a_{\varphi\varphi},\\
&a_3{}^3 = a^3{}_3 = a_{\varphi\varphi}.
\end{aligned}
\tag{B.32}
$$

B.2.4 δ-Tensor and ε-Tensor

The only nonzero coordinates are

$$
\begin{aligned}
&g_{11} = g^{11} = 1, \qquad g_{22} = r^2, \qquad g^{22} = \frac{1}{r^2},\\
&g_{33} = r^2 \sin^2\vartheta, \qquad g^{33} = \frac{1}{r^2\sin^2\vartheta};
\end{aligned}
\tag{B.33}
$$

$$
\begin{aligned}
&e_{123} = e^1{}_{23} = r^2 \sin\vartheta, \qquad e_1{}^2{}_3 = e^{12}{}_3 = \sin\vartheta,\\
&e_{12}{}^3 = e^1{}_2{}^3 = \frac{1}{\sin\vartheta}, \qquad e_1{}^{23} = e^{123} = \frac{1}{r^2\sin^2\vartheta},
\end{aligned}
\tag{B.34}
$$

and the corresponding permutations of the ε-tensor. Furthermore

$$
g := \det g_{ij} = r^4 \sin^2\vartheta. \tag{B.35}
$$

B.2.5 Christoffel Symbols

The only nonzero Christoffel symbols are

$$
\begin{aligned}
\Gamma^1_{22} &= -r, \qquad \Gamma^1_{33} = -r\sin^2\vartheta,\\
\Gamma^2_{12} &= \Gamma^2_{21} = \Gamma^3_{13} = \Gamma^3_{31} = \tfrac{1}{r},\\
\Gamma^2_{33} &= -\sin\vartheta\cos\vartheta, \qquad \Gamma^3_{23} = \Gamma^3_{32} = \cot\vartheta.
\end{aligned}
\tag{B.36}
$$

B.2.6 Differential Operators

For each operator the physical coordinates are given:

$$\operatorname{grad} a = \left\{\frac{\partial a}{\partial r}, \frac{1}{r}\frac{\partial a}{\partial \vartheta}, \frac{1}{r\sin\vartheta}\frac{\partial a}{\partial \varphi}\right\}; \tag{B.37}$$

$$\operatorname{div} \underline{a} = \frac{\partial a_r}{\partial r} + \frac{2a_r}{r} + \frac{1}{r}\frac{\partial a_\vartheta}{\partial \vartheta} + \frac{\cot\vartheta\, a_\vartheta}{r} + \frac{1}{r\sin\vartheta}\frac{\partial a_\varphi}{\partial \varphi}; \tag{B.38}$$

$$\begin{aligned}
(\operatorname{curl} \underline{a})_r &= \frac{1}{r}\frac{\partial a_\varphi}{\partial \vartheta} + \frac{\cot\vartheta\, a_\varphi}{r} - \frac{1}{r\sin\vartheta}\frac{\partial a_\vartheta}{\partial \varphi},\\
(\operatorname{curl} \underline{a})_\vartheta &= \frac{1}{r\sin\vartheta}\frac{\partial a_r}{\partial \varphi} - \frac{\partial a_\varphi}{\partial r} - \frac{a_\varphi}{r},\\
(\operatorname{curl} \underline{a})_\varphi &= \frac{\partial a_\vartheta}{\partial r} + \frac{a_\vartheta}{r} - \frac{1}{r}\frac{\partial a_r}{\partial \vartheta};
\end{aligned} \tag{B.39}$$

$$\begin{aligned}
(\operatorname{grad} \underline{a})_{rr} &= \frac{\partial a_r}{\partial r},\\
(\operatorname{grad} a)_{r\vartheta} &= \frac{1}{r}\frac{\partial a_r}{\partial \vartheta} - \frac{a_\vartheta}{r},\\
(\operatorname{grad} \underline{a})_{r\varphi} &= \frac{1}{r\sin\vartheta}\frac{\partial a_r}{\partial \varphi} - \frac{a_\varphi}{r},\\
(\operatorname{grad} a)_{\vartheta r} &= \frac{\partial a_\vartheta}{\partial r},\\
(\operatorname{grad} \underline{a})_{\vartheta\vartheta} &= \frac{1}{r}\frac{\partial a_\vartheta}{\partial \vartheta} + \frac{a_r}{r},\\
(\operatorname{grad} \underline{a})_{\vartheta\varphi} &= \frac{1}{r\sin\vartheta}\frac{\partial a_\vartheta}{\partial \varphi} - \frac{\cot\vartheta\, a_\varphi}{r},\\
(\operatorname{grad} a)_{\varphi r} &= \frac{\partial a_\varphi}{\partial r},\\
(\operatorname{grad} \underline{a})_{\varphi\vartheta} &= \frac{1}{r}\frac{\partial a_\varphi}{\partial \vartheta},\\
(\operatorname{grad} \underline{a})_{\varphi\varphi} &= \frac{1}{r\sin\vartheta}\frac{\partial a_\varphi}{\partial \varphi} + \frac{a_r}{r} + \frac{\cot\vartheta\, a_\vartheta}{r};
\end{aligned} \tag{B.40}$$

$$\begin{aligned}
[(\operatorname{grad} \underline{a})\cdot \underline{b}]_r &= \frac{\partial a_r}{\partial r} b_r + \frac{\partial a_r}{\partial \vartheta}\frac{b_\vartheta}{r} + \frac{\partial a_r}{\partial \varphi}\frac{b_\varphi}{r\sin\vartheta}\\
&\quad - \frac{a_\vartheta b_\vartheta + a_\varphi b_\varphi}{r},\\
[(\operatorname{grad} \underline{a})\cdot \underline{b}]_\vartheta &= \frac{\partial a_\vartheta}{\partial r} b_r + \frac{\partial a_\vartheta}{\partial \vartheta}\frac{b_\vartheta}{r} + \frac{\partial a_\vartheta}{\partial \varphi}\frac{b_\varphi}{r\sin\vartheta}\\
&\quad + \frac{a_r b_\vartheta - \cot\vartheta\, a_\varphi b_\varphi}{r},\\
[(\operatorname{grad} \underline{a})\cdot \underline{b}]_\varphi &= \frac{\partial a_\varphi}{\partial r} b_r + \frac{\partial a_\varphi}{\partial \vartheta}\frac{b_\vartheta}{r} + \frac{\partial a_\varphi}{\partial \varphi}\frac{b_\varphi}{r\sin\vartheta}\\
&\quad + \frac{a_r b_\varphi + \cot\vartheta\, a_\vartheta b_\varphi}{r};
\end{aligned} \tag{B.41}$$

$$
\begin{aligned}
(\operatorname{div} \underline{\underline{a}})_r &= \frac{\partial a_{rr}}{\partial r} + \frac{2\,a_{rr}}{r} + \frac{1}{r}\frac{\partial a_{r\vartheta}}{\partial \vartheta} + \frac{1}{r\sin\vartheta}\frac{\partial a_{r\varphi}}{\partial \varphi} \\
&\quad - \frac{a_{\vartheta\vartheta} + a_{\varphi\varphi}}{r} + \frac{\cot\vartheta\, a_{r\vartheta}}{r}, \\
(\operatorname{div} \underline{\underline{a}})_\vartheta &= \frac{\partial a_{\vartheta r}}{\partial r} + \frac{2\,a_{\vartheta r}}{r} + \frac{1}{r}\frac{\partial a_{\vartheta\vartheta}}{\partial \vartheta} \\
&\quad + \frac{1}{r\sin\vartheta}\frac{\partial a_{\vartheta\varphi}}{\partial \varphi} + \frac{\cot\vartheta(a_{\vartheta\vartheta} - a_{\varphi\varphi})}{r} + \frac{a_{r\vartheta}}{r}, \\
(\operatorname{div} \underline{\underline{a}})_\varphi &= \frac{\partial a_{\varphi r}}{\partial r} + \frac{2\,a_{\varphi r}}{r} + \frac{1}{r}\frac{\partial a_{\varphi\vartheta}}{\partial \vartheta} + \frac{1}{r\sin\vartheta}\frac{\partial a_{\varphi\varphi}}{\partial \varphi} \\
&\quad + \frac{\cot\vartheta(a_{\vartheta\varphi} + a_{\varphi\vartheta})}{r} + \frac{a_{r\varphi}}{r};
\end{aligned} \tag{B.42}
$$

$$
\Delta a = \frac{\partial^2 a}{\partial r^2} + \frac{2}{r}\frac{\partial a}{\partial r} + \frac{1}{r^2}\frac{\partial^2 a}{\partial \vartheta^2} + \frac{\cot\vartheta}{r^2}\frac{\partial a}{\partial \vartheta} + \frac{1}{r^2\sin^2\vartheta}\frac{\partial^2 a}{\partial \varphi^2}; \tag{B.43}
$$

$$
\begin{aligned}
(\Delta \underline{a})_r &= \frac{\partial^2 a_r}{\partial r^2} + \frac{2}{r}\frac{\partial a_r}{\partial r} + \frac{1}{r^2}\frac{\partial^2 a_r}{\partial \vartheta^2} + \frac{\cot\vartheta}{r^2}\frac{\partial a_r}{\partial \vartheta} \\
&\quad + \frac{1}{r^2\sin^2\vartheta}\frac{\partial^2 a_r}{\partial \varphi^2} - \frac{2\,a_r}{r^2} - \frac{2}{r^2}\frac{\partial a_\vartheta}{\partial \vartheta} - \frac{2\cot\vartheta\, a_\vartheta}{r^2} \\
&\quad - \frac{2}{r^2\sin\vartheta}\frac{\partial a_\varphi}{\partial \varphi}, \\
(\Delta \underline{a})_\vartheta &= \frac{\partial^2 a_\vartheta}{\partial r^2} + \frac{2}{r}\frac{\partial a_\vartheta}{\partial r} + \frac{1}{r^2}\frac{\partial^2 a_\vartheta}{\partial \vartheta^2} + \frac{\cot\vartheta}{r^2}\frac{\partial a_\vartheta}{\partial \vartheta} \\
&\quad + \frac{1}{r^2\sin^2\vartheta}\frac{\partial^2 a_\vartheta}{\partial \varphi^2} + \frac{2}{r^2}\frac{\partial a_r}{\partial \vartheta} - \frac{a_\vartheta}{r^2\sin^2\vartheta} - \frac{2\cot\vartheta}{r^2\sin\vartheta}\frac{\partial a_\varphi}{\partial \varphi}, \\
(\Delta \underline{a})_\varphi &= \frac{\partial^2 a_\varphi}{\partial r^2} + \frac{2}{r}\frac{\partial a_\varphi}{\partial r} + \frac{1}{r^2}\frac{\partial^2 a_\varphi}{\partial \vartheta^2} + \frac{\cot\vartheta}{r^2}\frac{\partial a_\varphi}{\partial \vartheta} \\
&\quad + \frac{1}{r^2\sin^2\vartheta}\frac{\partial^2 a_\varphi}{\partial \varphi^2} + \frac{2}{r^2\sin\vartheta}\frac{\partial a_r}{\partial \varphi} + \frac{2\cot\vartheta}{r^2\sin\vartheta}\frac{\partial a_\vartheta}{\partial \varphi} - \frac{a_\varphi}{r^2\sin^2\vartheta}.
\end{aligned} \tag{B.44}
$$

B.2.7 Curve-, Area-, and Volume Elements

The elements are

$$
\begin{aligned}
\mathrm{d}u^i &= (\mathrm{d}r,\ \mathrm{d}\vartheta,\ \mathrm{d}\varphi), \\
\mathrm{d}A_i &= (r^2\sin\vartheta\,\mathrm{d}\vartheta\,\mathrm{d}\varphi,\ r^2\sin\vartheta\,\mathrm{d}\varphi\,\mathrm{d}r,\ r^2\sin\vartheta\,\mathrm{d}r\,\mathrm{d}\vartheta), \\
\mathrm{d}V &= r^2\sin\vartheta\,\mathrm{d}r\,\mathrm{d}\vartheta\,\mathrm{d}\varphi.
\end{aligned} \tag{B.45}
$$

In (B.46) the physical coordinates are given:

$$
\begin{aligned}
\mathrm{d}\underline{x} &= (\mathrm{d}r,\ r\,\mathrm{d}\vartheta,\ r\sin\vartheta\,\mathrm{d}\varphi), \\
\mathrm{d}\underline{A} &= (r^2\sin\vartheta\,\mathrm{d}\vartheta\,\mathrm{d}\varphi,\ r\sin\vartheta\,\mathrm{d}\varphi\,\mathrm{d}r,\ r\,\mathrm{d}r\,\mathrm{d}\vartheta).
\end{aligned} \tag{B.46}
$$

Index

www.ingramcontent.com/pod-product-compliance
Lightning Source LLC
Chambersburg PA
CBHW080916220326
41598CB00034B/5588
9783110404258